NONLINEAR PROCESSES IN GEOPHYSICAL FLUID DYNAMICS

Nonlinear Processes in Geophysical Fluid Dynamics

A tribute to the scientific work of Pedro Ripa

Edited by

O.U. Velasco Fuentes
Departamento de Oceanografía Física, CICESE,
Ensenada, México

J. Sheinbaum
Departamento de Oceanografía Física, CICESE,
Ensenada, México

and

J. Ochoa
Departamento de Oceanografía Física, CICESE,
Ensenada, México

SPRINGER-SCIENCE+BUSINESS MEDIA, B.V.

A C.I.P. Catalogue record for this book is available from the Library of Congress.

ISBN 978-94-010-3996-3 ISBN 978-94-010-0074-1 (eBook)
DOI 10.1007/978-94-010-0074-1

Printed on acid-free paper

Cover photo by Pedro Ripa.

A la familia Ripa:
Andrea, Camila, Luisa, Natalia, Camilita y Damián

TABLE OF CONTENTS

PREFACE

This volume contains a collection of papers by international experts in geophysical fluid dynamics, based upon presentations at a colloquium held in memory of Pedro Ripa on the first anniversary of his untimely death. They review or present recent developments in hydrodynamic stability theory, Hamiltonian fluid mechanics, balanced dynamics, waves, vortices, general oceanography and the physical oceanography of the Gulf of California; all of them subjects in which Professor Ripa made important contributions. His work, but also his friendly spirit and kindness were highly regarded and appreciated by colleagues and students alike around the world. This book is a tribute to his scientific legacy and constitutes a valuable reference for researchers and graduate students interested in geophysical and general fluid mechanics.

Early in his career as a physical oceanographer, Pedro Ripa made two landmark contributions to geophysical fluid dynamics. In 1981, he showed that the conservation of the potential vorticity is related to the invariance of the equations of motion under the symmetry transformations of the labels that identify the fluid particles. That is, potential vorticity conservation is a consequence, via Noether's theorem, of the particle re-labelling symmetry. Two years later he published a paper entitled "General stability conditions for zonal flows in a one-layer model on the beta-plane or the sphere", where he established necessary conditions for stability in the shallow water equations, nowadays known as "Ripa's Theorem." This is one of the very few Arnol'd-like stability conditions that goes beyond two-dimensional or quasi-geostrophic flow, and stands alongside other famous stability criteria in making the foundations of the field.

It is then almost natural to begin this volume with Theodore Shepherd's review of Ripa's theorem, in which he vividly illustrates the state of the subject in the early eighties, its relation to Hamiltonian methods, and the role it played in the development of stability theory. The following four contributions are related to Hamiltonian formalism, variational methods and stability. Josefina Olascoaga *et al.* study the role of the deep ocean in the baroclinic instability of upper-ocean currents. Javier Beron-Vera's contribution develops a balanced model using modern mathematical techniques and Lagrangian-Eulerian variational principles. The paper by Philip Morrison shows how to find action-angle variables in non-canonical Hamiltonian systems with continuous spectra, ubiquitous in plasma and fluid mechanics, and put them into normal form. The last paper on this subject is from Jonas Nycander who investigates the nonlinear stability of vortices using isovortical (dynamically accessible) perturbations, and explains observations regarding the stability of anticyclones above ocean mountains.

The next five papers deal with studies of ocean processes and the general circulation of the atmosphere and oceans. Doron Nof's contribution discusses the dynamical fate of a southward outflow on a rotating basin on the β plane, whilst Gordon Swaters analyzes the generation of gravity waves by unstable dense overflows and their importance for deep ocean currents.

The third paper in this section is from Pedro Ripa himself, and deserves special mention. It investigates the impact upon sea-level of heat and fresh-water fluxes in mass- and volume-conserving models. Pedro was about to submit the manuscript for publication when he passed away. We decided to include the manuscript here so that it should not be left unpublished, but especially because it illustrates the insightful and fresh approach he would take on a relatively well-studied subject.

On general circulation issues, Andrew Bush looks at baroclinic waves shaping the mean atmospheric circulation in past and present climates using a coupled ocean-atmosphere model; and Geoffrey Vallis reviews recent thermocline theories and the role of eddies in defining the vertical density structure of the ocean.

Pedro's other scientific passions were coastal oceanography and the oceanography of the Gulf of California. Among his numerous contributions we should mention his important paper "Towards a physical explanation of the seasonal dynamics and thermodynamics of the Gulf of California" (1997), in which he proposes that the Pacific Ocean influences the seasonal variability of the Gulf of California by way of a baroclinic Kelvin wave. Miguel Lavín and Guido Marinone review the state of knowledge of the physical oceanography of the Gulf of California and highlight the important contributions made by Pedro for its understanding. Antoine Badan reviews recent ideas about the atmosphere over the Gulf, which has received a lot of attention being the core region of the North-American Monsoon system. Observational evidence of semi-permanent geostrophic gyres in the southern part of the Gulf is discussed by Manuel Figueroa *et al.*, whilst Guido Marinone and Miguel Lavín use a numerical model to study mixing processes and residual circulation in the central part of the Gulf. The paper by José Ochoa *et al.* discusses questions of optimal observation-array design for ocean transport calculations, using data from the Yucatan Channel in the Mexican Caribbean; Anatoliy Filonov studies internal tide generation and propagation in a coastal region of México's western Pacific coast.

A constant in Pedro's work was the search for simplified or approximate models that contained the essential physics of the problem under study. Many of his latest papers dealt with models with simplified vertical structure and their dynamical properties. James McWilliams contribution reviews current understanding of Balanced Equations models and highlights Pedro's contribution to the subject.

The final four papers are related to vortex dynamics, another subject in which Pedro made interesting contributions. GertJan van Heijst *et al.* analyze the effect of solid boundaries on decaying two dimensional turbulence in laboratory and numerical simulations. George Carnevale *et al.* discuss the effect of rotation in suppressing Rayleigh-Taylor instability and generation of vortex filaments. Using dynamical systems methods, Oscar Velasco Fuentes looks at the process of vortex merger on a β plane, whilst Rudolf Kloosterziel and George Carnevale derive a simplified model to study the formation and nonlinear dynamics of tripolar structures.

It has always been an issue whether or not science is a form of art. We believe it is, if only from the pleasure our own discoveries or those of others provide, of similar nature to the experience felt when reading a good book, admiring a beautiful painting or hearing a wonderful piece of music. Pedro Ripa was an outstanding performer of the scientific artistry. It was a joy to have had him as a colleague, teacher and friend, and we are certain we speak in the name of all the participants to the colloquium and those contributing in this book.

We would like to thank all the people who collaborated in the success of the colloquium and the edition of its proceedings. Especially Lupita Rodríguez, Roberto Soto, and Julio Figueroa, for their secretarial and computer assistance at CICESE; and Petra van Steenbergen and Mieke van der Fluit, at Kluwer. Our special thanks also to the reviewers who generously offered their time to referee the papers. Finally, the Academia Mexicana de Ciencias, México's CONACyT (Grants 28137-T and 39016-T), and CICESE, provided the financial support that made all of this possible.

Oscar Velasco Fuentes
Julio Sheinbaum
José Ochoa

Ensenada, México
May 2003

RIPA'S THEOREM AND ITS RELATIVES *

THEODORE G. SHEPHERD
Department of Physics
University of Toronto
60 St. George Street
Toronto, M5S IA7, Canada

Abstract. In 1983, Pedro Ripa proved a stability theorem for parallel flows in rotating shallow-water dynamics which has since come to be known as Ripa's theorem. It is singular for being the only known Arnol'd-type stability theorem (meaning that it is based on conservation laws) for non-resting basic states and non-symmetric unbalanced dynamics. It was also pioneering in its simultaneous use of momentum as well as energy conservation, together with the Casimir invariants which are crucial to obtaining such results. This turned out to be a fortuitous choice on Pedro's part, because the apparent generalization to non-parallel flows offered by energy-Casimir conservation alone is less than meets the eye, due to Andrews' theorem. Ripa's theorem has a clear physical interpretation, and has been found to be very useful in anticipating stability boundaries for coupled vortical/gravity-wave instabilities. It has also recently emerged in the context of balanced models that support coastal Kelvin waves, which brings together two of Pedro's loves: stability theory and coastal oceanography.

Key words: Hamiltonian fluid dynamics, conservation laws, stability theorems

1. Introduction

Pedro Ripa's career overlapped quite strongly with mine for a period of about 10 years, during which time we were both actively exploiting the Hamiltonian structure of geophysical fluid dynamics to study disturbance conservation laws and the stability theorems which follow from them. When I entered this field in the mid-1980s, Pedro was already famous for a number of seminal contributions made in the early 1980s. This field was perhaps his first love in physical oceanography, and it continued to occupy his attention until the very end of his life. A large fraction of his papers are devoted to this subject, including his most cited paper of all — Ripa (1983), in which he proves what has now come to be called *Ripa's theorem.*

* In memory of Pedro Ripa, 1946-2001

O.U. Velasco Fuentes et al. (eds.), Nonlinear Processes in Geophysical Fluid Dynamics, 1-14.

2. Hamiltonian structure

In the absence of forcing and dissipation, good models in geophysical fluid dynamics can be written in the Hamiltonian form

$$\boldsymbol{u}_t = J\frac{\delta\mathcal{H}}{\delta\boldsymbol{u}}, \tag{1}$$

where $\boldsymbol{u}$ are the dependent variables, $\mathcal{H}$ is the Hamiltonian (energy), and J is an operator with certain properties (e.g. Benjamin, 1984; Shepherd, 1990; Morrison, 1998; Salmon, 1998). The fundamental equations of geophysical fluid dynamics — the compressible, stratified Euler equations — have this property, as do most approximations to them, including the hydrostatic, the Boussinesq, the anelastic, the quasi-geostrophic, and the barotropic equations. Representation (1) is called the "symplectic" formulation; for canonical dynamics (Hamilton's equations) one has

$$\begin{pmatrix} q_t \\ \\ p_t \end{pmatrix} = \begin{pmatrix} \partial\mathcal{H}/\partial p \\ \\ -\partial\mathcal{H}/\partial q \end{pmatrix} = \underbrace{\begin{pmatrix} 0 & 1 \\ \\ -1 & 0 \end{pmatrix}}_{J} \begin{pmatrix} \partial\mathcal{H}/\partial q \\ \\ \partial\mathcal{H}/\partial p \end{pmatrix}, \tag{2}$$

but the representation (1) is considerably more general.

It is arguable that Pedro's favorite model system was the shallow-water equations (SWE), which on an f-plane take the form (Salmon, 1998)

$$\boldsymbol{v}_t + (f\hat{\boldsymbol{z}} + \boldsymbol{\nabla}\times\boldsymbol{v})\times\boldsymbol{v} + \boldsymbol{\nabla}(\frac{1}{2}|\boldsymbol{v}|^2) = -g\boldsymbol{\nabla}h, \quad h_t + \boldsymbol{\nabla}\cdot(h\boldsymbol{v}) = 0. \tag{3}$$

The SWE turn out to be Hamiltonian, with $\boldsymbol{u} = (\boldsymbol{v}, h)^T$,

$$\mathcal{H} = \frac{1}{2}\iint (h|\boldsymbol{v}|^2 + gh^2)\,\mathrm{d}x\mathrm{d}y, \quad J = \begin{pmatrix} 0 & q & -\partial_x \\ \\ -q & 0 & -\partial_y \\ \\ -\partial_x & -\partial_y & 0 \end{pmatrix}, \tag{4}$$

where $q \equiv [f + \hat{\boldsymbol{z}}\cdot(\boldsymbol{\nabla}\times\boldsymbol{v})]/h$ is the potential vorticity. To see this, the functional derivatives of $\mathcal{H}$ are evidently

$$\frac{\delta\mathcal{H}}{\delta\boldsymbol{v}} = h\boldsymbol{v}, \quad \frac{\delta\mathcal{H}}{\delta h} = \frac{1}{2}|\boldsymbol{v}|^2 + gh, \tag{5}$$

whence substitution of (4) and (5) into (1) yields

$$J\frac{\delta\mathcal{H}}{\delta\boldsymbol{u}} = \begin{pmatrix} 0 & q & -\partial_x \\ \\ -q & 0 & -\partial_y \\ \\ -\partial_x & -\partial_y & 0 \end{pmatrix} \begin{pmatrix} hu \\ \\ hv \\ \\ \frac{1}{2}|\boldsymbol{v}|^2 + gh \end{pmatrix}$$

$$= \begin{pmatrix} qhv - \boldsymbol{v}\cdot\boldsymbol{v}_x - gh_x \\ -qhu - \boldsymbol{v}\cdot\boldsymbol{v}_y - gh_y \\ -\boldsymbol{\nabla}\cdot(h\boldsymbol{v}) \end{pmatrix} = \begin{pmatrix} u_t \\ v_t \\ h_t \end{pmatrix} = \boldsymbol{u}_t. \tag{6}$$

For any functional $\mathcal{F}$,

$$\frac{\mathrm{d}\mathcal{F}}{\mathrm{d}t} = \iint \frac{\delta\mathcal{F}}{\delta\boldsymbol{u}}\boldsymbol{u}_t \,\mathrm{d}x\mathrm{d}y = \iint \frac{\delta\mathcal{F}}{\delta\boldsymbol{u}} J \frac{\delta\mathcal{H}}{\delta\boldsymbol{u}} \,\mathrm{d}x\mathrm{d}y \equiv [\mathcal{F},\mathcal{H}]. \tag{7}$$

This defines the skew-symmetric Poisson bracket $[\,\cdot\,,\,\cdot\,]$.

3. Symmetries and conservation laws

Skew-symmetry of the Poisson bracket implies that energy is conserved:

$$\frac{\mathrm{d}\mathcal{H}}{\mathrm{d}t} = [\mathcal{H},\mathcal{H}] = 0. \tag{8}$$

More generally, consider

$$\delta\mathcal{H} = \iint \frac{\delta\mathcal{H}}{\delta\boldsymbol{u}}\,\delta\boldsymbol{u}\,\mathrm{d}x\mathrm{d}y. \tag{9}$$

A functional $\mathcal{M}$ defines the infinitesimal variation

$$\delta_{\mathcal{M}}\boldsymbol{u} \equiv \varepsilon J\,\frac{\delta\mathcal{M}}{\delta\boldsymbol{u}}, \tag{10}$$

where ε is an infinitesimal parameter. It follows that

$$\delta_{\mathcal{M}}\mathcal{H} = \iint \frac{\delta\mathcal{H}}{\delta\boldsymbol{u}}\,\delta_{\mathcal{M}}\boldsymbol{u}\,\mathrm{d}x\mathrm{d}y = \varepsilon \iint \frac{\delta\mathcal{H}}{\delta\boldsymbol{u}} J \frac{\delta\mathcal{M}}{\delta\boldsymbol{u}}\,\mathrm{d}x\mathrm{d}y = \varepsilon\,[\mathcal{H},\mathcal{M}] = -\varepsilon\,\frac{\mathrm{d}\mathcal{M}}{\mathrm{d}t}. \tag{11}$$

This illustrates *Noether's theorem*: the Hamiltonian has a symmetry with respect to the variation induced by $\mathcal{M}$ (i.e. $\delta_{\mathcal{M}}\mathcal{H} = 0$) if and only if $\mathcal{M}$ is a conserved quantity (i.e. $\mathrm{d}\mathcal{M}/\mathrm{d}t = 0$). Conservation of energy results from symmetry in time; the spatial translational symmetries give conservation of the various momenta, linear and angular.

Consider translation in x (with $\delta\boldsymbol{u} = -\varepsilon\boldsymbol{u}_x$) for the SWE; solving

$$\begin{pmatrix} 0 & q & -\partial_x \\ -q & 0 & -\partial_y \\ -\partial_x & -\partial_y & 0 \end{pmatrix} \begin{pmatrix} \delta\mathcal{M}^x/\delta u \\ \delta\mathcal{M}^x/\delta v \\ \delta\mathcal{M}^x/\delta h \end{pmatrix} = \begin{pmatrix} -u_x \\ -v_x \\ -h_x \end{pmatrix} \tag{12}$$

for $\mathcal{M}^x$ gives

$$\mathcal{M}^x = \iint h(u - fy)\,\mathrm{d}x\mathrm{d}y, \tag{13}$$

with

$$\frac{\delta \mathcal{M}^x}{\delta u} = h, \quad \frac{\delta \mathcal{M}^x}{\delta v} = 0, \quad \frac{\delta \mathcal{M}^x}{\delta h} = u - fy. \tag{14}$$

This is the absolute zonal momentum. In a similar fashion, symmetries in y and θ (rotation) give conservation of

$$\mathcal{M}^y = \iint h(v + fx)\,\mathrm{d}x\mathrm{d}y, \quad \mathcal{M}^\theta = \iint h\Big(\hat{\boldsymbol{z}} \cdot (\boldsymbol{r} \times \boldsymbol{v}) + \frac{fr^2}{2}\Big)\,\mathrm{d}x\mathrm{d}y. \tag{15}$$

A special class of functionals $\mathcal{C}$, known as *Casimirs*, satisfy

$$J\frac{\delta \mathcal{C}}{\delta \boldsymbol{u}} = 0. \tag{16}$$

Since $\delta_{\mathcal{C}}\boldsymbol{u} \equiv 0$ they correspond to invisible symmetries; they are conserved since

$$\frac{\mathrm{d}\mathcal{C}}{\mathrm{d}t} = [\mathcal{C}, \mathcal{H}] = -[\mathcal{H}, \mathcal{C}] = 0. \tag{17}$$

Symmetry-related invariants are only defined to within a Casimir. For the SWE the Casimirs are

$$\mathcal{C} = \iint hC(q)\,\mathrm{d}x\mathrm{d}y \tag{18}$$

for arbitrary functions $C(\,\cdot\,)$. They are linked with Lagrangian conservation of q,

$$\frac{Dq}{Dt} = 0, \tag{19}$$

and associated with the *particle relabelling symmetry* (e.g. Ripa, 1981*a*).

I am sure that when presenting this sort of theory Pedro must have been, as I often am, asked what the point is of Hamiltonian structure. (He may have been particularly asked this by his Director!) It is true that most results derived from Hamiltonian structure follow from the conservation laws themselves, and can always be derived in a direct fashion from the equations of motion. However, Hamiltonian structure explains why this all works, and that it must do so generally. It also serves to unify theory across different models. Hamiltonian structure would have been second nature to a theoretical physicist like Pedro. In fact, he did not make a big fanfare about Hamiltonian structure in his papers, and almost certainly did not even know how to express it for Eulerian fluid models in the early 1980s when he made his most seminal contributions. But he knew that fluid models somehow *had* to be Hamiltonian, and thus possessed the connection between symmetries and conservation laws expressed in Noether's theorem. This was enough to give him the confidence to proceed.

Circa 1980, Hamiltonian structure for fluids was generally understood in terms of Hamilton's principle $\delta \int \mathcal{L}\,\mathrm{d}t = 0$, where $\mathcal{L}$ is the Lagrangian functional, in the Lagrangian (or particle-following) description — which is canonical. The connection between symmetries and conservation laws was understood in this context (Andrews & McIntyre, 1978; Ripa, 1981*a*). Yet Ripa (1981*b*) nevertheless defined energy and pseudomomentum *in Eulerian variables*, relative to a resting basic state, for both barotropic Rossby waves and internal gravity waves.[1] Re-reading these early papers, and trying to imagine what Pedro knew at the time, I find myself in wonder at how he made the connection in 1981 between pseudomomentum and the Eulerian conservation laws involving arbitrary functions of vorticity $F(\zeta)$ (for barotropic Rossby waves) and of density $F(\rho)$ (for internal gravity waves). Around the same time, Holliday & McIntyre (1981) had defined a finite-amplitude available potential energy for stratified flow using conservation of $F(\rho)$, but saw this as a special trick for a particular case and did not recognize that they had stumbled on a general way of constructing the pseudoenergy. [The same trick was then applied to barotropic Rossby waves to derive what might be called an "available enstrophy" by Killworth & McIntyre (1985), using conservation of $F(\zeta)$.] At the time, it was felt by many that the existence of finite-amplitude Eulerian disturbance conservation laws was a special property of those systems for which the Lagrangian information was somehow encoded through invariant functions such as $F(\zeta)$ or $F(\rho)$. This comes through quite clearly in Ripa (1981*a*), which makes the conceptual leap of Ripa (1981*b*) all the more remarkable. Of course, the general Eulerian Hamiltonian theory was already known to some workers at that time, and all would become clear shortly thereafter (Morrison, 1982; Benjamin, 1984; Holm *et al.*, 1985).

4. Stability

Returning to the SWE, consider a steady, parallel basic flow $U(y)$, $H(y)$ with $fU = -gH_y$. Define the disturbance by $u = U + u'$, $v = v'$, and $h = H + h'$. The disturbance energy

$$\Delta\mathcal{H} \equiv \mathcal{H}(\boldsymbol{v}, h) - \mathcal{H}(U, H) = \iint \Big\{ \frac{1}{2}(H + h')|\boldsymbol{v}'|^2 + Uh'u' + HUu' + \frac{1}{2}U^2 h' + \frac{1}{2}g(h')^2 + gHh' \Big\} \mathrm{d}x\mathrm{d}y \tag{20}$$

[1] Ripa's "energy" was already quadratic, but for internal gravity waves is actually the pseudoenergy (Shepherd, 1993).

is evidently not quadratic to leading order in the disturbance. This cannot happen in canonical systems, for which

$$\boldsymbol{U}_t = 0 \quad \Longrightarrow \quad J\,\frac{\delta\mathcal{H}}{\delta\boldsymbol{u}}\bigg|_{\boldsymbol{u}=\boldsymbol{U}} = 0 \quad \Longrightarrow \quad \frac{\delta\mathcal{H}}{\delta\boldsymbol{u}}\bigg|_{\boldsymbol{u}=\boldsymbol{U}} = 0. \tag{21}$$

Thus for canonical systems, steady states $\boldsymbol{U}$ are conditional extrema of $\mathcal{H}$, i.e. $\Delta\mathcal{H}$ is locally quadratic about $\boldsymbol{U}$. In contrast, with a non-invertible J the second implication of (21) is not true; however

$$J\,\frac{\delta\mathcal{H}}{\delta\boldsymbol{u}}\bigg|_{\boldsymbol{u}=\boldsymbol{U}} = 0 \quad \Longrightarrow \quad \frac{\delta\mathcal{H}}{\delta\boldsymbol{u}}\bigg|_{\boldsymbol{u}=\boldsymbol{U}} = -\frac{\delta\mathcal{C}}{\delta\boldsymbol{u}}\bigg|_{\boldsymbol{u}=\boldsymbol{U}} \tag{22}$$

for some Casimir $\mathcal{C}$. This defines a new invariant $\mathcal{H}+\mathcal{C}$ [or rather $\Delta(\mathcal{H}+\mathcal{C})$] which is quadratic to leading order in the disturbance, and which is known as the *pseudoenergy*.

Since the basic flow is symmetric in x (i.e. $\boldsymbol{U}_x = 0$), we have

$$J\,\frac{\delta\mathcal{M}^x}{\delta\boldsymbol{u}}\bigg|_{\boldsymbol{u}=\boldsymbol{U}} = 0 \tag{23}$$

and one can include $\mathcal{M}^x$ as well. Hence we can write

$$\mathcal{A} \equiv \Delta(\mathcal{H} - \alpha\mathcal{M}^x + \mathcal{C}) \tag{24}$$

for some arbitrary constant α, with $\mathcal{C}$ determined by

$$\frac{\delta\mathcal{H}}{\delta\boldsymbol{u}}\bigg|_{\boldsymbol{u}=\boldsymbol{U}} - \alpha\frac{\delta\mathcal{M}^x}{\delta\boldsymbol{u}}\bigg|_{\boldsymbol{u}=\boldsymbol{U}} + \frac{\delta\mathcal{C}}{\delta\boldsymbol{u}}\bigg|_{\boldsymbol{u}=\boldsymbol{U}} = 0. \tag{25}$$

$\mathcal{A}$ is guaranteed to be conserved, and quadratic to leading order in the disturbance, for any choice of α.

To work this out for the SWE, note that

$$\begin{aligned} \delta\mathcal{C} &= \iint \Big\{ C(q)\,\delta h + hC'(q)\underbrace{\Big[\frac{1}{h}(\delta v_x - \delta u_y) - \frac{q}{h}\,\delta h\Big]}_{\delta q} \Big\}\,\mathrm{d}x\mathrm{d}y \\ &= \iint \Big\{ \underbrace{[C(q) - qC'(q)]}_{\delta\mathcal{C}/\delta h}\,\delta h \underbrace{-C''(q)q_x}_{\delta\mathcal{C}/\delta v}\,\delta v + \underbrace{C''(q)q_y}_{\delta\mathcal{C}/\delta u}\,\delta u \Big\}\,\mathrm{d}x\mathrm{d}y \end{aligned} \tag{26}$$

(after integrating by parts). With $q = Q(y)$, *etc.*, this gives (Ripa, 1983)

$$C(Q) - QC'(Q) = \alpha(U - fy) - \frac{1}{2}U^2 - gH, \quad C''(Q)Q_y = (\alpha - U)H. \tag{27}$$

The right-hand sides of these expressions are functions of Q via y. The exact finite-amplitude pseudoenergy is then given by

$$\mathcal{A} = \iint \Big\{\frac{1}{2}(H+h')|\boldsymbol{v}'|^2 + (U-\alpha)h'u' + \frac{1}{2}g(h')^2 \\ + (H+h')\int_0^{q'} [C'(Q+\xi) - C'(Q)]\,\mathrm{d}\xi\Big\}\,\mathrm{d}x\mathrm{d}y, \tag{28}$$

which after completing the square becomes

$$\mathcal{A} = \iint \Big\{\frac{1}{2(H+h')}|(H+h')\boldsymbol{v}' + (U-\alpha)h'\hat{\boldsymbol{x}}|^2 + \frac{1}{2}\Big(g - \frac{(U-\alpha)^2}{H+h'}\Big)(h')^2 \\ + (H+h')\int_0^{q'} [C'(Q+\xi) - C'(Q)]\,\mathrm{d}\xi\Big\}\,\mathrm{d}x\mathrm{d}y. \tag{29}$$

Conservation of $\mathcal{A}$ has implications for stability and instability. If $\mathcal{A}$ is sign-definite, it defines a norm and the flow is stable. Normal-mode instabilities must have $\mathcal{A} = 0$. Many normal-mode stability theorems are just conservation of $\mathcal{A}$, and take the form $c_i \int \{\ldots\}\,\mathrm{d}\boldsymbol{x} = 0$, where c_i is the imaginary part of the phase speed.

5. Ripa's theorem

The term in (29) involving the Casimir is positive definite provided

$$C''(Q) = \frac{(\alpha-U)H}{Q_y} > 0 \quad \Longleftrightarrow \quad (\alpha - U)Q_y > 0. \tag{30}$$

This is the analogue of the *Fjørtoft-Pedlosky theorem* for quasi-geostrophic (QG) dynamics. The other terms are positive definite *at small amplitude only* provided

$$(U-\alpha)^2 < gH. \tag{31}$$

This so-called "subsonic" condition states that the Froude number must be less than unity in the reference frame for which the Fjørtoft-Pedlosky condition is satisfied. This is *Ripa's theorem* (Ripa, 1983): a flow is stable if there exists an α for which these conditions hold for all y. Note that the terms involving α cannot be inferred from a Galilean boost, because the SWE are not Galilean invariant (Ripa, 1983).

Ripa's theorem is the only known Arnol'd-type stability theorem (meaning that it is based on conservation laws) for non-resting basic states and non-symmetric unbalanced dynamics. The Fjørtoft-Pedlosky, Rayleigh-Kuo, Charney-Stern, and Arnol'd first and second theorems are for balanced dynamics; static stability is for a resting basic state; centrifugal and symmetric stability are for symmetric disturbances; and the Miles-Howard theorem[2] is for linearized normal-mode disturbances.

Once again it is interesting to consider what was known at that time. Virtually all stability theory was for normal modes. Drazin & Howard (1966) combined the disturbance energy and enstrophy equations to derive a non-parallel version of Fjørtoft's theorem for incompressible 2D flow. Blumen (1970) extended this result to compressible 2D parallel flow — which is isomorphic to the non-rotating SWE — and obtained a forerunner of Ripa's theorem.[3] Neither of these results used conservation laws, although Fjørtoft (1950), in his remarkable paper, did use circulation constraints in deriving his original variational result and was thus implicitly using Casimir invariants. However, it is not at all obvious how to extend Fjørtoft's analysis to finite amplitude or to non-circulation-preserving disturbances. Arnol'd (1965) used conservation laws, including Casimirs, to obtain the Drazin & Howard result, and in that sense anticipated Pedro's insight.

According to the Acknowledgements in Ripa (1983), Pedro learned of Arnol'd (1965) — but not Arnol'd (1966) — from a reviewer. The same reviewer also drew Pedro's attention to Blumen (1968), which he clearly studied because Ripa (1983) quotes Blumen's (1968) erroneous pagination of Arnol'd (1965)! But Blumen evidently had not made the connection between Arnol'd (1965) and conservation laws for general systems, because Blumen (1970) used a completely different technique. Arnol'd (1966), published in an obscure journal, was the paper which actually proved that Arnol'd (1965) is a finite-amplitude result. Holm *et al.* (1983), following Arnol'd (1965, 1966), also published Ripa's theorem — in ignorance of both Ripa (1983) and Blumen (1970). [It is important to emphasize that although both papers were published in the same year, Ripa (1983) was submitted more than 12 months before Holm *et al.* (1983) was submitted.] But a notable difference is that while Arnol'd and followers used energy-Casimir conservation plus (when appropriate) Galilean boosts, Pedro was unique in using momentum conservation as well.

The apparent generalization of Fjørtoft's theorem to non-parallel flows offered by Drazin & Howard (1966) and Arnol'd (1965) turned out to be less than meets the eye. This is because of *Andrews' theorem* (Andrews, 1984), which shows that no Arnol'd-stable non-parallel flow (meaning one whose stability follows from the energy-Casimir invariant) can exist in a domain with translational symmetry. Pedro admired Andrews' theorem enough to put it on a par with Noether's theorem and Arnol'd's theorem, two results which underpinned much of his work (Ripa, 1992). The situation turns out to be even worse than imagined by Andrews: Wirosoetisno & Shepherd

[2] *Pace* Abarbanel *et al.* (1986).

[3] So it is arguable that Ripa's theorem should be called the *Blumen-Ripa* theorem.

(2000) have recently shown that on a boundariless compact domain (e.g. a spheroid), or in a simply connected bounded domain with no net circulation, there can be no Arnol'd-stable flow in the generic case with no symmetry. Since with symmetry Andrews' theorem comes into play, this means that Arnol'd-stable flows are largely restricted to parallel flows, for which the momentum invariants are essential.

One thing I cannot understand is why Pedro did not generalize his wave-wave interaction theory (Ripa, 1981*b*) to non-resting basic states, since in Ripa (1983) he developed the tools to deal with such states. It was unusual for him to miss this kind of connection. Instead, this development was left to Vanneste & Vial (1994).

Returning to Ripa's theorem, the subsonic condition (31) is natural; it prohibits the direct coupling of the balanced (vortical) and unbalanced (gravity wave) components of the flow (cf. Saujani & Shepherd, 2002). When this condition is violated, these two components can couple through phase-locking. Indeed, that is exactly what happens in the case of a linear potential-vorticity jump, which permits one vortical wave and is nonlinearly stable under balanced dynamics, but for which the threshold for a coupled vortical/gravity-wave instability in the SWE is precisely predicted by Ripa's theorem (Ford, 1993). [See also Nore & Shepherd (1997) for the treatment of this instability, and Ripa's theorem, for a weak-wave approximation to the SWE.]

Ripa (1987) derived a circular analogue of Ripa's theorem: a flow is stable if there exists an α such that

$$(U - \alpha r)Q_r > 0, \qquad (U - \alpha r)^2 < gH. \tag{32}$$

However its utility is restricted; for a localized vortex in an unbounded domain, one must take $\alpha = 0$. It follows that no flow can satisfy these two conditions simultaneously (Nore & Shepherd, 1997, Appendix). This is related to Andrews' theorem, because with $\alpha = 0$ only the energy-Casimir invariant is used. It is ironic that Pedro, who had such an admiration for Andrews' theorem, did not notice this. The non-existence result is not hypothetical, as Ford (1994) showed that a circular potential-vorticity jump is unstable for any Froude number.

Ripa (1991) extended Ripa's theorem to multi-layer primitive-equations models. He showed that the subsonic condition becomes more difficult to satisfy with more layers, and that there is no stability theorem in the continuum limit.

One nagging open end concerning Ripa's theorem is that unlike so many Arnol'd-type stability theorems, it does not extend to finite amplitude. There is no problem with condition (30), which is the condition that arises for balanced models and does extend to finite amplitude, but condition (31) seems from (29) to be inherently limited to small amplitude. It remains an open question whether there is a Ripa-stable flow that exhibits nonlinear instability, or whether this loose end is merely technical.

6. Ripa's theorem in balanced models with coasts

It would seem to be a defensible proposition that any shallow-water flow that is stable by Ripa's theorem should also be stable under a reasonable model of balanced dynamics. We might call this the "Ripa test".

Salmon (1983) derived a "nearly geostrophic" balanced model (a.k.a. L_1 dynamics, or the HP model). It is a special case of a general class of constrained Hamiltonian balanced models of the form

$$\frac{\partial \boldsymbol{v}_C}{\partial t} + qh\,\hat{\boldsymbol{z}} \times \boldsymbol{v} = -\boldsymbol{\nabla} b, \qquad \frac{\partial h}{\partial t} + \boldsymbol{\nabla} \cdot (h\boldsymbol{v}) = 0 \tag{33}$$

(Salmon, 1988; Allen & Holm, 1996; McIntyre & Roulstone, 2002). Here $\boldsymbol{v}_C[h]$ is the constraint velocity [Salmon (1983) takes $\boldsymbol{v}_C = \boldsymbol{v}_G$]; $\boldsymbol{v} \to \boldsymbol{v}_C$ in the acceleration term, in $q \equiv [f + \hat{\boldsymbol{z}} \cdot (\boldsymbol{\nabla} \times \boldsymbol{v}_C)]/h$, and in all conservation laws; the generalized Bernoulli function b is determined by the Hamiltonian structure; the boundary conditions are $\boldsymbol{v} \cdot \hat{\boldsymbol{n}} = 0$ plus *another* lateral boundary condition determined by the Hamiltonian structure; and $\boldsymbol{v}$ is determined by the consistency of $\partial \boldsymbol{v}_C/\partial t$ and $\partial h/\partial t$, a procedure which requires the extra lateral boundary condition. For Salmon's model the extra lateral boundary condition is $\boldsymbol{v}_{AG} \times \hat{\boldsymbol{n}} = 0$. This is a semi-geostrophic approximation, valid for weakly-curved boundaries (Ren & Shepherd, 1997). Salmon's model possesses an analogue of Ripa's theorem, with U replaced by U_G (Ren & Shepherd, 1997). This is no surprise, since Ripa's theorem follows from the conservation laws of the model. The subsonic condition is still relevant, because of coastal Kelvin waves (this is not so in the QG model).

The same thing applies for the general class of constrained Hamiltonian balanced models, with U replaced by U_C (Ren & Shepherd, 1997). However the semi-geostrophic (SG) model does not appear to possess such a theorem. Constant-shear (uniform Q) anti-cyclonic SG flow in a channel is unstable for any Froude number, due to interacting coastal Kelvin waves (Kushner, McIntyre & Shepherd 1998). This presumably has something to do with the phase speeds of coastal Kelvin waves in these various balanced models (Figure 1, left panel).

Consider coastal Kelvin waves in fluid of uniform depth H (Allen, Barth & Newberger, 1990). For shallow-water dynamics, $c = \sqrt{gH}$. For Salmon's model,

$$c = \sqrt{gH}\Big(1 + \frac{gHk^2}{f^2}\Big)^{1/2}, \tag{34}$$

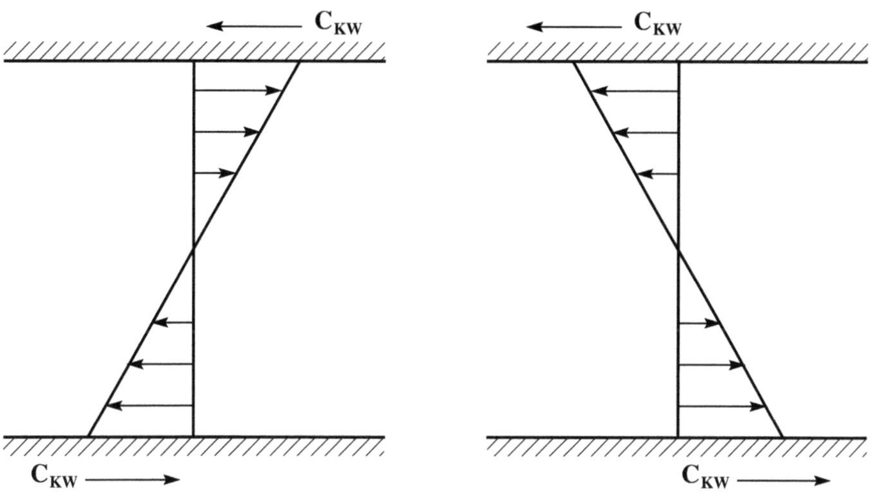

Figure 1. Left panel: Anti-cyclonic shear flow. The coastal Kelvin waves both propagate against the flow. When the Froude number is less than unity, the two coastal Kelvin waves can only phase lock if their phase speed is less than $\sqrt{gH}$, as in the SG model. For models (including the SWE) where the phase speed is $\geq \sqrt{gH}$, phase locking requires a Froude number of unity or greater, which corresponds to the violation of Ripa's stability criterion (31). Right panel: Cyclonic shear flow. The coastal Kelvin waves propagate with the flow. Phase locking of the two coastal Kelvin waves is now impossible no matter what their phase speed is.

which represents a "supersonic" distortion in the large-k limit. For virtually all other known balanced models, including SG,

$$c = \sqrt{gH}\Big(1 + \frac{gHk^2}{f^2}\Big)^{-1/2}, \tag{35}$$

which represents a "subsonic" distortion in the large-k limit. For these latter models, the "subsonic" condition of Ripa's theorem becomes impossible to satisfy! In particular, there is no separation between the vortical and gravity-wave components of the flow in the large-k limit, even for Fr $\ll 1$.

For the general constrained Hamiltonian balanced models, taking zero-PV disturbances (to isolate coastal Kelvin waves), the pseudoenergy $\mathcal{E}$ and pseudomomentum $\mathcal{P}$ are given by

$$\mathcal{E} = \iint \frac{1}{2}\Big\{H|\boldsymbol{v}'_C|^2 + g(h')^2\Big\}\mathrm{d}x\mathrm{d}y, \quad \mathcal{P} = \iint h'u'_C\,\mathrm{d}x\mathrm{d}y. \tag{36}$$

It follows that the phase speed c satisfies

$$c = \frac{\mathcal{E}}{\mathcal{P}} = \frac{\iint \frac{1}{2}\Big\{H(v'_C)^2 + (\sqrt{H}u'_C \pm \sqrt{g}h')^2\Big\}\mathrm{d}x\mathrm{d}y}{\mathcal{P}} \mp \sqrt{gH}, \tag{37}$$

and one sees that $|c| > \sqrt{gH}$ (Ren & Shepherd, 1997). This then provides a physical explanation for the existence of an analogue of Ripa's theorem for any such system, irrespective of its actual form. It is amusing to note that the same simple trick of completing the square, which was the key to Ripa's theorem, is the key step to obtaining this result.

For the SG model there is a stability theorem analogous to the Fjørtoft-Pedlosky theorem, but with the additional condition that $(U - \alpha)$ be cyclonic on lateral boundaries (Kushner & Shepherd, 1995). In that case, the coastal Kelvin waves cannot interact with each other (or with the interior flow), no matter what their phase speed is (Figure 1, right panel). Such flows are also stable in Salmon's model (Ren & Shepherd, 1997).

On the basis of (35), and the results of Kushner *et al.* (1998) for SG dynamics, one may conjecture that for anti-cyclonic shear flows which are provably stable by Ripa's theorem (for shallow water), most known balanced models may be prone to a physically spurious instability at large k involving coastal Kelvin waves.

7. Summary

Hamiltonian fluid dynamics, conservation laws, and stability theory represented a major part of Pedro Ripa's research interests. This was natural, given his background in theoretical physics. His contributions to this field in the early 1980s were at the leading edge of concurrent developments at this exciting time. In particular, his connection between symmetries and Eulerian pseudoenergy and pseudomomentum was without apparent precedent. Ripa's theorem (1983) remains the only known Arnol'd-type stability theorem for non-resting basic states and non-symmetric unbalanced dynamics. Its predictions for coupled vortical/gravity-wave instabilities have been well vindicated. Pedro's use of momentum conservation was also unique at that time. It turns out that stability results without momentum constraints are exceedingly limited, so this was a prescient decision on his part! Ripa's theorem has recently reared its head for balanced models that support coastal Kelvin waves.

Acknowledgements

My research in this area is supported by the Natural Sciences and Engineering Research Council of Canada. I am grateful to the organizers of the Symposium for inviting me to what was a delightful celebration of the achievements of a wonderful scientist and human being.

References

Abarbanel, H. D. I., D. D. Holm, J. E. Marsden and T. Ratiu. Nonlinear stability analysis of stratified fluid equilibria. *Phil. Trans. Roy. Soc. Lond. A*, 318:349–409, 1986.

Allen, J. S., J. A. Barth and P. A. Newberger. On intermediate models for barotropic continental shelf and slope flow fields. Part I: Formulation and comparison of exact solutions. *J. Phys. Oc.*, 20:1017–1042, 1990.

Allen, J. S. and D. D. Holm. Extended-geostrophic Hamiltonian models for rotating shallow water motion. *Physica D*, 98:229–248, 1996.

Andrews, D. G. On the existence of nonzonal flows satisfying sufficient conditions for stability. *Geophys. Astrophys. Fluid Dyn.*, 28:243–256, 1984.

Andrews, D. G. and M. E. McIntyre. On wave-action and its relatives. *J. Fluid Mech.*, 89:647–664, 1978. (See also *Corrigenda*, 95:796.)

Arnol'd, V. I. Conditions for nonlinear stability of stationary plane curvilinear flows of an ideal fluid. *Dokl. Akad. Nauk. SSSR*, 162:975–978. [English translation: *Sov. Math.*, 6:773–777 (1965).]

Arnol'd, V. I. On an a priori estimate in the theory of hydrodynamical stability. *Izv. Vyssh. Uchebn. Zaved. Mat.*, 54(5):3–5, 1966. [English translation: *Amer. Math. Soc. Transl., Ser. 2*, 79:267–269 (1969).]

Benjamin, T. B. Impulse, flow force and variational principles. *IMA J. Appl. Math.*, 32:3–68, 1984.

Blumen, W. On the stability of quasi-geostrophic flow. *J. Atmos. Sci.*, 25:929–931, 1968.

Blumen, W. Shear layer instability of an inviscid compressible fluid. *J. Fluid Mech.*, 40:769–781, 1970.

Drazin, P. G. and L. N. Howard. Hydrodynamic stability of parallel flow of inviscid fluid. *Adv. Appl. Mech.*, 9:1–89, 1966.

Fjørtoft, R. Application of integral theorems in deriving criteria of stability for laminar flows and for the baroclinic circular vortex. *Geofys. Publ.*, 17(6):1–52, 1950.

Ford, R. Gravity wave generation by vortical flows in a rotating frame. Ph.D. thesis, Department of Applied Mathematics and Theoretical Physics, Cambridge University, 1993.

Ford, R. The instability of an axisymmetric vortex with monotonic potential vorticity in rotating shallow water. *J. Fluid Mech.*, 280:303–334, 1994.

Holliday, D. and M. E. McIntyre. On potential energy density in an incompressible, stratified fluid. *J. Fluid Mech.*, 107:221–225, 1981.

Holm, D. D., J. E. Marsden, T. Ratiu and A. Weinstein. Nonlinear stability conditions and a priori estimates for barotropic hydrodynamics. *Phys. Lett.*, 98A:15–21, 1983.

Holm, D. D., J. E. Marsden, T. Ratiu and A. Weinstein. Nonlinear stability of fluid and plasma equilibria. *Phys. Reports*, 123:1–116, 1985.

Killworth, P. D. and M. E. McIntyre. Do Rossby-wave critical layers absorb, reflect, or over-reflect? *J. Fluid Mech.*, 161:449–492, 1985.

Kushner, P. J., M. E. McIntyre and T. G. Shepherd. Coupled Kelvin-wave and mirage-wave instabilities in semi-geostrophic dynamics. *J. Phys. Oc.*, 28:513–518, 1998.

Kushner, P. J. and T. G. Shepherd. Wave-activity conservation laws and stability theorems for semi-geostrophic dynamics. Part 2: Pseudoenergy-based theory. *J. Fluid Mech.*, 290:105–129, 1995.

McIntyre, M. E. and I. Roulstone. Are there higher-accuracy analogues of semi-geostrophic theory? In *Large-Scale Atmosphere-Ocean Dynamics I* (J. Norbury and I. Roulstone, eds.), Cambridge University Press, pp. 300–363, 2002.

Morrison, P. J. Poisson brackets for fluids and plasmas. In *Mathematical Methods in Hydrodynamics and Integrability in Dynamical Systems* (M. Tabor and Y. M. Treve, eds.), AIP Conf. Proc., 88:13–46, 1982.
Morrison, P. J. Hamiltonian description of the ideal fluid. *Rev. Mod. Phys.*, 70:467–521, 1998.
Nore, C. and T. G. Shepherd. A Hamiltonian weak-wave model for shallow-water flow. *Proc. Roy. Soc. Lond. A*, 453:563–580, 1997.
Ren, S. and T. G. Shepherd. Lateral boundary contributions to wave-activity invariants and nonlinear stability theorems for balanced dynamics. *J. Fluid Mech.*, 345:287–305, 1997.
Ripa, P. Symmetries and conservation laws for internal gravity waves. In *Nonlinear Properties of Internal Waves* (B. J. West, ed.), AIP Conf. Proc., 76:281–306, 1981*a*.
Ripa, P. On the theory of nonlinear wave-wave interactions among geophysical waves. *J. Fluid Mech.*, 103:87–115, 1981*b*.
Ripa, P. General stability conditions for zonal flows in a one-layer model on the β-plane or the sphere. *J. Fluid Mech.*, 126:463–489, 1983.
Ripa, P. On the stability of elliptical vortex solutions of the shallow water equations. *J. Fluid Mech.*, 183:343–363, 1987.
Ripa, P. General stability conditions for a multi-layer model. *J. Fluid Mech.*, 222: 119–137, 1991.
Ripa, P. A tale of three theorems. *Rev. Mex. Fís.*, 38:229–242, 1992.
Salmon, R. Practical use of Hamilton's principle. *J. Fluid Mech.*, 132:431–444, 1983.
Salmon, R. Semigeostrophic theory as a Dirac-bracket projection. *J. Fluid Mech.*, 196:345–358, 1988.
Salmon, R. *Lectures on Geophysical Fluid Dynamics.* Oxford University Press, 1998.
Saujani, S. and T. G. Shepherd. Comments on "Balance and the slow quasimanifold: some explicit results". *J. Atmos. Sci.*, 59:2874–2877, 2002.
Shepherd, T. G. Symmetries, conservation laws, and Hamiltonian structure in geophysical fluid dynamics. *Adv. Geophys.*, 32:287–338, 1990.
Shepherd, T. G. A unified theory of available potential energy. *Atmos.-Ocean*, 31:1–26, 1993.
Vanneste, J. and F. Vial. On the nonlinear interactions of geophysical waves in shear flows. *Geophys. Astrophys. Fluid Dyn.*, 78:121–144, 1994.
Wirosoetisno, D. and T. G. Shepherd. On the existence of two-dimensional Euler flows satisfying energy-Casimir stability criteria. *Phys. Fluids*, 12:727–730, 2000.

DEEP OCEAN INFLUENCE ON UPPER OCEAN BAROCLINIC INSTABILITY SATURATION

M. J. OLASCOAGA AND F. J. BERON-VERA
RSMAS, University of Miami
4600 Rickenbacker Causeway
Miami, FL 33149-1098, U.S.A.

J. SHEINBAUM
Departamento de Oceanografía Física, CICESE, México

Abstract. In this paper we extend earlier results regarding the effects of the lower layer of the ocean (below the thermocline) on the baroclinic instability within the upper layer (above the thermocline). We confront quasigeostrophic baroclinic instability properties of a 2.5-layer model with those of a 3-layer model with a very thick deep layer, which has been shown to predict spectral instability for basic state parameters for which the 2.5-layer model predicts nonlinear stability. We compute and compare maximum normal-mode perturbation growth rates, as well as rigorous upper bounds on the nonlinear growth of perturbations to unstable basic states, paying particular attention to the region of basic state parameters where the stability properties of the 2.5- and 3-layer model differ substantially. We found that normal-mode perturbation growth rates in the 3-layer model tend to maximize in this region. We also found that the size of state space available for eddy-amplitude growth tends to minimize in this same region. Moreover, we found that for a large spread of parameter values in this region the latter size reduces to only a small fraction of the total enstrophy of the system, thereby allowing us to make unambiguous assessments of the significance of the instabilities.

Key words: layer model, reduced-gravity, stability, instability saturation

1. Introduction

Observations indicate that most of the world oceans variability is confined in a thin layer limited from below by the permanent thermocline. There, the density is approximately uniform in the vertical but has important horizontal gradients. The latter imply the existence of a considerable reservoir of potential energy within this layer, stored in the isopycnals tilt and available to feeding baroclinic instability processes (Gill *et al.*, 1974). Haine and Marshall (1998) have argued that these processes are of outmost importance for the dynamics of the upper ocean. These authors pointed out

O.U. Velasco Fuentes et al. (eds.), Nonlinear Processes in Geophysical Fluid Dynamics, 15-28.

that baroclinic instability waves can be efficient transport agents capable of stopping convective processes, thereby exerting a large influence in the thermodynamic state of the ocean.

Because the total depth of the ocean is much larger than that of the upper thermocline layer, the reduced-gravity setting has been commonly adopted to studying the upper ocean baroclinic instability (Fukamachi *et al.*, 1995; Ripa, 1995; Young and Chen, 1995; Beron-Vera and Ripa, 1997; Ripa, 1999b; Olascoaga and Ripa, 1999; Ripa, 1999c; Ripa, 1999a; Ripa, 2000a; Ripa, 2000b; Ripa, 2001). In this setting the active fluid layer is considered as floating on top of a quiescent, infinitely deep layer. Olascoaga (2001) showed, however, that a thick—but finite—abyssal active layer can substantially alter the stability properties of the upper ocean for certain baroclinic zonal flows, such as the Atlantic North Equatorial Current (ANEC) (Beron-Vera and Olascoaga, 2003). Olascoaga (2001) considered spectral (i.e. linear, normal-mode), formal (or Arnold), and nonlinear (or Lyapunov) stability (Holm et al., 1985; cf. also McIntyre and Shepherd, 1987) in a 3-layer quasigeostrophic (QG) model. Primary attention was given to the limit of a very thick bottom layer. The stability results were compared with those from a reduced-gravity 2-layer (or 2.5-layer) model (Olascoaga and Ripa, 1999), and assessments were made of the influence of the deep ocean on upper ocean baroclinic instability.

To make further assessments, in this paper we turn our attention to baroclinic instability saturation. Employing Shepherd's (1988) method, we establish and confront rigorous bounds on nonlinear instability saturation in 2.5- and 3-layer models. This method, which builds on the existence of a nonlinear stability theorem, has been previously used to compute saturation bounds in 2.5- (Olascoaga and Ripa, 1999) and 3-layer (Paret and Vanneste, 1996) models. In addition to considering more general model configurations than in these earlier works, we focus on the size of state space available for the growth of eddies in the region of parameter space where the models present discrepancies in their stability properties. Also, unlike Paret and Vanneste (1996), who computed numerical energy-norm bounds based on both Arnold's first and second theorems, we derive analytical expressions for enstrophy-norm bounds based on Arnold's first theorem. Maximum normal-mode perturbation growth rates in both 2.5- and 3-layer models are also calculated and contrasted.

The reminder of the paper has the following organization. Section 2 presents the 3-layer model, from which the 2.5-layer model follows as a limiting case. Normal-mode perturbation growth rates are computed in § 3, along with an exposition of the main results of formal and nonlinear stability analyses. Nonlinear saturation bounds are then derived in § 4. We want to remark that the number of basic state parameters that define a 3-

layer flow is too large to be explored in full detail. To facilitate the analysis we reduce in some cases this number by fixing certain parameters to values that can be taken as "realistic," because of being found appropriate for a region of the ocean mainly dominated by the ANEC, which is a good example of a major zonal current. Section 5 presents a discussion and the conclusions. Appendices A and B are reserved for mathematical details concerning the computation of the saturation bounds in the 3- and 2.5-layer models, respectively.

2. The Layer Models

Let $\mathbf{x}$ denote the horizontal position with Cartesian coordinates x (eastward) and y (northward), let t be the time, and let D be an infinite (or periodic) zonal channel domain on the β plane with coasts at $y = \pm\frac{1}{2}W$. The unforced, inviscid evolution equations for QG motions in a ***3-layer model***, with rigid surface and flat bottom, are given by (cf. e.g. Ripa, 1992)

$$\partial_t q_i = \hat{\mathbf{z}} \cdot \nabla q_i \times \nabla \psi_i, \quad \dot{\gamma}_i^{\pm} = 0, \tag{1a}$$

where ψ_i, being a nonlocal function of $\mathbf{q} := (q_i)^{\mathrm{T}}$ and $\boldsymbol{\gamma} := (\gamma_i^{\pm})^{\mathrm{T}}$, is uniquely determined by

$$\nabla^2 \psi_i - \sum_j \mathsf{R}_{ij} \psi_j = q_i - f \tag{1b}$$

on D, where

$$\mathsf{R} := \frac{1}{(1+r_1)R^2} \begin{bmatrix} 1 & -1 & 0 \\ -r_1 & (1+s)r_1 & -sr_1 \\ 0 & -s\frac{r_1 r_2}{1+r_1} & s\frac{r_1 r_2}{1+r_1} \end{bmatrix}, \tag{1c}$$

and

$$\int \mathrm{d}x \, \partial_y \psi_i = \gamma_i^{\pm}, \quad \partial_x \psi_i = 0 \tag{1d}$$

at $y = \pm\frac{1}{2}W$. Here, $q_i(\mathbf{x}, t)$, $\psi_i(\mathbf{x}, t)$, and $\gamma_i^{\pm} = \text{const.}$ denote the QG potential vorticity, streamfunction, and Kelvin circulation along the boundaries of the channel, respectively, in the top ($i = 1$), middle ($i = 2$) and bottom ($i = 3$) layers. The Coriolis parameter is represented as $f = f_0 + \beta y$, the Nabla operator $\nabla = (\partial_x, \partial_y)$, and $\hat{\mathbf{z}}$ denotes the vertical unit vector. The quantities

$$R^2 := \frac{g_1 \bar{H}_1 \bar{H}_2}{f_0^2 \bar{H}}, \quad s := \frac{g_1}{g_2}, \quad r_1 := \frac{\bar{H}_1}{\bar{H}_2}, \quad r_2 := \frac{\bar{H}}{\bar{H}_3}, \tag{2}$$

where g_i is the buoyancy jump at the interface of the i-th and $(i+1)$-th layers, and $\bar{H} := \bar{H}_1 + \bar{H}_2$ with $\bar{H}_i$ the i-th layer reference thickness.

The ***2.5-layer model*** follows from (1) in the limit of infinitely thick $(r_2 \to 0)$ and quiescent $(\psi_3 \to 0)$ lower layer. In the latter case, $(1 + r_1)(r_1/s)^{1/2}R$ and R are equal to the first (equivalent barotropic) and second (baroclinic) deformation radius, respectively, in the limit of weak internal stratification $(s \to 0)$.

The evolution of system (1) is constrained by the conservation of ***energy***, ***zonal momentum***, and an infinite number of vorticity-related ***Casimirs***, which are given by

$$\mathcal{E} := -\tfrac{1}{2}\langle \psi_i q_i \rangle, \quad \mathcal{M} := \langle y q_i \rangle, \quad \mathcal{C} := \langle C_i(q_i) \rangle \tag{3}$$

(modulo Kelvin circulations along the boundaries), where $C_i(\cdot)$ is an arbitrary function and $\langle \cdot \rangle := \sum_i \bar{H}_i \int_D \mathrm{d}^2\mathbf{x}\,(\cdot)$.

3. Spectral, Formal, and Nonlinear Stability

In this paper we deal with the stability of a ***basic state***, i.e. equilibrium or steady solution of (1), of the form

$$\Psi_i = -U_i y, \tag{4}$$

which represents a baroclinic zonal flow. Here, $U_i = \sum_{i_1=i}^{2} g_{i_1} \sum_{i_2=1}^{i_1} H_{i_2,y}/f_0$ = const., where $H_i(y)$ is the thickness of the i-th layer in the basic state, whereas U_3 is an arbitrary constant (set here to zero with no loss of generality). The following six parameters are enough to characterize the solutions of the 3-layer model stability problem:

$$\kappa := \sqrt{k^2 + l^2}R,\ s,\ b := \frac{\beta \bar{H}_1 \bar{H}_2}{f_0 \bar{H} H_{1,y}} \equiv \frac{\beta R^2}{U_\mathrm{s}},\ b_\mathrm{T} := \frac{H_{,y}}{H_{1,y}} \equiv \frac{sU_2}{U_\mathrm{s}},\ r_1,\ r_2. \tag{5}$$

The first parameter, κ, is a nondimensional ***wavenumber*** of the perturbation, where k and l are the zonal and meridional wavenumbers, respectively. The second parameter, s, is a nondimensional measure of the ***stratification***. The third parameter, b, is a ***planetary Charney number***, namely the ratio of the planetary β effect and the topographic β effect due to the geostrophic slope of the upper interface. Here, $U_\mathrm{s} := U_1 - U_2$ is the velocity jump at the interface (i.e. the current vertical "shear"). The fourth parameter, b_T, is a ***topographic Charney number*** given by the ratio of the topographic β effects due to the geostrophic slopes of the lower and upper interfaces. Finally, the fifth (resp., sixth), r_1 (resp., r_2), parameter is the ***aspect ratio*** of the upper to intermediate (resp., upper-plus-intermediate to lower) reference layer thicknesses. The problem considered by Olascoaga

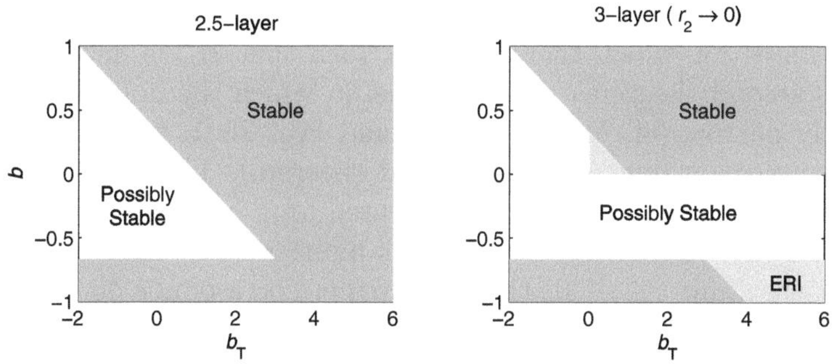

Figure 1. Stability/instability regions in the planetary b vs. topographic b_{T} Charney numbers space. Dark-shaded regions are the locus of positive-definite pseudo energy–momentum integrals. In the blank regions a pseudo energy–momentum integral can be found to be negative definite if the zonal channel flow is narrow enough. In the light-shaded regions no pseudo energy–momentum integral can be proved to be sign definite.

(2001) had $r_1 = 1$. In turn, the 2.5-layer problem treated by Olascoaga and Ripa (1999), which can be recovered upon making $r_2 \to 0$ and $\psi_3 \to 0$, also had $r_1 = 1$.

Choosing a Casimir such that $\delta(\mathcal{E} + \mathcal{C} - \alpha\mathcal{M}) = 0$ for any constant α, the ***pseudo energy–momentum***,

$$\mathcal{H}_\alpha[\delta\mathbf{q}] := (\Delta - \delta)(\mathcal{E} + \mathcal{C} - \alpha\mathcal{M}) = \mathcal{E}[\delta\boldsymbol{\psi}] + \tfrac{1}{2}\langle C_{i,QQ}\delta q_i^2\rangle, \tag{6}$$

where $\boldsymbol{\psi} := (\psi_i)^{\mathrm{T}}$, is an exact finite-amplitude invariant, quadratic in the ***perturbation*** $\delta q_i(\mathbf{x}, t)$ on the basic state potential vorticity $Q_i(y)$. Here, the symbols Δ and δ stand for total and first variations of a functional, respectively, and $C_i(Q_i) = \int \mathrm{d}Q_i\,(\alpha - U_i)Y(Q_i)$ [Y is the meridional coordinate of an isoline of Q_i], where

$$Q_1 = f_0 + (b + \rho)\, yU_{\mathrm{s}}/R^2, \tag{7}$$
$$Q_2 = f_0 + [b + \rho r_1 (b_{\mathrm{T}} - 1)]\, yU_{\mathrm{s}}/R^2, \tag{8}$$
$$Q_3 = f_0 + \left(b - \rho^2 r_1 r_2 b_{\mathrm{T}}\right) yU_{\mathrm{s}}/R^2, \tag{9}$$

with $\rho := (1 + r_1)^{-1}$. Arnold's (1965; 1966) method for proving ***formal stability*** of Q_i relies upon the sign-definiteness of $\mathcal{H}_\alpha$. For evaluating the latter, it is useful to make the Fourier expansion $\delta\mathbf{q} = \sum_{k,l}\hat{\mathbf{q}}(t)\mathrm{e}^{\mathrm{i}kx}\sin ly$, which implies $\mathcal{H}_\alpha = \frac{1}{2}\sum_{k,l}(\hat{\mathbf{q}}^*)^{\mathrm{T}}\mathbf{H}_\alpha\hat{\mathbf{q}}$ for certain matrix $\mathbf{H}_\alpha(\kappa, s, b, b_{\mathrm{T}}, r_1, r_2)$ (cf. Beron-Vera and Olascoaga, 2003, § 2.2.1), so that the sign of $\mathcal{H}_\alpha$ is determined from the inspection of the elements of $\mathbf{H}_\alpha$ (cf. Mu et al., 1994; Paret and Vanneste, 1996; Ripa, 2000a).

In Figure 1 the regions of the (b, b_{T})-space labeled "Stable" correspond to basic states for which there exists α such that $\mathcal{H}_\alpha$ is positive definite

(Arnold's first theorem). The regions labeled "Possibly Stable" are locus of basic states for which there exists α such that $\mathcal{H}_\alpha$ is negative definite (Arnold's second theorem) if the channel in which the flow is contained is sufficiently narrow (cf. Mu, 1998; Mu and Wu, 2001, for details on optimality issues regarding Arnold's second theorem). The results, which are independent of the choice of s, are presented for $r_1 = 0.5$, a value estimated for the ANEC. The r.h.s. panel in this figure corresponds to the 3-layer model in the limit $r_2 \to 0$; the l.h.s. panel corresponds to the 2.5-layer model. Clearly, as $r_2 \to 0$ the 3-layer model stable region does not reduce to that of the 2.5-layer model; it also requires $\delta\psi_3 \to 0$ (Olascoaga, 2001). In the regions labeled "ERI" no sign-definite $\mathcal{H}_\alpha$ can be found. Consequently, the corresponding states are always unstable either through normal-mode perturbations or explosive resonant interaction (ERI) (Vanneste, 1995). By contrast, in the 2.5-layer instability problem all basic states subject to ERI are spectrally unstable. Finally, ***nonlinear stability*** can be proven for all formally stable states. Namely, the departure from these basic states can be bounded at all times by a multiple of the initial distance.

For ***spectral stability*** a perturbation is assumed to be infinitesimal and with the structure of a normal mode, i.e. $\hat{\mathbf{q}} = \varepsilon\tilde{\mathbf{q}}e^{-ikct} + O(\varepsilon^2)$, where $\varepsilon \to 0$. Nontrivial solutions for $\tilde{\mathbf{q}}$, which satisfies $\mathbf{H}_c\tilde{\mathbf{q}} = 0$, require condition $\det \mathbf{H}_c = 0$ to be fulfilled. This implies the eigenvalue $c(\kappa; s, b, b_T, r_1, r_2)$ to satisfy $P(c) = 0$, where $P(\cdot)$ is a cubic characteristic polynomial (cf. Beron-Vera and Olascoaga, 2003; appendix B).

Figure 2 shows the 3-layer model maximum perturbation growth rate, $\max_\kappa\{\kappa \operatorname{Im} c\}$, for $r_1 = 0.5$, and different values of parameters r_2 and s in the planetary b vs. topographic b_T Charney numbers space. (In constructing this figure and the following figures one is of course assuming that the channel in which the flow evolves is sufficiently wide.) In general, the maximum perturbation growth rate increases with increasing r_2 and decreasing s. As b_T increases, the maximum perturbation growth rate tends to achieve the largest values in the region where the 2.5-layer model is nonlinearly stable as a consequence of Arnold's first theorem, even for (realistically) small values of r_2 as depicted in the bottom panels of the figure.

Figure 3 shows instability regions in (κ, b_T)-space for $b = -0.35$, a value estimated for the ANEC, and the same values of parameters r_1, r_2, and s as in figure 2. The area of the region of possible wavenumbers for destabilizing perturbations in the 3-layer model increases with increasing r_2 and decreases with decreasing s. Thus the likelihood of these instabilities, which are not present in the 2.5-layer model, appear to be quite limited because they are confined only to small bands of wavenumbers for small s and r_2. Yet the perturbation growth rates in these bands are not negligible even for very small values of r_2 according to Olascoaga (2001), who explained these

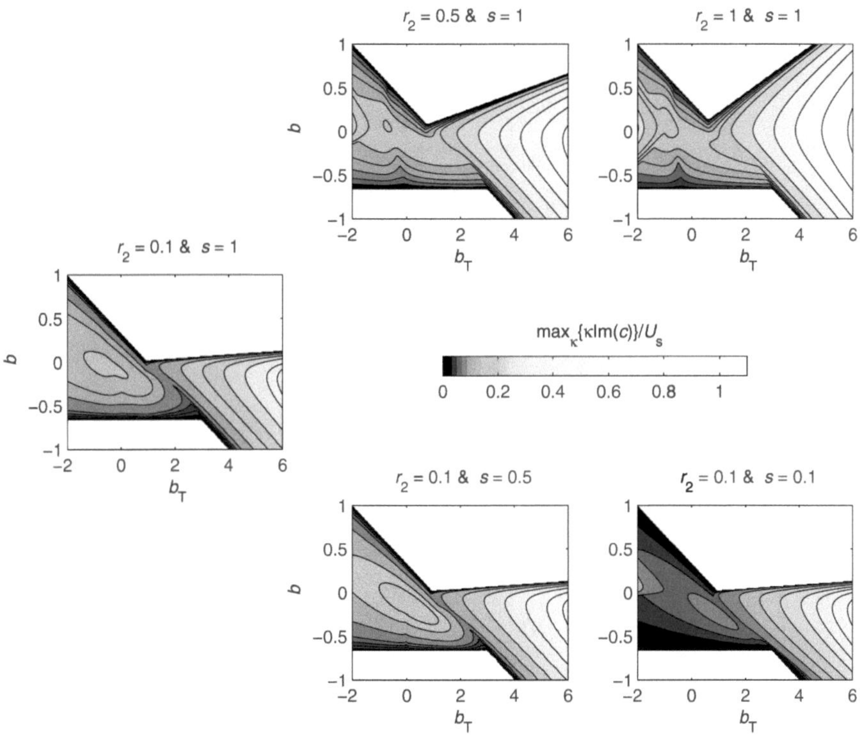

Figure 2. Maximum 3-layer model perturbation growth rate in the planetary b vs. topographic b_T Charney numbers space for a fixed value of the aspect ratio $r_1(= 0.5)$ of the upper to intermediate reference layer thicknesses, and different values of the aspect ratio r_2 of the upper-plus-intermediate to lower reference layer thicknesses and the stratification parameter s.

instabilities as a result of the resonant interaction between a neutral mode of the 2.5-layer model instability problem and a short Rossby wave in the bottom layer. In the next section we will see, however, how the existence of nonlinearly stable states contributes to arrest the eddy-amplitude growth, restricting the significance of these instabilities, at least for certain basic state parameters.

Let us finally turn our attention to the region of (b, b_T)-space where the two models allow for the possibility of instability. The 2.5-layer model maximum perturbation growth rates acquire their largest values for $b = \frac{1}{2}\rho[(1 - b_T)r_1 - 1] = -\frac{1}{6}(1 + b_T)$ and $b_T < 1 + r_1^{-1} = 3$ as $s \to 0$, which corresponds to an exact cancellation of the planetary and topographic β effects (i.e. $Q_1 + Q_2 = 0$). This result does not hold for the 3-layer model because of the presence of instabilities not present in the 2.5-layer problem. The latter instabilities are confined to very narrow branches in the (κ, b_T)-space

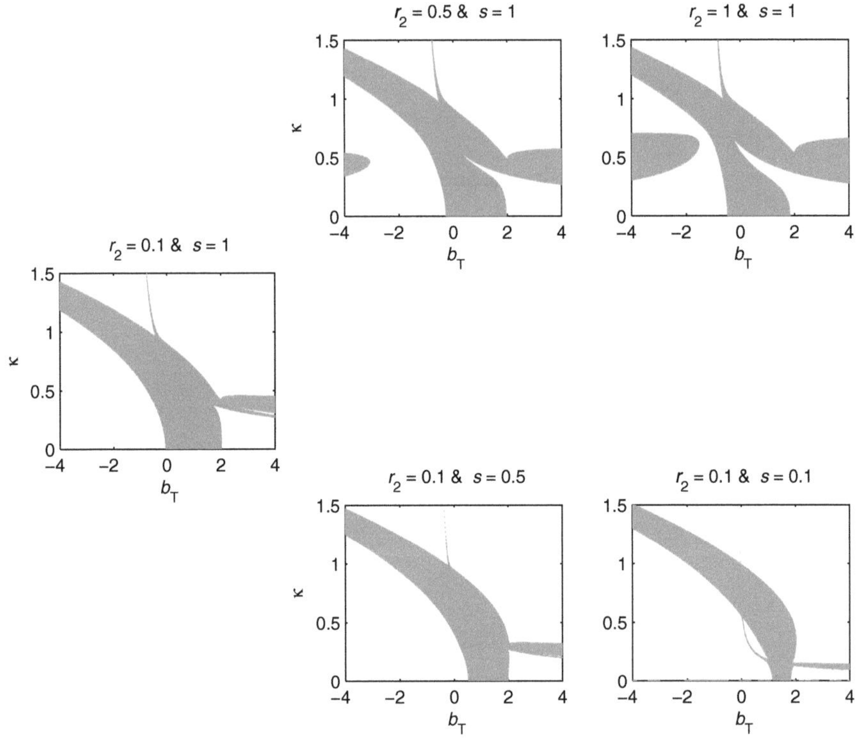

Figure 3. Three-layer model instability regions in the nondimensional wavenumber κ vs. topographic Charney number b_T space for a fixed value of the planetary Charney number $b(= -0.35)$ and the same values of the aspect ratio parameters r_1 and r_2, and the stratification parameter s as in the previous figure.

and were also explained by Olascoaga (2001) as a result of the resonant interplay of a neutral mode in the 2.5-layer model instability problem and a short Rossby wave in the abyssal layer. The magnitude of the maximum perturbation growth rates associated with these instabilities is larger than that of the 2.5-layer model maximum perturbation growth rates. However, we will see that the fraction of the total enstrophy of the system available for eddy-amplitude growth can be much smaller in the 3-layer model than in the 2.5-layer model for certain unstable basic states.

4. Upper Bounds on Instability Saturation

When a basic state is unstable, a priori upper bounds on the finite-amplitude growth of the perturbation to this state can be obtained using Shepherd's (1988) method. This method relies upon the existence of a nonlinear stability theorem, and the bounds are given in terms of the "distance" between the unstable basic state, Q_i^{U} say, and the nonlinearly stable state, Q_i^{S} say, in the infinite-dimensional phase space.

Let $\delta q_i'(\mathbf{x}, t)$ be that part of the perturbation representing the "waves" or "eddies," which result upon subtracting from the perturbation its zonal (i.e. along-channel) average. Let $\mathbb{S}$ denote the space of all possible nonlinearly stable basic states, let $\|\mathbf{a}\|^2 := \langle a_i^2 \rangle / \mathcal{Z}$, where $\mathcal{Z} := \langle (Q_i^{\mathrm{U}})^2 \rangle$, and assume $q_i \approx Q_i^{\mathrm{U}}$ at $t = 0$ so that $\mathcal{Z}$ corresponds to the ***total enstrophy*** of the system. According to Shepherd (1988), a rigorous ***enstrophy-norm upper bound on eddy-amplitude growth***, based on Arnold's first theorem, must have the form

$$\|\delta \mathbf{q}'\|^2 \leq \frac{1}{\mathcal{Z}} \min_{Q_i^{\mathrm{S}} \in \mathbb{S}} \left\{ \frac{\max Q_{i,y}^{\mathrm{S}}}{Q_{i,y}^{\mathrm{S}}} \left\langle \left(Q_i^{\mathrm{U}} - Q_i^{\mathrm{S}} \right)^2 \right\rangle \right\}. \tag{10}$$

We want to mention that bounds—not treated here—on the zonal-mean perturbation, $\delta \bar{q}_i(y, t) := \delta q_i - \delta q_i'$, or the total perturbation can also be derived as δq_i, $\delta \bar{q}_i$, and $\delta q_i'$ satisfy the Pythagorean relationship $\|\delta \mathbf{q}\|^2 = \|\delta \bar{\mathbf{q}}\|^2 + \|\delta \mathbf{q}'\|^2$ (cf. Ripa, 1999c; Ripa, 2000b).

Figure 4 shows the tightest bound on instability saturation corresponding to the 3-layer model (thick curves, cf. appendix A) as a function of the planetary Charney number b, for aspect ratios $r_1 = 0.5$ and $r_2 = 0.1$, and various values of the topographic Charney number b_{T}. The focus on the small value $r_2 = 0.1$ will allow us to make comparisons with the stability properties of the 2.5-layer model. The latter model's bound (cf. appendix B) is also plotted in the figure (dashed lines) assuming $r_1 = 0.5$ and the same values of b_{T}. It is important to remark that the 2.5-layer bound does not follow from that of the 3-layer model in the limit $r_2 \to 0$; it also requires $\psi_3 = 0$ (Olascoaga, 2001). Both the 2.5- and 3-layer model bounds are independent of the stratification parameter s. The 2.5-layer model bound curves are only present in the upper three panels of the figure because the 2.5-layer model predicts nonlinear stability as a consequence of Arnold's first theorem for $b_{\mathrm{T}} > 1 + r_1^{-1} = 3$ and any value of b (cf. also Figure 2, lower-right panel).

Vertical dashed lines in each panel of Figure 4 indicate the values of b for marginal stability, denoted by $b_\pm$. In the 3-layer model, $b_- = \min\{-\rho, \rho r_1(1 - b_{\mathrm{T}})\}$ and $b_+ = \max\{\rho r_1(1 - b_{\mathrm{T}}), \rho^2 r_1 r_2 b_{\mathrm{T}}\}$, whereas for the 2.5-layer model, $b_- = -\rho$ and $b_+ = \rho r_1(1 - b_{\mathrm{T}})$. Note that in the 2.5-layer model, while the b_- marginal stability value remains fixed at $b \approx -0.66667$, the b_+ moves toward smaller values as b_{T} increases, until it collapses with b_- at $b_{\mathrm{T}} = 3$ (not shown in the figure). For $b > b_+$ and $b < b_-$ the basic flow in both the 2.5- and 3-layer models is nonlinearly stable. For $b_+ < b < b_-$ the basic flow is unstable unless the zonal channel flow is narrow enough for Arnold's second stability theorem to be fulfilled. The latter is not always

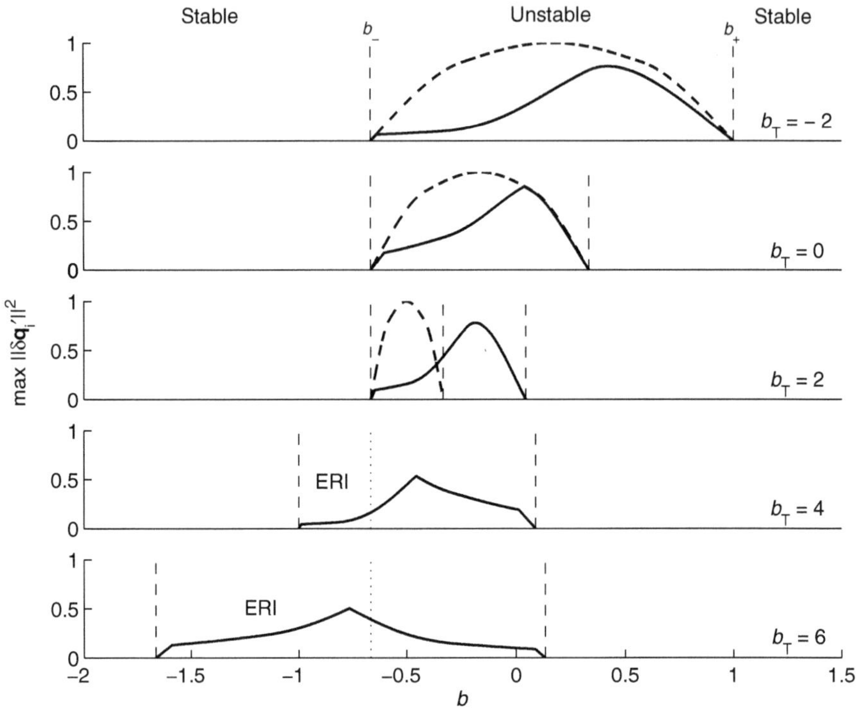

Figure 4. Fraction of the total potential enstrophy available for eddy-amplitude growth in the 2.5- (dashed lines) and 3- (solid lines) layer models as a function of the planetary Charney number b, with aspect ratios $r_1 = 0.5$ and $r_2 = 0.1$ (the 2.5-layer model has $r_2 \to 0$), for different values of the topographic Charney number b_{T}.

true in the 3-layer model case, however, since there is a possibility that a spectrally stable basic state could become unstable through ERI.

Three-layer surface-confined flows are susceptible to suffer more destabilization than 2.5-layer flows. However, the state space available, (determined by the fraction of total potential enstrophy), for eddy-amplitude growth in the 3-layer model tends to be smaller than the space available in the 2.5-layer model, at least for certain basic state parameters. This is evident in the upper three panels of Figure 4. There is an overall tendency of the 3-layer model bound to decrease as b_{T} increases. Moreover, for a large set of parameters this bound reduces to only a small fraction of the total enstrophy of the system. In these cases, the significance of the associated instabilities is relative. On the other hand, there are basic state parameters for which this fraction is not negligible. As an example not shown in the figure, for $b = -0.35$ and $b_{\mathrm{T}} = 2.5$, which are appropriate for a region similar to the ANEC, the fraction of total enstrophy is about 45%, which is not negligible. Of course, when the upper bounds are not small enough, no unambiguous conclusion can be drawn about the significance of an instability.

Figure 4 also shows that the 3-layer model bound can be significantly small for certain potentially ERI unstable flows (cf. lower two panels in the figure). This also allows us to make an unambiguous assessment of the significance of these type of instabilities in the sense that they can be certainly negligible for some basic state parameters.

Before closing this section, two points deserve additional discussion. First, the result that the bounds for the 3-layer model with a very thick deep layer are smaller than the 2.5-layer model bounds in the region of parameters where the two models share similar instability properties might seem at odds with the fact that the 3-layer model is less constrained than the 2.5-layer model, which allows for the development of more unstable states. However, we believe that this result should not be surprising inasmuch as the space over which the minimization is carried out is larger in the 3-layer model than in the 2.5-layer model, which offers the possibility of finding tighter bounds (cf. Olascoaga, 2001). Second, Paret and Vanneste (1996) were not able to draw a conclusion on the significance of ERI instability as in the present paper. These authors computed energy-norm saturation bounds, according to both Arnold's first and second theorems, using numerical minimization algorithms. These bounds, whose analytical computation appears to be too difficult, were not found to minimize at basic state parameters for which ERI instability is possible. The analytical minimization involved in the derivation of the enstrophy-norm of this paper has shown that the tightest bounds are obtained using stable basic states whose parameters have quite spread numerical values. The minimization thus requires to search for a solution in a considerably large space, making numerical computations extremely expensive. This might explain the difficulty of Paret and Vanneste (1996) to find tighter bounds for potentially ERI unstable basic flows.

5. Concluding Remarks

A previous study showed that the quasigeostrophic baroclinic instability properties of a surface-confined zonal current may differ substantially between a 2.5-layer model and a 3-layer, if the former is considered to be a simplified 3-layer model with a very thick deep layer. For certain basic state parameters, the 2.5-layer model predicts nonlinear stability whereas the 3-layer model spectral instability. That study thus suggested that the effects of the deep ocean on the baroclinic instability of the upper thermocline layer of the ocean may be important for certain currents.

In this paper we have made further assessments of the importance of the deep ocean on upper baroclinic instability. We have achieved this by analyzing (i) maximum normal-mode perturbation growth rates and (ii) rigorous enstrophy-norm upper bounds on the growth of perturbations to unstable basic states, in both 2.5- and 3-layer models of baroclinic instability. The new results show that instabilities, which the 3-layer model predicts in the region of basic state parameters where the 2.5-layer model predicts nonlinear stability, appear to maximize their growth rates. At the same time, however, the saturation bounds tend to minimize in this same region of basic state parameters, thereby reducing the size of state space available for eddy-amplitude growth. Moreover, for a large subset of parameters in the region, the latter reduces to only a small fraction of the total enstrophy of the system. In these cases we have been able to make unambiguous assessments of the significance of the associated instabilities in the sense that they can be certainly negligible.

We close remarking that the important issue of making assessments of the accuracy of the saturation bounds as predictors of equilibrated eddy amplitudes is still largely open. This cannot be addressed without performing direct numerical simulations. The importance of this subject relies upon the potential of the bounds in the architecture of transient-eddy parametrization schemes. The treatment of these issues will be the subject of future investigations.

Acknowledgements

We thank Ted Shepherd and an anonymous reviewer for helpful comments. M.J.O. and F.J.B.V. were supported by NSF (USA). J.S. was supported by CICESE's core funding and by CONACyT (México).

A. Three-Layer Model Bounds

Upon minimizing the r.h.s. of (10) over all stable states, we have been able to find, in addition to the trivial bound $\max \|\delta \mathbf{q}'\|^2 = 1$, various sets of possible bounds. A first set involves 9 possibilities, for which $Q^{\mathrm{S}}_{I,y} = \max\{Q^{\mathrm{S}}_{i,y}\}$ and is given by

$$\max \|\delta \mathbf{q}'\|^2 = \begin{cases} -Q^{\mathrm{U}}_{i,y}(Q^{\mathrm{U}}_{I,y} + Q^{\mathrm{U}}_{i,y}) \\ -\sum_i Q^{\mathrm{U}}_{i,y} \sum_j Q^{\mathrm{U}}_{j,y} \end{cases} \tag{A.1}$$

$\div\frac{1}{4}\sum_j(Q^{\mathrm{U}}_{j,y})^2$, for $i \neq I = 1,2,3$. A second set involves other 9 possibilities, for which $\max\{Q^{\mathrm{S}}_{i,y}\} = Q^{\mathrm{S}}_{I_1,y} = Q^{\mathrm{S}}_{I_2,y}$ and is given by

$$\max\|\delta\mathbf{q}'\|^2 = \begin{cases} (Q^{\mathrm{U}}_{I_1,y})^2 + (Q^{\mathrm{U}}_{I_2,y})^2 \\ \frac{1}{2}(Q^{\mathrm{U}}_{I_1,y} - Q^{\mathrm{U}}_{I_2,y})^2 \\ \frac{1}{2}(Q^{\mathrm{U}}_{I_1,y} - Q^{\mathrm{U}}_{I_2,y})^2 - 2Q_{i,y}\sum_j Q^{\mathrm{U}}_{j,y} \end{cases} \quad \text{(A.2)}$$

$\div\frac{1}{4}\sum_j(Q^{\mathrm{U}}_{j,y})^2$, for $i \neq I_1, I_2$, where $\{I_1, I_2\} = \{1,2\},\{2,3\},\{1,3\}$. Another possibility finally results for $Q^{\mathrm{S}}_{1,y} = Q^{\mathrm{S}}_{2,y} = Q^{\mathrm{S}}_{3,y}$, and is given

$$\max\|\delta\mathbf{q}'\|^2 = \frac{2}{3}\left[1 - \frac{Q^{\mathrm{U}}_{1,y}Q^{\mathrm{U}}_{2,y} + Q^{\mathrm{U}}_{1,y}Q^{\mathrm{U}}_{3,y} + Q^{\mathrm{U}}_{2,y}Q^{\mathrm{U}}_{3,y}}{\sum_j(Q^{\mathrm{U}}_{j,y})^2}\right]. \quad \text{(A.3)}$$

The tightest bound follows as the least continuous bound of the above 20 possible bounds in the 4-dimensional space of unstable basic state parameters, with coordinates $(b, b_{\mathrm{T}}, r_1, r_2)$ (the bounds are independent of s).

B. Two-and-a-Half-Layer Model Bounds

In the 2.5-layer model ($r_2 \to 0$ and $\psi_3 \to 0$) the least bound in the 3-dimensional space of unstable basic state parameters, with coordinates (b, b_{T}, r_1), is given by

$$\max\|\delta\mathbf{q}'\|^2 = \begin{cases} -4Q^{\mathrm{U}}_{1,y}(Q^{\mathrm{U}}_{1,y} + Q^{\mathrm{U}}_{2,y}) & \text{if } -\rho < b < b_1 \\ \frac{1}{2}(Q^{\mathrm{U}}_{2,y} - Q^{\mathrm{U}}_{1,y})^2 & \text{if } b_1 \leq b \leq b_2 \\ -4Q^{\mathrm{U}}_{2,y}(Q^{\mathrm{U}}_{1,y} + Q^{\mathrm{U}}_{2,y}) & \text{if } b_2 < b < -r_1\rho(b_{\mathrm{T}} - 1) \end{cases} \quad \text{(B.1)}$$

$\div\ [(Q^{\mathrm{U}}_{1,y})^2 + (Q^{\mathrm{U}}_{2,y})^2]$, where $b_1 := -\frac{1}{4}\rho\,[r_1(b_{\mathrm{T}} - 1) + 3]$ and $b_2 := -\frac{1}{4}\rho \times [3r_1(b_{\mathrm{T}} - 1) + 1]$. This result extends to arbitrary r_1 that of Olascoaga and Ripa (1999).

References

Arnold, V. Condition for Nonlinear Stability of Stationary Plane Curvilinear Flows of an Ideal Fluid. *Dokl. Akad. Nauk. USSR*, 162:975–978, 1965. Engl. transl. *Sov. Math.*, 6:773-777, 1965.

Arnold, V. On an Apriori Estimate in the Theory of Hydrodynamical Stability. *Izv. Vyssh. Uchebn. Zaved Mat.*, 54:3–5, 1966. Engl. transl. *Am. Math. Soc. Transl. Series II*, 79, 267-269, 1969.

Beron-Vera, F. J. and M. J. Olascoaga. Spectral, Formal, and Nonlinear Stability in a Layered Quasigeostrophic Model with Application to the Atlantic North Equatorial Current. In: P. Malanotte-Rizzoli and G. J. Goni (eds.): *Interhemispheric Water Exchange in the Atlantic Ocean*, Elsevier Oceanography Series. Elsevier Science, in press, 2003.

Beron-Vera, F. J. and P. Ripa. Free Boundary Effects on Baroclinic Instability. *J. Fluid Mech.*, 352:245–264, 1997.

Fukamachi, Y., J. McCreary, and J. Proehl. Instability of Density Fronts in Layer and Continuously Stratified Models. *J. Geophys. Res.*, 100:2559–2577, 1995.

Gill, A., J. Green, and A. Simmons. Energy Partition in the Large-Scale Ocean Circulation and the Production of Mid-Ocean Eddies. *Deep Sea Res.*, 21:499–528, 1974.

Haine, T. W. and J. Marshall. Gravitational, Symmetric, and Baroclinic Instability of the Ocean Mixed Layer. *J. Phys. Oceanogr.*, 28:634–658, 1998.

Holm, D. D., J. E. Marsden, T. Ratiu, and A. Weinstein. Nonlinear Stability of Fluid and Plasma Equilibria. *Phys. Rep.*, 123:1–116, 1985.

McIntyre, M. and T. Shepherd. An Exact Local Conservation Theorem for Finite-Amplitude Disturbances to Non-Parallel Shear Flows, with Remarks on Hamiltonian Structure and on Arnol'd's Stability Theorems. *J. Fluid Mech.*, 181:527–565, 1987.

Mu, M. Optimality of a nonlinear stability of two-layer Phillips model. *Chinese Science Bulletin*, 43:656–659, 1998.

Mu, M. and Y. Wu. Arnold nonlinear stability theorems and their applications to the atmosphere and oceans. *Surveys in Geophysics*, 22:383–426, 2001.

Mu, M., Q. C. Zeng, T. G. Shepherd, and Y. Liu. Nonlinear stability of multilayer quas-geostrophic flow. *J. Fluid Mech.*, 264:165–184, 1994.

Olascoaga, M. J. Deep Ocean Influence on Upper Ocean Baroclinic Instability. *J. Geophys. Res.*, 106:26,863–26,877, 2001.

Olascoaga, M. J. and P. Ripa. Baroclinic Instability in a Two-Layer Model with a Free Boundary and β Effect. *J. Geophys. Res.*, 104:23,357–23,366, 1999.

Paret, J. and J. Vanneste. Nonlinear Saturation of Baroclinic Instability in a Three-Layer Model. *J. Atmos. Sci.*, 53:2905–2917, 1996.

Ripa, P. Wave Energy-Momentum and Pseudo Energy-Momentum Conservation for the Layered Quasi-Geostrophic Instability Problem. *J. Fluid Mech.*, 235:379–398, 1992.

Ripa, P. On Improving a One-Layer Ocean model With Thermodynamics. *J. Fluid Mech.*, 303:169–201, 1995.

Ripa, P. A Minimal Nonlinear Model of Free Boundary Baroclinic Instability. In: *Proceedings of the 12th Conference on Atmospheric and Oceanic Fluid Dynamics*. pp. 249–252, American Meteorological Society, 1999a.

Ripa, P. On the Validity of Layered Models of Ocean Dynamics and Thermodynamics with Reduced Vertical Resolution. *Dyn. Atmos. Oceans*, 29:1–40, 1999b.

Ripa, P. On Upper Ocean Baroclinic Instability. In: J. Ramos-Mora and J. Herrera (eds.): *Escuela de Turbulencia (School of Turbulence)*. Sociedad Mexicana de Física, 1999c.

Ripa, P. Baroclinic Instability in a Reduced Gravity, Three-Dimensional, Quasi-Geostrophic Model'. *J. Fluid Mech.*, 403:1–22, 2000a.

Ripa, P. On the Generation of Turbulence by Baroclinic Instability in the Upper Ocean. In: C. Dopazo et al. (ed.): *Advances in Turbulence VIII. Proceedings of the 8th European Turbulence Conference*, pp. 371–374, Kluwer Academic, 2000b.

Ripa, P. Waves and Resonance in Free-Boundary Baroclinic Instability. *J. Fluid Mech.*, 428:387–408, 2001.

Shepherd, T. Nonlinear Saturation of Baroclinic Instability. Part I: The two-layer model. *J. Atmos. Sci.*, 45:2014–2025, 1998.

Vanneste, J. Explosive Resonant Interaction of Rossby Waves and Stability of Multilayer Quasi-Geostrophic Flow. *J. Fluid Mech.*, 291:83–107, 1995.

Young, W. and L. Chen. Baroclinic Instability and Thermohaline Gradient Alignment in the Mixed Layer. *J. Phys. Oceanogr.*, 25:3172–3185, 1995.

CONSTRAINED-HAMILTONIAN SHALLOW-WATER DYNAMICS ON THE SPHERE

F. J. BERON-VERA
RSMAS, University of Miami
4600 Rickenbacker Causeway
Miami, FL 33149-1098, U.S.A.

Abstract. Salmon's nearly geostrophic model for rotating shallow-water flow is derived in full spherical geometry. The model, which results upon constraining the velocity field to the height field in Hamilton's principle for rotating shallow-water dynamics, constitutes an important prototype of Hamiltonian balanced models. Instead of Salmon's original approach, which consists in taking variations of particle paths at fixed Lagrangian labels and time, Holm's approach is considered here, namely variations are taken on Lagrangian particle labels at fixed Eulerian positions and time. Unlike the classical quasigeostrophic model, Salmon's is found to be sensitive to the differences between geographic and geodesic coordinates. One consequence of this result is that the β plane approximation, which is included in Salmon's original derivation, is not consistent for this class of model.

Key words: Hamilton's principle, shallow water, balance, sphere

1. Introduction

The rotating shallow-water (SW) equations constitute a paradigm for geophysical fluid motions ranging from fast timescale dynamics, associated with inertia-gravity waves, to slow advective-timescale dynamics, associated with nonlinear vortical motions and Rossby waves (cf. Gill, 1982; Pedlosky, 1987). This set of equations constitute the "primitive" equations on which different approximations are usually performed. In this paper I deal with those approximations which involve the introduction of balance relations or constraints that lead to filtering out the fast degrees of freedom. Terms commonly used to denote the resulting models are "balanced," "constrained," or "intermediate;" the latter, in particular, reflects the fact of being at a level which is in between the primitive equations and the equations for geostrophic motion. For an extensive review on the wide variety of balanced models that exists in the literature the reader is referred to Allen *et al.* (1990a).

O.U. Velasco Fuentes et al. (eds.), Nonlinear Processes in Geophysical Fluid Dynamics, 29-51.

Of particular interest are those balanced models derived by performing approximations directly in Hamilton's principle (HP) for SW dynamics as proposed by Salmon(1983, hereafter referred to as S83). This procedure allows the fundamental symmetry-based conservation laws of the underlying primitive system to be preserved. The approach consists in substituting leading order balance relations and asymptotic expansions into HP before taking variations. In particular, S83's model is derived by constraining the velocity field to the height field in the form of a geostrophic balance relation, i.e. between the pressure gradient and the Coriolis force. This so-called L1 model, however, was shown to produce less accurate solutions to the SW equations than those produced by other non-Hamiltonian intermediate models (Allen *et al.*, 1990a; Allen *et al.*, 1990b; Barth *et al.*, 1990). This is indicative of the known fact that possession of Hamiltonian structure is no guarantee of model's accuracy. Nevertheless, other balance relation choices—potentially more accurate than that considered by S83—are possible (Allen and Holm, 1996; Allen *et al.*, 2002). This fact makes the L1 model an important prototype of constrained Hamiltonian models, and thus motivates the present study.

The L1 nearly geostrophic model, as well as its relatives the extended-geostrophic Hamiltonian models of Allen and Holm (1996) and Allen *et al.* (2002), have been derived in the Cartesian coordinates of the β plane approximation. Such approximation relies upon expansion of the equations of motion with respect to geographic (e.g. spherical longitude and latitude) coordinates about some fixed point on the surface of the planet, in inverse powers of the (mean) radius of the planet. The expansion is then truncated at first order but retaining only the first order variation of the Coriolis parameter (the so called β term) and neglecting all metric terms, which are of the same order as the β term! Consequently, the β plane approximation is only valid locally and in geodesic coordinates (Phillips, 1973; Verkley, 1990). These coordinate systems are such that all the derivatives of the metric tensor vanishes identically at the origin and thus locally look like Cartesian coordinates. Geographic coordinates are not geodesic in general, except at the equator where coordinate curves are geodesic curves, e.g. great circles in spherical geometry. Consequently, only at the equator the β plane approximation is valid when written in geographic coordinates, but this region is forbidden for the L1 model.

Remarkable is the fact that the quasigeostrophic (QG) model—perhaps the most exploited model of (slow advective-timescale) intermediate dynamics—has the property of being insensitive to differences between geographic and geodesic coordinates, namely the β plane approximation gives the right QG equations (Pedlosky, 1987; Ripa, 1997a, hereafter referred to as R97). Even though the QG system does not fit within the frame of models of the L1 class, i.e. it does not follow from an approximation made in HP for SW motion, it can be derived from HP but for stationary variations of a particularly chosen action (Virasoro, 1981; Holm and Zeitlin, 1998).

The goal of this paper is to derive an L1 model using non-Cartesian geometry in order to make an assessment of the sensitivity of this model to the difference between geographic and geodesic coordinates. I am not aware of a similar development within the Hamiltonian framework except for the works of Shutts (1989) and Verkley (2001). Shutts (1989) derived a modified version of the Hoskins' (1975) semigeostrophic equations, which are another type of intermediate equations that can be derived from the L1 model through a transformation into "geostrophic coordinates" (Salmon, 1985). Verkley (2001), in turn, presented a derivation of an isentropic L1-type model for application to atmospheric flows; the model derived here is based on SW dynamics. Unlike both Shutts' (1989) and Verkley's (2001) derivations, in this paper I use tools from non-Cartesian tensor algebra, which leads to an invariant formulation for the dynamical equations of the L1 model.

The reminder of the paper is organized as follows. In § 2 I set up a mathematical model for the Earth's surface that defines the space in which the analysis is carried out. Section 3 includes a derivation of the general equations for a free particle on the smooth surface of the Earth in invariant form. This is done from HP for a general spheroidal Earth in § 3.1. The usual spherical approximation is then applied to the resulting motion equations, which, in particular, are written in geographic coordinates (§ 3.2). Section 3.3 presents a discussion of the consistency of the so-called planar approximations, which include the classical f and β. Section 4 is devoted to extending into non-Cartesian geometry Holm's (1996, hereafter referred to as H96) general HP for variations of Lagrangian particle labels at fixed Eulerian positions and time. The SW and L1 model equations are derived in §§ 4.4 and 4.5, respectively, using the spherical Earth's model. The equations are written in a coordinate-invariant fashion on the sphere and then particularized to the common geographic coordinate system. Concluding remarks are given in § 5. Appendix A presents various relationships involved in the derivation of the equations. Appendix B is reserved for the discussion and comparison of alternative HPs.

2. Earth's Shape Model

I consider here some basic geophysical facts that relate to the shape of the Earth and the forces acting on its equilibrium surface (e.g. Stommel and Moore, 1989; Ripa, 1995; Ripa, 1997b; R97). The mathematical framework on which the invariant formulation of the equations derived in this paper is based involve concepts from non-Cartesian tensor algebra (e.g. Abraham *et al.*, 1988; Dubrovin *et al.*, 1992) that I start by reviewing first.

2.1. NON-CARTESIAN TENSOR ALGEBRA BACKGROUND

Let S be a two-dimensional manifold, coordinatized by $\mathbf{x} := (x^1, x^2)$. Two-dimensional intrinsic vectors on S at any point $\mathbf{x}$ define the tangent space, $T_\mathbf{x}S$. The disjoint union of tangent spaces constitute the tangent bundle, TS. Let $\{e_i\}$ be a basis for $T_\mathbf{x}S$ and $\{e^i\}$ for the dual space, $(T_\mathbf{x}S)^*$, namely

$$e^i(e_j) = \delta^i_j, \tag{1}$$

where δ^i_j is the Kroenecker symbol which equals 1 if $i = j$ and 0 otherwise. Let $T^m_n(T_\mathbf{x}S)$ be the space of m-contravariant and n-covariant real valued tensors or, simply, (m, n)-tensors. Vectors $a \in T^1_0(T_\mathbf{x}S) = T_\mathbf{x}S$ are expressed as $a = a^i e_i$ and covectors $\alpha \in T^0_1(T_\mathbf{x}S) = (T_\mathbf{x}S)^*$ as $\alpha = \alpha_i e^i$; the quantities $a^i = a(e^i)$ and $\alpha_i = \alpha(e_i)$ are the components of a and α, respectively. (N.B. The convention of summation over repeated lower and upper indices is understood.) In general, a (m, n)-tensor A expresses as

$$A = A^{i_1\cdots i_m}_{j_1\cdots j_n} e_{i_1} \otimes \cdots \otimes e_{i_m} \otimes e^{j_1} \otimes \cdots e^{j_n}, \tag{2}$$

where $A^{i_1\cdots i_m}_{j_1\cdots j_n} = A(e^{i_1}, \cdots, e^{i_m}, e_{j_1}, \cdots, e_{j_n})$ and $\otimes$ denotes the tensor product.

Assume now that S is endowed with a Riemannian metric, namely a symmetric, positive definite, bilinear form

$$\langle\langle\cdot,\cdot\rangle\rangle := m_{ij}\, e^i \otimes e^j, \tag{3}$$

where $m_{ij}(\mathbf{x}) := \langle\langle e_i, e_j\rangle\rangle$. The inner product of two vectors $a, b \in T_\mathbf{x}S$ is computed with respect to the metric, i.e.

$$\langle\langle a, b\rangle\rangle = \left(m_{ij}\, e^i \otimes e^j\right) \cdot (a, b) = m_{ij}\, e^i(a)\, e^j(b) = m_{ij}\, a^i b^j. \tag{4}$$

In particular, the square of the distance between two nearby positions on S, $\mathbf{x}$ and $\mathbf{x} + \mathrm{d}\mathbf{x}$, is given by

$$\mathrm{d}s^2 = \langle\langle \mathrm{d}\mathbf{x}, \mathrm{d}\mathbf{x}\rangle\rangle = \|\mathrm{d}\mathbf{x}\|^2 = m_{ij}\, \mathrm{d}x^i \mathrm{d}x^j. \tag{5}$$

Let $^\flat$ be the index lowering operator, and $^\natural$, its inverse, be the index raising operator, which are defined by

$$^\flat : T_\mathbf{x}S \to (T_\mathbf{x}S)^*;\ a \mapsto \langle\langle a, \cdot\rangle\rangle \quad \text{and} \quad ^\natural : (T_\mathbf{x}S)^* \to T_\mathbf{x}S, \tag{6}$$

respectively. The matrix of $^\flat$ is $[m_{ij}]$, i.e. $(a^\flat)_i = m_{ij}\, a^j =: a_i$, whereas that of $^\natural$ is $[m_{ij}]^{-1} = m^{-1}\,\mathrm{adj}[m_{ij}] =: [m^{ij}]$, i.e. $(\alpha^\natural)^i = m^{ij}\alpha_j =: \alpha^i$. Here, $m := \det[m_{ij}]$ and adj denotes adjoint (transpose cofactor).

Let, in addition, reserve the symbol $\mathbf{d}$ to denote the exterior derivative (or generalized gradient operator), whose action on a skew-symmetric $(0,k)$-tensor or k-form α, i.e.

$$\alpha = \alpha_{i_1 \cdots i_k}\, e^{i_1} \wedge \cdots \wedge e^{i_k}, \tag{7}$$

where $\wedge$ denotes the exterior product, is defined by

$$\mathbf{d}\alpha := \sum_{j,i_1 < \cdots < i_k} \partial_j \alpha_{i_1 \cdots i_k}\, e^j \wedge e^{i_1} \wedge \cdots \wedge e^{i_k}. \tag{8}$$

Notice that, in particular, if $k = 0$ then α is simply a scalar and, hence, $\mathbf{d}\alpha = \alpha_{,i} e^i =: \operatorname{grad} \alpha$. N.B. The shorthand notations $\partial_i(\cdot)$ and $(\cdot)_{,i}$ for partial differentiation $\partial(\cdot)/\partial x^i$ are in use.

Finally, let $\mathbb{P}$ be a linear map, with matrix elements $\mathbb{P}_{ij} = \sqrt{m_{ij}}$ for $i = j$ and $\mathbb{P}_{ij} = 0$ otherwise. Then $\mathbb{P} \cdot a$ (resp., $\mathbb{P}^{-1} \cdot a^\flat$) denotes the physical—nontensorial—contravariant (resp., covariant) counterpart of vector a. For orthogonal coordinates, i.e. with $m_{ij} = 0$ for $i \neq j$, physical contravariant and covariant counterparts coincide, namely $\mathbb{P} \cdot a \equiv \mathbb{P}^{-1} \cdot a^\flat$.

2.2. GENERAL ASSUMPTIONS ON S

Two main assumptions make the two-dimensional manifold S an idealized model of the surface of the (solid) Earth. First, S is assumed to be embedded in a three-dimensional Euclidean space which rotates steadily, with spinning frequency Ω, with respect to a Newtonian inertial space. Second, S is assumed to be a geopotential surface. Namely the projections onto $T_{\mathbf{x}}S$ of the ***centrifugal force*** (due to the spinning of the planet with respect to an inertial reference frame) and the ***gravitational attraction*** (due to the deviation of the shape of the planet from a perfect sphere and to inhomogeinities in the mass distribution within the planet) are assumed to balance one another exactly.

As a consequence of the second assumption it follows that

$$\Phi := V + V_{\mathrm{C}} = \text{const.} \tag{9}$$

on S, where V is the ***gravitational potential***, V_{C} stands for the ***centrifugal potential***, and their sum, Φ, defines the ***geopotential***. (The constant in the above expression is arbitrary and can be freely set to zero.) The centrifugal potential (per unit mass) can be expressed in invariant form as

$$V_{\mathrm{C}} = -\tfrac{1}{2} \left\| \sigma \right\|^2, \tag{10}$$

where σ is the velocity of the $\mathbf{x}$-system with respect to a suitable inertial frame.

Finally, the ***acceleration of gravity*** is defined as the minus gradient of Φ, thereby determining the vertical direction at each point $\mathbf{x}$ on S. Its magnitude is thus given by

$$g(\mathbf{x}) = \|\mathrm{grad}\,\Phi\| = \sqrt{m^{ij}\Phi_{,i}\Phi_{,j}}. \tag{11}$$

2.3. SPHERICAL MODEL

It is convenient—and quite accurate—to consider S as a (two-dimensional) sphere of radius R, say, but keeping the main effect of the gravitational force. Namely that it can sustain a steady rotation, relative to an inertial frame, in any point on S. Thus let the coordinates on S be given by

$$x^1 = (\lambda - \lambda_0)\, R \cos\vartheta_0, \quad x^2 = (\vartheta - \vartheta_0)\, R, \tag{12}$$

which are rescaled longitude, λ, and latitude, ϑ, that will be referred here to as ***geographic coordinates***. In this case one can introduce the usual notation (x, y) for (x^1, x^2) and $(\hat{\mathbf{x}}, \hat{\mathbf{y}})$ for (e^1, e^2). The corresponding metric matrix, velocity of the $\mathbf{x}$-system, and centrifugal potential, respectively, read:

$$[m_{ij}] = \begin{bmatrix} \gamma^2 & 0 \\ 0 & 1 \end{bmatrix}, \quad \sigma = \frac{f}{2\tau\gamma}\hat{\mathbf{x}}, \quad V_{\mathrm{C}} = -\frac{f^2}{8\tau^2}. \tag{13}$$

Here,

$$f := 2\Omega \sin\vartheta, \quad \gamma := \sec\vartheta_0 \cos\vartheta, \quad \tau := R^{-1}\tan\vartheta; \tag{14}$$

the first parameter is the Coriolis parameter whereas the other two are the geometric coefficients as defined by Ripa (2000a,b). Consistently with this spherical approximation, the acceleration of gravity is taken as a constant, namely $g \approx 9.8\ \mathrm{m^2\,s^{-1}}$.

More accurate models (not treated here) should account explicitly for the flattening of the planet at the poles. For instance, although still crude, next in accuracy can be mentioned one that has the form of an axisymmetric spheroid of revolution (Chandrasekhar, 1969; cf. also R97).

3. Particle Dynamics

In this section the manifold S is assumed to represent a smooth and frictionless Earth's surface on which a particle moves freely. The derivation of the particle's equations of motion is instructive inasmuch as it sets the grounds for tackling the more complicated problem of the following section. In particular, it shows clearly how the Coriolis force—which finds its origin in the gravitational force—arises directly from a HP with an action appropriate for an inertial observer, but written in coordinates fixed to the planet. The method is in essence the same as the one used by Pierre Simon de Laplace (1749–1827) to introduce this force over quarter a century before Gaspard Gustave de Coriolis (1792–1843) was born (cf. R95, R96). The analysis of the particle's equations allows, in addition, one to simplify the discussion on the consistency of the so-called planar approximations (cf. R97).

3.1. GENERAL EQUATIONS

From an inertial observer viewpoint, the only force acting on the particle is the gravitational one. The particle's kinetic and potential energies (per unit mass) as measured by this observer are given by

$$T(\mathbf{x},\dot{\mathbf{x}}) := \tfrac{1}{2}\left\|\dot{\mathbf{x}}+\sigma\right\|^2, \quad V(\mathbf{x}) = -V_{\mathrm{C}} = \tfrac{1}{2}\left\|\sigma\right\|^2, \tag{15}$$

respectively, where the overdot denotes time differentiation and a zero value of the geopotential has been assigned to the Earth's surface. The Lagrangian function, $L : TS \to \mathbb{R}$, is constructed in the usual way, i.e.

$$L(\mathbf{x},\dot{\mathbf{x}}) := T - V = \tfrac{1}{2}\left\|\dot{\mathbf{x}}\right\|^2 + \langle\langle \dot{\mathbf{x}}, \sigma\rangle\rangle. \tag{16}$$

Let δt be a time displacement and $\delta\mathbf{x} := \mathrm{d}/\mathrm{d}\varepsilon|_{\varepsilon=0}\,\mathbf{x}(t+\varepsilon\delta t)$ a variation of the curve $\mathbf{x} : [t_0, t_1] \to T_{\mathbf{x}}S$. Let, in addition,

$$\mathcal{S}[\mathbf{x}] := \int_{t_0}^{t_1} \mathrm{d}t\, L : \mathcal{F}([t_0,t_1]) \to \mathbb{R} \tag{17}$$

be the action functional, where $\mathcal{F}([t_0,t_1])$ denotes the set of sufficiently smooth real valued functions on $[t_0,t_1]$. Subject to fixed endpoint conditions, i.e. $\delta\mathbf{x}(t_0) = 0 = \delta\mathbf{x}(t_1)$, the first variation of $\mathcal{S}$, defined as $\delta\mathcal{S} := \mathrm{d}/\mathrm{d}\varepsilon|_{\varepsilon=0}\,\mathcal{S}[\mathbf{x}+\varepsilon\delta\mathbf{x}]$, is given by

$$\begin{aligned}\delta\mathcal{S} &= \int_{t_0}^{t_1}\mathrm{d}t\,\left(L_{,i}\delta x^i + \frac{\partial L}{\partial \dot{x}^i}\delta\dot{x}^i\right)\\ &= \int_{t_0}^{t_1}\mathrm{d}t\,\left(L_{,i} - \frac{\mathrm{d}}{\mathrm{d}t}\frac{\partial L}{\partial \dot{x}^i}\right)\delta x^i\\ &= \int_{t_0}^{t_1}\mathrm{d}t\,\left[\tfrac{1}{2}\left(m_{jk,i} - m_{ki,j} - m_{ji,k}\right)\dot{x}^j\dot{x}^k - \ddot{x}_i - \left(\sigma_{i,j}-\sigma_{j,i}\right)\dot{x}^j\right]\delta x^i. \end{aligned}\tag{18}$$

HP ($\delta\mathcal{S} = 0$) then yields the Newton's law for the particle in covariant form

$$\boxed{\mathrm{D}\dot{\mathbf{x}}^\flat/\mathrm{d}t + \mathbf{d}\sigma^\flat\cdot\dot{\mathbf{x}} = 0.} \tag{19}$$

In this equation, the coordinate representation of the object $\mathrm{D}\alpha/\mathrm{d}t$, for any covector α, is given by $(\mathrm{D}\alpha/\mathrm{d}t)_i = \dot{\alpha}_i - \Gamma^k_{ij}\dot{x}^j\alpha_k$, where $\Gamma^k_{ij}(\mathbf{x}) := \frac{1}{2}m^{kl}(m_{il,j}+m_{jl,i}-m_{ij,l})$ are the Christoffel symbols (of second kind), which establish the (Levi–Civita) connection on the Riemannian manifold S. In addition,

$$\mathbf{d}\sigma^\flat = \sigma_{i,j}\, e^i \wedge e^j = (\sigma_{i,j} - \sigma_{j,i})\; e^i \otimes e^j, \tag{20}$$

which can be regarded as the ***Coriolis two-form***. Notice that $\mathbf{d}\sigma^\flat \cdot \dot{\mathbf{x}} = [(\sigma_{i,j} - \sigma_{j,i})\; e^i \otimes e^j] \cdot \dot{\mathbf{x}} = (\sigma_{i,j} - \sigma_{j,i})\; e^i(\cdot)\, e^j(\dot{\mathbf{x}}) = (\sigma_{i,j} - \sigma_{j,i})\, \dot{x}^j\, e^i$. The operator $^\natural$ transforms (19) into its contravariant counterpart

$$\mathrm{D}\dot{\mathbf{x}}/\mathrm{d}t + (\mathbf{d}\sigma^\flat \cdot \dot{\mathbf{x}})^\natural = 0; \tag{21}$$

here, $\mathrm{D}a/\mathrm{d}t$, for any covector a, in components reads $(\mathrm{D}a/\mathrm{d}t)^i = \dot{a}^i + \Gamma^i_{jk}\dot{x}^j a^k$.

Equations (19) or (21) are invariant under general coordinate transformations on S, which in this case is not restricted to the spherical Earth model. In particular, these equations nicely show that the Coriolis term is responsible for the particle's trajectory to depart from a geodesic curve on S, i.e. a pure Galilean inertial motion. The latter is only consistent with motions with sufficiently large initial kinetic energy as shown by R97, who described all possible solutions on a sphere, namely the so-called inertial oscillations.

3.2. EQUATIONS ON THE SPHERE IN GEOGRAPHIC COORDINATES

In the geographic coordinate system (12) of the spherical Earth's model the only nonzero Christoffel symbols are $\Gamma^2_{11} = \gamma^2\tau$ and $\Gamma^1_{12} = \Gamma^1_{21} = -\tau$. In turn, the matrix of the Coriolis two-form takes the form

$$[\sigma_{i,j} - \sigma_{j,i}] = \begin{bmatrix} 0 & -\gamma f \\ \gamma f & 0 \end{bmatrix}. \tag{22}$$

Thus equation (19), with the spherical approximation and particularized to geographic coordinates, takes the following component representation:

$$\left.\begin{aligned} \ddot{x}_1 - \gamma^2\tau\dot{x}^1\dot{x}_2 - (\gamma f + \tau\dot{x}_1)\dot{x}^2 = 0, \\ \ddot{x}_2 + (\gamma f + \tau\dot{x}_1)\dot{x}^1 = 0. \end{aligned}\right\} \tag{23}$$

Let $\mathbf{u} := (u, v) := \mathbb{P} \cdot \dot{\mathbf{x}}$ ($\equiv \mathbb{P}^{-1} \cdot \dot{\mathbf{x}}^\flat$ since the coordinates are orthogonal; cf. § 2.1). Application of $\mathbb{P}^{-1}$ transforms set (23) into the more familiar form (e.g. R97)

$$\left.\begin{aligned} \dot{u} - (f + \tau u)v = 0, \\ \dot{v} + (f + \tau u)u = 0, \end{aligned}\right\} \tag{24}$$

which can be written in vector notation as well, i.e.

$$\dot{\mathbf{u}} + (f + \tau u)\,\hat{\mathbf{z}} \times \mathbf{u} = 0, \tag{25}$$

where $\hat{\mathbf{z}}$ is the vertical unit vector and $\times$ denotes the cross product of vectors.

3.3. "PLANAR" APPROXIMATIONS

In addition to the spherical approximation, other standard approximations introduced in the equations are the "planar" approximations. These approximations, which are meant to be valid locally at a point on the sphere in geographic coordinates, are obtained by expanding the equations in inverse powers of the radius of the sphere R. The most common approximations being the f and β. The former is a consistent zeroth-order approximation. The latter, however, is an inconsistent first-order approximation, except at the equator. A consistent nth-order approximation is understood as one that produces $O(R^{-n-1})$ errors in the integrals of motion associated with the equations on the sphere. These integrals are the (kinetic) energy of the particle as measured by a terrestrial observer,

$$E := \tfrac{1}{2}\mathbf{u}^2, \tag{26}$$

and the absolute angular momentum (with respect to the center of the planet and in the direction of the axis of rotation), which, up to some constants, is given by

$$M := \gamma u - \Omega R\,(\cos\vartheta_0 - \gamma\cos\vartheta)\,. \tag{27}$$

R97 showed that a consistent first-order "planar" approximation must have

$$\textbf{\textit{Ripa "plane"}} : \gamma = 1 - \tau_0 y,\ \tau = \tau_0/\gamma,\ f = f_0 + \beta y/\gamma \tag{28}$$

where $\tau_0 := R^{-1}\tan\vartheta_0$, $f_0 := 2\Omega\sin\vartheta_0$, and $\beta := 2\Omega R^{-1}\cos\vartheta_0$. With this approximation the equations of motion conserve $\frac{1}{2}\dot{\mathbf{x}}^2 - \tau_0 y\dot{x}^2 = E - O(R^{-2})$ and $(1-\tau_0 y)u - f_0 y - \frac{1}{2}\beta(1 - R^2\tau_0^2)y^2 = M - O(R^{-2})$. The f plane approximation has

$$\textbf{\textit{f plane}} : \gamma = 1,\ \tau = 0,\ f = f_0, \tag{29}$$

which consistently implies conservation of $\frac{1}{2}\dot{\mathbf{x}}^2 = E - O(R^{-1})$ and $u - f_0 y = M - O(R^{-1})$. The β plane approximation, in turn, has

$$\boldsymbol{\beta}\ \textbf{\textit{plane}} : \gamma = 1,\ \tau = 0,\ f = f_0 + \beta y \tag{30}$$

and implies conservation of $\frac{1}{2}\dot{\mathbf{x}}^2$ and $u - f_0 y - \frac{1}{2}\beta y^2$, which produce $O(R^{-1})$ errors to E and M, respectively, everywhere except at $\vartheta_0 = 0$ where these errors are $O(R^{-2})$ because $\tau_0 \equiv 0$.

It is thus clear that a consistent first-order approximation must include, in general, non-Cartesian terms in order to correctly reproduce the conservation laws of the system. (That is the reason for the quotation marks in this section's title.) It is worthwhile remarking that this is no longer necessary for motions around the equator. Geographic coordinates at the equator are ***geodesic coordinates*** because all the derivatives of the metric vanish there. For this reason locally at the equator the geometry in geographic coordinates looks like Cartesian and, hence, the β plane is a consistent approximation there. In general, for any point of a space with a symmetric affine connection coordinatized by x^i, $i = 1, 2, \cdots$, say, there exists a coordinate system x'^i, $i = 1, 2, \cdots$, say, such that the coefficients of the connection vanish identically. Such a system can be defined implicitly by $x^i = x'^i - \frac{1}{2}\Gamma^i_{jk}(0)\, x'^j x'^k$ which can be readily seen to result in $\Gamma'^i_{jk}(0) \equiv 0$. For geographic coordinates the transformation $(x', y') \mapsto (x, y)$ reads $(x, y) = (x', y') + \tau_0 x'(y', -\frac{1}{2}x')$, which reduces to the identity at $\vartheta = \vartheta_0$. Of course, the practical use of geodesic coordinates (away from the equator) is questionable (cf. Phillips, 1973; Verkley, 1990).

4. Fluid Dynamics

In this section I derive from HP the equations of motion for (inviscid, unforced) SW and L1 dynamics on the spherical model for the Earth's surface. The derivation makes use of H96's approach but extended to non-Cartesian geometry. In this approach variations of Lagrangian particle labels are performed at fixed Eulerian positions and time. One advantage of H96's approach is that the equations result directly in Eulerian coordinates.

4.1. LAGRANGIAN AND EULERIAN COORDINATES

Identification of fluid particles in a SW motion requires two-dimensional labels $\mathfrak{x} := (\mathfrak{x}^1, \mathfrak{x}^2)$, say, which are defined in certain affine (metricless) space $\mathfrak{S}$, say. Let

$$\varphi \times \mathrm{id} : \mathfrak{S} \times \mathbb{R} \to S \times \mathbb{R};\ (\mathfrak{x}, t) \mapsto (\mathbf{x}, t) = (\varphi(\mathfrak{x}, t), t) \tag{31}$$

be the map that relates the Lagrangian labels with the Eulerian two-dimensional positions at time t, and consider its inverse:

$$\varphi^{-1} \times \mathrm{id} : S \times \mathbb{R} \to \mathfrak{S} \times \mathbb{R};\ (\mathbf{x}, t) \mapsto (\mathfrak{x}, t) = (\varphi^{-1}(\mathbf{x}, t), t). \tag{32}$$

Let now J and $\mathfrak{J}$ be the Jacobians of these maps, respectively, which are defined by

$$J := \det[J^i_{\mathfrak{i}}], \quad J^i_{\mathfrak{i}} := \partial x^i / \partial \mathfrak{x}^{\mathfrak{i}}, \tag{33a}$$

$$\mathfrak{J} := \det[\mathfrak{J}^{\mathfrak{i}}_i], \quad \mathfrak{J}^{\mathfrak{i}}_i := \partial \mathfrak{x}^{\mathfrak{i}} / \partial x^i. \tag{33b}$$

The time derivative of a Lagrangian label, following a fluid particle, is zero by construction. Consequently, $\dot{\mathbf{x}} = \partial_t \varphi + \varphi_{,\mathfrak{i}} \dot{\mathfrak{x}}^{\mathfrak{i}} \equiv \partial_t \varphi$. The latter defines the ***Lagrangian or material velocity***

$$\mathfrak{v}(\mathfrak{x}, t) := \partial_t \varphi; \tag{34}$$

the ***Eulerian or spatial velocity***, in turn, is defined by

$$\mathbf{v}(\mathbf{x}, t) := \mathfrak{v}(\mathfrak{x}, t). \tag{35}$$

Finally, the time derivative of any scalar function $a(\mathbf{x}, t)$ is $\dot{a} = (\partial_t + v^i \partial_i) a =: \mathrm{D}a/\mathrm{D}t$, where $v^i = \mathbf{v}(e^i)$.

4.2. VOLUME CONSERVATION

Let $R(t) \subset S$ be a material spherical cap (made of the same fluid particles) and let $h(\mathbf{x}, t)$ be the depth of the fluid. Let, in addition, $h_0(\mathfrak{x})$ be the density of Lagrangian labels in container $R(t)$. Since $R(t)$ is material, the Lagrangian labels are defined in certain fixed region $\mathfrak{R} \subset \mathfrak{S}$. As a consequence of the metricless nature of $\mathfrak{S}$, the following equality holds:

$$\int_{R(t)} \mathrm{d}^2\mathbf{x}\, \sqrt{m} h = \int_{\mathfrak{R}} \mathrm{d}^2\mathfrak{x}\, h_0. \tag{36}$$

The latter implies

$$\boxed{\sqrt{m} h J = h_0,} \tag{37}$$

which is the Lagrangian form of the volume conservation law. In order to obtain the Eulerian counterpart of this law, one needs to take the time derivative of the l.h.s. of (36), i.e.

$$\begin{aligned} \frac{\mathrm{d}}{\mathrm{d}t} \int_{R(t)} \mathrm{d}^2\mathbf{x}\, \sqrt{m} h &= \int_{\mathfrak{R}} \mathrm{d}^2\mathfrak{x} \left[\frac{\mathrm{d}J}{\mathrm{d}t} \sqrt{m} h + J \frac{\mathrm{d}}{\mathrm{d}t} (\sqrt{m} h) \right] \\ &= \int_{\mathfrak{R}} \mathrm{d}^2\mathfrak{x}\, \sqrt{m} J \left[\partial_t h + \mathrm{div}(h\mathbf{v}) \right], \end{aligned} \tag{38}$$

where the relationships of appendix A have been used. The conservation law follows upon setting to zero the latter result:

$$\boxed{\partial_t h + \mathrm{div}(h\mathbf{v}) = 0,} \tag{39}$$

where $\operatorname{div}(h\mathbf{v}) := \partial_i\left(\sqrt{m}hv^i\right)/\sqrt{m}$. Notice that in geographic coordinates $\operatorname{div}(h\mathbf{v}) = \gamma^{-1}[\partial_x(hu) + \partial_y(\gamma hv)] =: \nabla\cdot(h\mathbf{u})$.

4.3. GENERAL HP IN EULERIAN COORDINATES

Following H96, I consider an action functional of the form

$$\mathcal{S}[\mathfrak{x}] := \int_{t_0}^{t_1} \mathrm{d}t\, L[\mathbf{v},\mathfrak{J}] = \int_{t_0}^{t_1} \mathrm{d}t \int_D \mathrm{d}^2\mathbf{x}\, l(\mathbf{v},\mathfrak{J};\mathbf{x}), \tag{40}$$

where D is a fixed region on S with solid boundary ∂D. Here, L is the Lagrangian functional and, unlike H96 who adopted Cartesian coordinates, $l/\sqrt{m}$ is the Lagrangian density. Variations of Lagrangian particle labels at fixed Eulerian positions and time result in

$$\begin{aligned}\delta\mathcal{S} &= \int\left(\frac{\delta L}{\delta v^i}\delta v^i + \frac{\delta L}{\delta\mathfrak{J}}\delta\mathfrak{J}\right)\\ &= \int \mathfrak{J}J^i_{\mathfrak{i}}\delta\mathfrak{x}^{\mathfrak{i}}\left[\frac{\mathrm{D}}{\mathrm{D}t}\left(J\frac{\delta L}{\delta v^i}\right) + J\frac{\delta L}{\delta v^j}\partial_i v^j - \partial_i\frac{\delta L}{\delta\mathfrak{J}}\right]\\ &\quad + \int\partial_i\left(\delta x^i\mathfrak{J}\frac{\delta L}{\delta\mathfrak{J}} - v^i J^j_{\mathfrak{i}}\delta\mathfrak{x}^{\mathfrak{i}}\frac{\delta L}{\delta v^j}\right)\\ &\quad - \int\partial_t\left(J^i_{\mathfrak{i}}\delta\mathfrak{x}^{\mathfrak{i}}\frac{\delta L}{\delta v^i}\right),\end{aligned} \tag{41}$$

where $\int(\cdot) := \int_{t_0}^{t_1}\mathrm{d}t\int_D\mathrm{d}^2\mathbf{x}\,(\cdot)$. The derivation of (41) involved the use of the relationships of appendix A. Fixed endpoint conditions, $\delta\mathfrak{x}(\mathbf{x},t_1) = 0 = \delta\mathfrak{x}(\mathbf{x},t_2)$, allows one to get rid of the last integral in (41). Then HP implies the motion equation

$$\boxed{\partial_t\left(J\frac{\delta L}{\delta\mathbf{v}}\right) + \boldsymbol{\pounds}_{\mathbf{v}}\left(J\frac{\delta L}{\delta\mathbf{v}}\right) - \mathbf{d}\frac{\delta L}{\delta\mathfrak{J}} = 0} \tag{42}$$

and the no-flow boundary condition

$$\boxed{\langle\!\langle\mathbf{v},n\rangle\!\rangle = 0 \quad @ \quad \partial D,} \tag{43}$$

where n is the external normal to the boundary. In (42), $\boldsymbol{\pounds}_a\alpha := \mathbf{d}\alpha\cdot a + \mathbf{d}\langle\!\langle a,\alpha\rangle\!\rangle$ is the Lie derivative of covector α along vector a; in components $(\boldsymbol{\pounds}_a\alpha)_i = a^j\alpha_{i,j} + \alpha_j a^j_{\ ,i}$. Result (43), in turn, made use of Gauss' theorem, namely $\int_D\mathrm{d}^2\mathbf{x}\,\sqrt{m}\operatorname{div}a = \int_D\mathrm{d}^2\mathbf{x}\,\partial_i(\sqrt{m}a^i) = \oint_{\partial D}\mathrm{d}s\,a^i n_i$ for all vector a.

Finally, it must be mentioned that the Euler–Poincaré formalism provides an alternative way to obtaining (42)–(43) (Holm *et al.*, 2002).

4.4. HP FOR SW DYNAMICS ON THE SPHERE

Under the assumption that the layer of fluid is thin enough so that it does not represent a source of gravitation, an appropriate Lagrangian density for a HP for SW dynamics on the sphere has

$$\boxed{l(\mathbf{v}, \mathfrak{J}; \mathbf{x}) := h_0 \mathfrak{J} \left(\tfrac{1}{2} \|\mathbf{v}\|^2 + \langle\!\langle \mathbf{v}, \sigma \rangle\!\rangle\right) - \tfrac{1}{2} g \sqrt{m} \left(\frac{h_0 \mathfrak{J}}{\sqrt{m}} - H\right)^2} \tag{44}$$

along with the definitions

$$\boxed{h := h_0 \mathfrak{J} / \sqrt{m}, \quad p := g\,(h - H)\,.} \tag{45}$$

Here, h_0 and g are both constants, and $p(\mathbf{x}, t)$ is the hydrostatic pressure, where $H(\mathbf{x})$ is the reference depth including the possibility of an irregular topography. The choice $h_0 = \text{const.}$ is necessary in order for the Lagrangian density to be independent of the Lagrangian labels. The assumption $g = \text{const.}$, in turn, is consistent with the spherical approximation for the Earth's surface. The last term on the r.h.s. of (44), which is not present in (16), relates to the gravitational potential of the fluid column due to the departure of the free surface from the resting position.

According to

$$\frac{\delta L}{\delta \mathbf{v}} = h_0 \mathfrak{J} \, (\mathbf{v} + \sigma)^\flat \,, \tag{46}$$

$$\frac{\delta L}{\delta \mathfrak{J}} = h_0 \left(\tfrac{1}{2} \|\mathbf{v}\|^2 + \langle\!\langle \mathbf{v}, \sigma \rangle\!\rangle - p\right), \tag{47}$$

equations (42) imply the following equivalent sets of equations:

$$\boxed{\begin{aligned} &\text{a. } \partial_t \, (\mathbf{v} + \sigma)^\flat + \boldsymbol{\pounds}_{\mathbf{v}} \, (\mathbf{v} + \sigma)^\flat + \mathbf{d}\left(p - \tfrac{1}{2} \|\mathbf{v}\|^2 - \langle\!\langle \mathbf{v}, \sigma \rangle\!\rangle\right) = 0, \\ &\text{b. } \partial_t \mathbf{v}^\flat + \mathbf{d}\,(\mathbf{v} + \sigma)^\flat \cdot \mathbf{v} + \mathbf{d}\left(p + \tfrac{1}{2} \|\mathbf{v}\|^2\right) = 0, \\ &\text{c. } (\partial_t + \nabla_{\mathbf{v}}) \mathbf{v}^\flat + \mathbf{d}\sigma^\flat \cdot \mathbf{v} + \mathbf{d}p = 0. \end{aligned}} \tag{48}$$

Equation (48b) involves the identity $\boldsymbol{\pounds}_a \alpha = \mathbf{d}\alpha \cdot a + \mathbf{d}\langle\!\langle a, \alpha \rangle\!\rangle$, particularized for $a = \mathbf{v}$ and $\alpha = (\mathbf{v} + \sigma)^\flat$. Equation (48c), in turn, $\mathbf{d}\langle\!\langle a, \alpha \rangle\!\rangle = \nabla_a \alpha + \nabla_{\alpha^\natural} a^\flat - \mathbf{d}\alpha \cdot a - \mathbf{d}a^\flat \cdot \alpha^\natural$, specialized for $a = \mathbf{v}$ and $\alpha = \mathbf{v}^\flat$. Here, $\nabla_a \alpha$ denotes the covariant derivative of covector α in the direction of vector a; in components $(\nabla_a \alpha)_i = a^k \alpha_{i,k} - \Gamma^j_{ik} \alpha_j a^k$. Any set selected from (48) together with the volume conservation equation (39), all subject to the no-flow boundary condition (43), constitute the covariant form of the SW equations on a region D defined on the sphere. These equations (or their contravariant counterpart via the metric) are invariant under general changes of coordinates on the sphere.

The SW system conserves energy and Casimirs, namely

$$\mathcal{E} := \tfrac{1}{2}\int_D \mathrm{d}^2\mathbf{x}\,\sqrt{m}\left(h\,\|\mathbf{v}\|^2 + p^2/g\right), \quad \mathcal{C} := \int_D \mathrm{d}^2\mathbf{x}\,\sqrt{m}hC(q), \tag{49}$$

for arbitrary $C(\cdot)$ and where

$$qh := \frac{1}{\sqrt{m}h_0}\varepsilon^{ij}\partial_i\left(J\frac{\delta L}{\delta v^j}\right) = \frac{1}{\sqrt{m}}\varepsilon^{ij}\partial_i\left(v_j + \sigma_j\right) \tag{50}$$

defines the potential vorticity q. The latter is conserved following fluid particles, i.e. $\mathrm{D}q/\mathrm{D}t = 0$, as readily follows upon noticing that

$$\begin{aligned}\frac{\mathrm{d}}{\mathrm{d}t}\oint_{\partial D}\mathrm{d}x^i\,J\frac{\delta L}{\delta v^i} &= \oint_{\partial D}\mathrm{d}x^i\left[\frac{\mathrm{D}}{\mathrm{D}t}\left(J\frac{\delta L}{\delta v^i}\right) + J\frac{\delta L}{\delta v^j}\partial_i v^j\right]\\ &= \oint_{\partial D}\mathrm{d}x^i\partial_i\frac{\delta L}{\delta \mathfrak{J}}\\ &\equiv 0.\end{aligned} \tag{51}$$

The physical counterpart of any of the equations in (48) follows from application of the inverse map $\mathbb{P}^{-1}$. In geographic coordinates, the physical counterpart of, for instance, set (48b), reads (e.g. R97)

$$\left.\begin{aligned}\partial_t u - (\xi + f)v + \gamma^{-1}\partial_x B = 0,\\ \partial_t v + (\xi + f)u + \partial_y B = 0,\end{aligned}\right\} \tag{52}$$

where $\xi := \gamma^{-1}\partial_x v - \partial_y u - \tau u$ and $B := p + \frac{1}{2}(u^2 + v^2)$ are the relative vorticity and Bernoulli head, respectively. System (52) can also be written in vector notation, i.e.

$$\partial_t\mathbf{u} + (\xi + f)\,\hat{\mathbf{z}}\times\mathbf{u} + \nabla B = 0, \tag{53}$$

with $\nabla a := (\gamma^{-1}\partial_x a, \partial_y a)$ the gradient of any scalar function $a(\mathbf{x})$ in geographic coordinates, and where $\xi = \nabla\cdot(\mathbf{u}\times\hat{\mathbf{z}})$ and $B = p + \frac{1}{2}\mathbf{u}^2$. Finally, the integrals of motion take the form

$$\mathcal{E} = \tfrac{1}{2}\int_D \mathrm{d}^2\mathbf{x}\,\gamma\left(h\mathbf{u}^2 + p^2/g\right), \quad \mathcal{C} = \int_D \mathrm{d}^2\mathbf{x}\,\gamma hC(q), \tag{54}$$

where $q = (\xi + f)/h$ satisfies

$$\mathrm{D}q/\mathrm{D}t = (\partial_t + u\gamma^{-1}\partial_x + v\partial_y)q = (\partial_t + \mathbf{u}\cdot\nabla)q = 0. \tag{55}$$

If D is a zonal channel and the topography has the same symmetry, i.e. $\partial_x H \equiv 0$, then the zonal momentum,

$$\mathcal{M} := \int_D \mathrm{d}^2\mathbf{x}\, \gamma h \left[\gamma u - \Omega R \left(\cos\vartheta_0 - \gamma \cos\vartheta\right)\right], \tag{56}$$

is also an integral of motion.

4.5. HP FOR L1 DYNAMICS ON THE SPHERE

The starting point of S83's method to derive approximate models by making approximations in the HP for SW consists in expanding the velocity field as

$$\begin{array}{ccccc} \mathbf{v} & = & \mathbf{v}_\mathrm{G} & + & \mathbf{v}_\mathrm{A}, \\ O & : & \varepsilon & & \varepsilon^2 \end{array} \tag{57}$$

where $\varepsilon \to 0$ is an appropriate Rossby number. The lowest-order contribution to the velocity is assumed to satisfy the geostrophic balance and thus is a function of the height (mass) field. In invariant form this reads

$$\mathbf{v}_\mathrm{G}(\mathfrak{J}) = -(\mathbf{d}\sigma^\flat)^{-1} \cdot \mathbf{d}p \tag{58}$$

(at least there where $\mathbf{d}\sigma^\flat$ is invertible). The Lagrangian density for L1 dynamics on the sphere is obtained from (44) after replacing $\mathbf{v}$ by (57), with $\mathbf{v}_\mathrm{G}$ given by (58), and by dropping the $O(\varepsilon^4)$-term $\frac{1}{2}\|\mathbf{v}_\mathrm{A}\|^2$ in the first parenthesis. Thus

$$\boxed{l_1(\mathbf{v},\mathfrak{J};\mathbf{x}) := h_0\mathfrak{J}\left(\langle\langle \mathbf{v}, \mathbf{v}_\mathrm{G} + \sigma\rangle\rangle - \tfrac{1}{2}\|\mathbf{v}_\mathrm{G}\|^2\right) - \tfrac{1}{2} g\sqrt{m}\left(\frac{h_0\mathfrak{J}}{\sqrt{m}} - H\right)^2,} \tag{59}$$

together with the definitions (45), gives the L1 model's Lagrangian, i.e. $L_1 := \int_D \mathrm{d}^2\mathbf{x}\, l_1$. (A notation more consistent with my dimensional approach should in fact be L_3 for this Lagrangian.) According to

$$\frac{\delta L_1}{\delta \mathbf{v}} = h_0\mathfrak{J}\left(\mathbf{v}_\mathrm{G} + \sigma\right)^\flat, \tag{60}$$

$$\frac{\delta L_1}{\delta \mathfrak{J}} = h_0\left(\langle\langle \mathbf{v}, \mathbf{v}_\mathrm{G} + \sigma\rangle\rangle - \tfrac{1}{2}\|\mathbf{v}_\mathrm{G}\|^2 - p_\mathrm{AG}\right), \tag{61}$$

where $p_{\rm AG} := p - \operatorname{div}[gh(\mathbf{d}\sigma^\flat)^{-1}\cdot \mathbf{v}_{\rm A}^\flat]$. HP implies the following equivalent equations:

$$\boxed{\begin{aligned}&\text{a. } \partial_t\left(\mathbf{v}_{\rm G}+\sigma\right)^\flat + \pounds_{\mathbf{v}}\left(\mathbf{v}_{\rm G}+\sigma\right)^\flat + \mathbf{d}\Big(p_{\rm AG} - \tfrac{1}{2}\left\|\mathbf{v}_G\right\|^2 - \langle\!\langle \mathbf{v}, \mathbf{v}_{\rm G}+\sigma\rangle\!\rangle\Big) = 0,\\ &\text{b. } \partial_t \mathbf{v}_{\rm G}^\flat + \mathbf{d}\left(\mathbf{v}_{\rm G}+\sigma\right)^\flat\cdot\mathbf{v} + \mathbf{d}\Big(p_{\rm AG}+\tfrac{1}{2}\left\|\mathbf{v}_{\rm G}\right\|^2\Big) = 0,\\ &\text{c. } (\partial_t + \nabla_{\mathbf{v}})\mathbf{v}_{\rm G}^\flat + \mathbf{d}\mathbf{v}_{\rm G}^\flat\cdot\mathbf{v}_{\rm A} + \mathbf{d}\sigma^\flat\cdot\mathbf{v} + \mathbf{d}p_{\rm AG} = 0.\end{aligned}} \tag{62}$$

Because of the presence of the term $\langle\!\langle \mathbf{v}, \mathbf{v}_{\rm G}\rangle\!\rangle$ in (59), in addition to the no-flow boundary condition (43), HP also implies the following condition:

$$\boxed{\langle\!\langle (\mathbf{d}\sigma^\flat)^{-1}\cdot\mathbf{v}_{\rm A}^\flat, n\rangle\!\rangle = 0 \quad @ \quad \partial D.} \tag{63}$$

Any set selected from (62) (or the corresponding contravariant counterpart through the metric) together with the volume conservation equation (39), all subject to boundary conditions (43) *and* (63), constitute the invariant form of the L1 model on a region D on the sphere. Since $\mathbf{v}_{\rm G}$ and h are not independent the L1 system has only one scalar prognostic equation; the other two scalar equations provide the constraints to determine $\mathbf{v}_{\rm A}$.

The L1 model conserves geostrophic versions of the SW energy and Casimirs, namely

$$\mathcal{E}_{\rm G} := \tfrac{1}{2}\int_D \mathrm{d}^2\mathbf{x}\,\sqrt{m}\left(h\left\|\mathbf{v}_{\rm G}\right\|^2 + p^2/g\right), \quad \mathcal{C}_{\rm G} := \int_D \mathrm{d}^2\mathbf{x}\,\sqrt{m}hC(q_{\rm G}) \tag{64}$$

for arbitrary $C(\cdot)$, where

$$q_{\rm G}h := \frac{1}{\sqrt{m}h_0}\varepsilon^{ij}\partial_i\left(J\frac{\delta L_1}{\delta v^j}\right) = \frac{1}{\sqrt{m}}\varepsilon^{ij}\partial_i\left(v_{{\rm G}j}+\sigma_j\right) \tag{65}$$

defines the geostrophic potential vorticity $q_{\rm G}$, which is materially conserved as is advected by the *total* flow (i.e. $\mathrm{D}q_G/\mathrm{D}t = 0$).

In geographic coordinates, the physical counterpart of, for instance, set (62b) is given by

$$\left.\begin{aligned}\partial_t u_{\rm G} - (\xi_{\rm G}+f)v + \gamma^{-1}\partial_x B_{\rm AG} = 0,\\ \partial_t v_{\rm G} + (\xi_{\rm G}+f)u + \partial_y B_{\rm AG} = 0,\end{aligned}\right\} \tag{66}$$

where $\xi_{\rm G} := \gamma^{-1}\partial_x v_{\rm G} - \partial_y u_{\rm G} - \tau u_{\rm G}$, $B_{\rm AG} := p_{\rm AG} + \frac{1}{2}(u_{\rm G}^2 + v_{\rm G}^2)$, and $p_{\rm AG} = p + \gamma^{-1}\partial_x\left(ghv_{\rm A}/f\right) - \partial_y\left(ghu_{\rm A}/f\right) - \tau ghu_{\rm A}/f$. In vector notation set (66) expresses as

$$\partial_t\mathbf{u}_{\rm G} + (\xi_{\rm G}+f)\,\hat{\mathbf{z}}\times\mathbf{u} + \nabla B_{\rm AG} = 0, \tag{67}$$

where $\xi_{\rm G} = \nabla\cdot(\mathbf{u}_{\rm G}\times\hat{\mathbf{z}})$, $B_{\rm AG} = p_{\rm AG}+\frac{1}{2}\mathbf{u}_{\rm G}^2$, and $p_{\rm AG} = p+\nabla\cdot(gh\mathbf{u}_{\rm A}\times\hat{\mathbf{z}}/f)$. Boundary condition (63), in turn, takes the form

$$\mathbf{u}_{\rm A}\cdot\hat{\mathbf{z}}\times\hat{\mathbf{n}} = 0 \quad @ \quad \partial D \tag{68}$$

(cf. Ren and Shepherd, 1997 for a physical interpretation of this condition). The set of diagnostic equations which determines $\mathbf{u}_{\rm A}$ is given by

$$\left.\begin{aligned} \mathsf{A}(hv_{\rm A}) + \mathsf{B}(hu_A) &= F_1, \\ \mathsf{A}((g/f)hv_{\rm A}) + (g/f)\mathsf{B}(hu_A) &= F_2, \end{aligned}\right\} \tag{69}$$

where the differential operators

$$\mathsf{A}(\cdot) := \nabla^2(\cdot) - R^{-2} - \tau^2 - (f/g)q_{\rm G}, \quad \mathsf{B}(\cdot) := (f'/f)\gamma^{-1}\partial_x(\cdot), \tag{70}$$

and the functions

$$F_1(h) := -\partial_y\nabla\cdot(h\mathbf{u}_{\rm G}) + (f/g)\left(hq_{\rm G}v_{\rm G} - \gamma^{-1}\partial_x B_{\rm G}\right), \tag{71}$$

$$F_2(h) := -(g/f)\gamma^{-1}\partial_x\nabla\cdot(h\mathbf{u}_{\rm G}) + hq_{\rm G}u_{\rm G} + \partial_y B_{\rm G}; \tag{72}$$

here, $\nabla^2 a := \nabla\cdot\nabla a = \gamma^{-2}\partial_{xx}a + \partial_{yy}a - \tau\partial_y a$ is the Laplacian of any scalar function $a(\mathbf{x})$ in geographic coordinates*. For completeness, from (69) it follows

$$hu_A = \left(\mathsf{A}(g/f) - (g/f)\mathsf{B}\mathsf{A}^{-1}\mathsf{B}\right)^{-1}\left(F_2 - (g/f)\mathsf{B}\mathsf{A}^{-1}F_1\right), \tag{73a}$$

$$hv_A = \left(\mathsf{A} - \mathsf{B}(f/g)\mathsf{A}^{-1}(g/f)\mathsf{B}\right)^{-1}\left(F_1 - \mathsf{B}(f/g)\mathsf{A}^{-1}F_2\right), \tag{73b}$$

which upon substitution in the volume conservation equation (39) results in a single evolution equation for the height field. The (Cartesian) f-plane version of the latter was derived by Vanneste and Bokhove (2002) using a Dirac-bracket approach. Finally, the integrals of motion of the L1 system read

$$\mathcal{E}_{\rm G} := \tfrac{1}{2}\int_D \mathrm{d}^2\mathbf{x}\,\gamma\left(h\mathbf{u}_{\rm G}^2 + p^2/g\right), \quad \mathcal{C}_{\rm G} := \int_D \mathrm{d}^2\mathbf{x}\,\gamma h C(q_{\rm G}), \tag{74}$$

where $q_{\rm G} = (\xi_{\rm G} + f)/h$; as before if D and H are zonally symmetric then

$$\mathcal{M}_{\rm G} := \int_D \mathrm{d}^2\mathbf{x}\,\gamma h\left[\gamma u_{\rm G} - \Omega R\left(\cos\vartheta_0 - \gamma\cos\vartheta\right)\right] \tag{75}$$

*Because $\gamma^2 > 0$ (excluding, of course, the poles) the elliptic problem (73) has a unique solution on D (bounded or periodic in one or both directions) provided that $q_{\rm G}f \geq -g(R^{-2} + \tau^2)$ (cf. Courant and Hilbert, 1962), which holds for all time because $q_{\rm G} \sim f/h$ as $\varepsilon \to 0$.

is also conserved.

Other decompositions, appart than (57), as well as other balance relationships, different than (58), are possible (Allen and Holm, 1996; Allen *et al.*, 2002). This freedom is what allows for the existence of approximate models which can be potentially more accurate than the L1 model.

5. Concluding Remarks

The scaling

$$\{\mathbf{u}, \partial_t, y/R, h - H(y), H'\} = O(\varepsilon) \tag{76}$$

implies, at $O(\varepsilon^2)$, the classical QG equation (cf. Pedlosky, 1987)

$$(\partial_t + \partial_x\psi\partial_y - \partial_y\psi\partial_x)\, q_{\mathrm{QG}} = 0, \tag{77a}$$

where

$$q_{\mathrm{QG}} := \left[\partial_{xx} + \partial_{yy} - f_0^2/(gH_0)\right]\psi + (\beta + \beta_{\mathrm{T}})\, y. \tag{77b}$$

Here, $H(y) = H_0(1 - \beta_{\mathrm{T}}y/f_0)$, with $H_0 = \mathrm{const}.$, and $\psi := g[h - H(y)]/f_0$ is the geostrophic streamfunction, i.e. $\mathbf{u} = (-\partial_y\psi, \partial_x\psi) + O(\varepsilon^2)$. (More complicated topographies can of course be considered.) Notice the absence of geometric coefficients in (77). Those terms, which *do* appear in the corresponding (diagnostic) momentum and volume conservation equations, have (fortuitously) cancelled out in the process of constructing the (prognostic) potential vorticity equation (77) [Pedlosky 1987; R97]. Consequently—and remarkably—QG flows develop *as if* the geometry were Cartesian, "feeling" the latitudinal variation of the Coriolis parameter as the *only* effect of the Earth's sphericity.

The L1 model shares a series of differences and similarities with the above QG model. Although both models are derivable from HP, the QG model's action is not seen to derive from approximations performed in SW's action. As QG motions, those governed by the L1 model are not allowed at the equator, i.e. where f vanishes. In addition to Rossby waves, the linear waves of the L1 model include (a form of) Kelvin waves, which are not supported by the QG model. Unlike QG motions, L1 motions are restricted neither to $O(\varepsilon)$ meridional excursions nor to $O(\varepsilon)$ displacements of the free surface from the position of equilibrium at rest, nor to the presence of $O(\varepsilon)$ topographic variations. In a reduced-gravity setting, the equations for both SW and L1 models have the same structure as those presented here, except that in that case g must be identified with the buoyancy jump at the interface between the active and the quiescent (infinitely deep) bottom layer, and $H(\mathbf{x})$ must be understood as the nonuniform thickness of the active layer at rest, including the possibility of a nonspherical rigid surface.

Consequently, in contrast to the QG model, the L1 model is able to describe the dynamics of frontal structures.

The integrals of motion of the L1 system, in geographic coordinates, expand in inverse powers of the radius of the spherical Earth R as

$$\mathcal{E}_{\mathrm{G}}=\left\langle\left(\frac{1}{2}-\frac{\beta y}{f_0}\right)\frac{(\partial_x p)^2+(\partial_y p)^2}{f_0^2}+\frac{\tau_0 y}{f_0^2}(\partial_x p)^2+\frac{p^2}{2gh}\right\rangle \tag{78}$$

$$\mathcal{M}_{\mathrm{G}}=\left\langle\left(\frac{2\tau_0 f_0+\beta}{f_0}y-1\right)\frac{\partial_y p}{f_0}-(1-\tau_0 y)\,f_0 y-\frac{1}{2}\beta\left(1-R^2\tau_0^2\right)y^2\right\rangle \tag{79}$$

$$\mathcal{C}_{\mathrm{G}}=\langle(1-\tau_0 y)C(q_0)+q_1 C'(q_0)\rangle \tag{80}$$

$+\, O(R^{-2})$. Here, $\langle\cdot\rangle := \int_D \mathrm{d}^2\mathbf{x}\, h\,(\cdot)$, and

$$q_0 h := \frac{\partial_{xx}p+\partial_{yy}p}{f_0}+f_0, \tag{81}$$

$$q_1 h := \frac{(2\tau_0 f_0-\beta)\partial_{xx}p-\beta\partial_{yy}p}{f_0^2}y+\frac{\tau_0 f_0-\beta}{f_0^2}\partial_y p+\beta y. \tag{82}$$

Clearly, a consistent (not necessarily the optimal, though) geometric approximation for L1 dynamics, which is first-order accurate in R, is given by the non-Cartesian Ripa "plane" and *not* by the standard β plane (recall that it has $\tau_0 = 0$). The latter is the one included in the original derivation of the L1 model. An important contribution of the present work to the above list of differences and similarities between the L1 and QG models is thus the sensitivity of the former to the differences between geographic and geodesic coordinates. This result confirms Ripa's (2000b) in the sense that Earth's curvature effects increase in importance as the motions deviate from strictly geostrophic (divergence-free) motions. The thorough evaluation of these effects, apart from checking that the equations have the right conservation laws, is a subject for futher research. The latter should involve direct numerical simulations in which predictions of the L1 model on the β-plane and the sphere (or the Ripa "plane") are compared.

Finally, Ripa (1983) showed that steady SW flows on the sphere posses a formal stability theorem. The latter involving an Arnold-like first theorem for the stability of QG flows, and a condition for the flow to be "subsonic" in the sense that the (geostrophic) basic flow must be everywhere slower than the slowest gravity-wave of the system. Ren and Shepherd (1997) showed, in turn, that steady L1 flows on the β plane posses a Ripa-like formal stability theorem, as well as a nonlinear (or Lyapunov) stability theorem in which the "subsonic" condition of Ripa's theorem is replaced by a condition that the flow be cyclonic along the lateral boundaries. The latter was shown by Ren and Shepherd (1997) to have an interpretation involving coastal Kelvin waves, which are not included in the QG model. Whether or not L1 flows on the sphere (or the Ripa "plane") enjoy similar stability properties is an issue that needs more investigation.

Acknowledgements

Part of this work was carried out while I was an ScD student at CICESE (México) under the supervision of Pedro Ripa. His untimely death will not prevent me from finding in him a source of inspiration. I have benefited from fruitful conversations with Alejandro Parés, Julio Sheinbaum, M. Josefina Olascoaga, and Oscar U. Velasco Fuentes. Corrections of the manuscript by M. Josefina Olascoaga and Michael G. Brown are sincerely appreciated. The comments of an anonymous reviewer lead to improvements in the paper. My work was partly supported by CICESE, CONACyT (México), and NSF (USA).

A. Useful Relations

The following relationships can be shown to hold:

$$J\mathfrak{J} = 1 \Longleftrightarrow J^i_{\mathrm{i}}\mathfrak{J}^{\mathrm{i}}_j = \delta^i_j, \tag{A.1a}$$

$$(\operatorname{adj} J)^{\mathrm{i}}_i = J\mathfrak{J}^{\mathrm{i}}_i \Longleftrightarrow J = J^i_{\mathrm{i}}(\operatorname{adj} J)^{\mathrm{i}}_i \Longleftrightarrow \partial J/\partial J^i_{\mathrm{i}} = J\mathfrak{J}^{\mathrm{i}}_i, \tag{A.1b}$$

$$(\operatorname{adj} \mathfrak{J})^i_{\mathrm{i}} = \mathfrak{J}J^i_{\mathrm{i}} \Longleftrightarrow \mathfrak{J} = \mathfrak{J}^{\mathrm{i}}_i(\operatorname{adj} \mathfrak{J})^i_{\mathrm{i}} \Longleftrightarrow \partial\mathfrak{J}/\partial\mathfrak{J}^{\mathrm{i}}_i = \mathfrak{J}J^i_{\mathrm{i}}, \tag{A.1c}$$

$$\partial_{\mathrm{i}}\left(J\mathfrak{J}^{\mathrm{i}}_i\right) = 0 = \partial_i\left(\mathfrak{J}J^i_{\mathrm{i}}\right), \tag{A.1d}$$

$$\delta J = J\partial_i\delta x^i, \ \delta\mathfrak{J} = \mathfrak{J}\partial_{\mathrm{i}}\delta\mathfrak{x}^{\mathrm{i}}, \tag{A.1e}$$

$$\partial_{\mathrm{i}} = J^i_{\mathrm{i}}\partial_i, \ \partial_i = \mathfrak{J}^{\mathrm{i}}_i\partial_{\mathrm{i}}. \tag{A.1f}$$

In deriving (A.1e) the following properties of the determinants were very helpful

$$\frac{\partial(a,b)}{\partial(\mathfrak{x}^1,\mathfrak{x}^2)} = \frac{\partial(a,b)}{\partial(x^1,x^2)}\frac{\partial(x^1,x^2)}{\partial(\mathfrak{x}^1,\mathfrak{x}^2)}, \quad \frac{\partial(a,x^2)}{\partial(x^1,x^2)} = \frac{\partial a}{\partial x^1} \tag{A.2}$$

for all scalar functions $a, b(\mathbf{x})$.

In addition, it can be shown that:

$$\dot{\mathfrak{x}}^{\mathrm{i}} = \partial_t\mathfrak{x}^{\mathrm{i}} + v^i\mathfrak{J}^{\mathrm{i}}_i = 0 \Longrightarrow v^i = -\mathfrak{J}^i_{\mathrm{i}}\partial_t\mathfrak{x}^{\mathrm{i}}, \tag{A.3a}$$

$$\partial_{\mathrm{i}}v^i = J^j_{\mathrm{i}}\partial_j v^i \Longrightarrow \dot{J}^i_{\mathrm{i}} = J^j_{\mathrm{i}}\partial_j v^i, \tag{A.3b}$$

$$\delta v^i = -J^i_{\mathrm{i}}(\partial_t + v^j\partial_j)\delta\mathfrak{x}^{\mathrm{i}}, \quad \delta\mathfrak{J} = \mathfrak{J}J^i_{\mathrm{i}}\partial_i\delta\mathfrak{x}^{\mathrm{i}}, \tag{A.3c}$$

$$\delta J = \mathfrak{J}\partial_i\delta x^i \Longrightarrow \partial_t\mathfrak{J} + \partial_i\left(\mathfrak{J}v^i\right) = 0. \tag{A.3d}$$

B. Alternative HPs

B.1. EULERIAN COORDINATES

The standard approach for fields $\mathfrak{x}(\mathbf{x},t)$ consists in considering an action functional of the form

$$\mathcal{S}[\mathfrak{x}] := \int_{t_0}^{t_1} \mathrm{d}t\, L[\mathfrak{x}, \partial_t\mathfrak{x}] = \int_{t_0}^{t_1} \mathrm{d}t \int_D \mathrm{d}^2\mathbf{x}\, l(\mathfrak{x}, \partial_t\mathfrak{x}, \mathfrak{J}^{\mathfrak{i}}_i; \mathbf{x}, t), \tag{B.1}$$

which, after invoking HP results in the familiar Euler–Lagrange equations

$$\partial_t \frac{\delta L}{\delta \mathfrak{x}^{\mathfrak{i}}_{,t}} - \frac{\delta L}{\delta \mathfrak{x}^{\mathfrak{i}}} = \partial_t \frac{\partial l}{\partial \mathfrak{x}^{\mathfrak{i}}_{,t}} + \partial_i \frac{\partial l}{\partial \mathfrak{J}^{\mathfrak{i}}_i} - \partial_{\mathfrak{i}} l = 0 \tag{B.2}$$

(plus boundary conditions). According to $\partial_i(\partial l/\partial \mathfrak{J}^{\mathfrak{i}}_i) = \partial_i(\mathfrak{J} J^i_{\mathfrak{i}} \partial l/\partial\mathfrak{J}) = \mathfrak{J} J^i_{\mathfrak{i}} \partial_i(\partial l/\partial \mathfrak{J})$ and (A.3a), equations (B.2) transform into

$$\partial_t \left(J^i_{\mathfrak{i}} \frac{\partial l}{\partial v^i} \right) - \mathfrak{J} J^i_{\mathfrak{i}} \partial_i \frac{\partial l}{\partial \mathfrak{J}} + \partial_{\mathfrak{i}} l = 0. \tag{B.3}$$

The latter can be shown to be equivalent to (42) only in the particular case $\partial_{\mathfrak{i}} l \equiv 0$.

B.2. LAGRANGIAN COORDINATES

In the variational approach for fields $\mathbf{x}(\mathfrak{x},t)$ the action is of the form

$$\mathcal{S}[\mathbf{x}] := \int_{t_0}^{t_1} \mathrm{d}t\, L[\mathbf{x}, \partial_t\mathbf{x}] = \int_{t_0}^{t_1} \mathrm{d}t \int_{\mathfrak{D}} \mathrm{d}^2\mathfrak{x}\; l(\mathbf{x}, \partial_t\mathbf{x}, J^i_{\mathfrak{i}}; \mathfrak{x}, t). \tag{B.4}$$

Upon variations of particle paths at fixed Lagrangian labels and time, the following Euler–Lagrange equations result from HP:

$$\partial_t \frac{\delta L}{\delta x^i_{,t}} - \frac{\delta L}{\delta x^i} = \partial_t \frac{\partial l}{\partial x^i_{,t}} + \partial_{\mathfrak{i}} \frac{\partial l}{\partial J^i_{\mathfrak{i}}} - \partial_i l = 0 \tag{B.5}$$

(plus appropriate boundary conditions). This is but the infinite-dimensional analogue of the particle's HP. For instance, S83's derivation of the SW and L1 systems in the Cartesian coordinates of the β plane and Van der Toorn's (1997) derivation of the SW equations on the sphere are based on this HP. One disadvantage of this variational approach, however, is that the resulting equations are in Lagrangian coordinates, which requires application of the inverse map φ^{-1} (32) to transform back to Eulerian coordinates.

References

Abraham, R., J. E. Marsden, and T. Ratiu. *Manifolds, Tensor Analysis, and Applications*, Second Edition, Applied Mathematical Sciences 75. Springer, 1998.

Allen, J. S., J. A. Barth, and P. A. Newberger. On Intermediate Models for Barotropic Continental Shelf and Slope Flow Fields. Part I: Formulation and Comparison of Exact Solutions. *J. Phys. Oceanogr.* 20:1017–1042, 1990a.

Allen, J. S., J. A. Barth, and P. A. Newberger. On Intermediate Models for Barotropic Continental Shelf and Slope Flow Field. Part III: Comparison of Numerical Model Solutions in Periodic Channels. *J. Phys. Oceanogr.* 20:1017–1042, 1990b.

Allen, J. S. and D. D. Holm. Extended-Geostrophic Hamiltonian Models for Rotating Shallow Water Motion. *Physica D* 98:229–248, 1996.

Allen, J. S., D. D. Holm, and P. A. Newberger. Toward an Extended-Geostrophic Euler–Poincaré Model for Mesoscale Oceanographic Flow. In: J. Norbury and I. Roulstone (eds.): *Large-Scale Atmosphere-Ocean Dynamics I: Analytical Methods and Numerical Models*, pp. 101–125, Cambridge University Press, 2002.

Barth, J., J. Allen, and P. Newberger. On Intermediate Models for Barotropic Continental Shelf and Slope Flow Fields. Part II: Comparison of Numerical Model Solutions in Doubly-Periodic Domains. *J. Phys. Oceanogr.* 20:1044–1076, 1990.

Chandrasekhar, S. *Ellipsoidal Figures of Equilibrium.* Yale University Press, 1969.

Courant, R. and D. Hilbert. *Methods of Mathematical Physics*, Vol. II. John Wiley & Sons, 1962.

Dubrovin, B. A., A. T. Fomenko, and S. P. Novikov. *Modern Geometry, Methods and Applications, Part I.* Springer, 1992.

Gill, A. E. *Atmosphere-Ocean Dynamics.* Academic, 1982.

Holm, D. and V. Zeitlin. Hamilton's Principle for Quasigeostrophic Motion. *Phys. Fluids* 10:800–806, 1998.

Holm, D. D. Hamiltonian Balance Equations. *Physica D* 98:379–414, 1996.

Holm, D. D., J. E. Marsden, and T. S. Ratiu. The Euler-Poincaré Equations in Geophysical Fluid Dynamics. In: J. Norbury and I. Roulstone (eds.): *Large-Scale Atmosphere-Ocean Dynamics II: Geometric Methods and Models*, pp. 251–299, Cambridge University Press, 2002.

Hoskins, B. J. The Geostrophic Momentum Approximation and the Semi-Geostrophic Equations. *J. Atmos. Sci.* 32:233–242, 1975.

Pedlosky, J. *Geophysical Fluid Dynamics*, Second Edition. Springer, 1987.

Phillips, N. A. Principles of Large Scale Numerical Weather Prediction. In: P. Morel (ed.): *Dynamic Meteorology*, pp. 3–96, Reidel, 1973.

Ren, S. and T. G. Shepherd. Lateral Boundary Contributions to Wave-Activity Invariants and Nonlinear Stability Theorems for Balanced Dynamics. *J. Fluid Mech.* 345:287–305, 1997.

Ripa, P. General Stability Conditions for Zonal Flows in a One-Layer Model on the Beta-Plane or the Sphere. *J. Fluid Mech.* 126:463–487, 1983.

Ripa, P. Caída Libre y la Figura de la Tierra. *Rev. Mex. Fís.* 41:106–127, 1995.

Ripa, P. "Inertial" Oscillations and the β-Plane Approximation(s). *J. Phys. Oceanogr.* 27:633–647, 1997a.

Ripa, P. *La Increíble Historia de la Malentendida Fuerza de Coriolis ("The Incredible Story of the Misunderstood Coriolis Force").* Fondo de Cultura Económica, 1997b.

Ripa, P. Effects of the Earth's Curvature on the Dynamics of Isolated Objects. Part I: The Disk. *J. Phys. Oceanogr.* 30:2,072–2,087, 2000a.

Ripa, P. Effects of the Earth's Curvature on the Dynamics of Isolated Objects. Part II: The Uniformly Translating Vortex. *J. Phys. Oceanogr.* 30:2504–2514, 2000b.

Salmon, R. Practical Use of Hamilton's Principle. *J. Fluid Mech.* 132:431–444, 1983.

Salmon, R. New Equations for Nearly-Geostrophic Flow. *J. Fluid Mech.* 153:461–477, 1985.

Shutts, G. Planetary Semi-Geostrophic Equations Derived from Hamilton's Principle. *J. Fluid Mech.* 208:545–573, 1989.

Stommel, H. M. and D. W. Moore. *An Introduction to the Coriolis Force.* Columbia University, 1989.

Van der Toorn, R. Geometry, Angular Momentum and the Intrinsic Drift of Oceanic Monopolar Vortices. Ph.D. thesis, Utrecht University, 1997.

Vanneste, J. and O. Bokhove. Dirac-Bracket Approach to Nearly Geostrophic Hamiltonian Balanced Models. *Physica D* 164:152–167, 2002.

Verkley, W. T. M. On the Beta Plane Approximation. *J. Atmos. Sci.* 47:2453–2459, 1990.

Verkley, W. T. M. Salmon's Hamiltonian Approach to Balanced Flow Applied to a One-Layer Isentropic Model of the Atmosphere. *Q. J. Meteorol. Soc.* 127:597–600, 2001.

Virasoro, M. A. Variational Principle for Two-Dimensional Incompressible Hydrodynamics and Quasigeostrophic Flows. *Phys. Rev. Lett.* 47:1181–1183, 1981.

HAMILTONIAN DESCRIPTION OF FLUID AND PLASMA SYSTEMS WITH CONTINUOUS SPECTRA

P. J. MORRISON
Department of Physics and Institute for Fusion Studies
University of Texas at Austin
Austin, TX 78712-1060, U.S.A.

In memory of Pedro Ripa 1946–2001

Abstract. We show how to transform a large class of infinite degree-of-freedom Hamiltonian systems into normal form. The energy-Casimir method that is widely used for ascertaining stability in Hamiltonian fluid and plasma systems is only the first step. A complete description involves changing to coordinates in which the energy is diagonal. This amounts to a transformation to action-angle variables. Because fluid and plasma systems typically have a continuous eigenspectrum, this transformation is nontrivial. It will be shown that a family of integral transforms, which is a generalization of the Hilbert transform, yields action-angle variables for a large class of fluid and plasma systems.

Key words: Hamiltonian fluid and plasma dynamics, normal forms, stability

1. Introduction

The goal of this paper is to show how to transform a class of infinite degree-of-freedom Hamiltonian systems, i.e. Hamiltonian field theories, into action-angle form. The class of systems is distinguished by two important features: first, it is Hamiltonian in the noncanonical sense of possessing a Lie-Poisson bracket description of the dynamics (e.g. Shepherd, 1990; Morrison, 1998; Marsden & Ratiu, 1999) and second, the class as a whole possesses a particular kind of continuous spectra that is akin to that discovered by Van Kampen (1955) in plasma physics. These two features present roadblocks to the usual construction of the transformation to action-angle form. Because of the noncanonical form one must first *canonize*, i.e. find a set of canonical variables. Because of the continuous spectrum the transformation constructed to *diagonalize* the Hamiltonian is novel and possesses some intricacies.

O.U. Velasco Fuentes et al. (eds.), Nonlinear Processes in Geophysical Fluid Dynamics, 53-69.

Finite linear Hamiltonian systems are simplified by transforming them into normal form (Birkhoff, 2002; Williamson, 1936; Moser, 1958), a form that is determined mostly by their eigenspectra. Action-angle form is the normal form that occurs when the system is stable, i.e. when the spectrum is neutral and nondegenerate. For finite systems the normal form for stable systems is given variously by one of the following:

$$H = \sum_{n=1}^{N} \frac{\omega_n}{2} \left(p_n^2 + q_n^2 \right) = i \sum_{n=1}^{N} \omega_n \, Q_n \, P_n = \sum_{n=1}^{N} \omega_n \, J_n \,, \tag{1}$$

where (q_n, p_n) and (Q_n, P_n) are canonical coordinates, and the last expression of (1) is the action-angle form, with J_n denoting the action variable. The frequency associated with the degree of freedom labeled by n is given by $\omega_n := \sigma_n |\omega_n|$ with $\sigma_n \in \{-1, 1\}$. The quantity σ_n is the signature that determines whether or not a stable oscillation possesses positive or negative energy and plays an important role in the bifurcation theory described by the Krein-Moser theorem (Kreĭn, 1950; Kreĭn & Jakubovič, 1980; Moser, 1958).

Infinite systems have the capacity for rich spectra composed of discrete, continuous, and residual components (e.g. Kato, 1966; Reed & Simon, 1980; Riesz & Nagy, 1955). If we restrict to the case of only a continuous spectrum we would expect the analog of the last term of (1) to be

$$H = \int \omega(u) \, \mathcal{J}(u) \, du \,, \tag{2}$$

where $\mathcal{J}(u)$ is a field action variable and the discrete sum over n is replaced by the integration over a continuum label u. Describing how to effect this transformation for the class of infinite dimensional Hamiltonian systems is the main result of this paper. This is a substantial task and we only attempt to sketch the basic ideas. (Greater detail for the cases of Vlasov-Poisson and shear flow dynamics can be found in Morrison & Pfirsch (1992), Morrison & Shadwick (1994), and Balmforth & Morrison (2002), with the most rigor given in Morrison (2000).)

The canonization and diagonalization procedure we describe here complements the stability arguments of Rayleigh (1896) and the energy-Casimir type of Kruskal & Oberman (1958), Arnol'd (1965), Ripa (1983), and others. (See e.g. Holm *et al.*, 1985; Morrison & Eliezer, 1986; Morrison, 1998 for more details.) The essence of the energy-Casimir method is to ascertain stability, as Dirichlet did for finite Hamiltonian systems, by using essentially the Hamiltonian as a Lyapunov fuction. Our diagonalization procedure completes the stability problem by finding the transformation to normal mode coordinates. Alternatively, one can view the procedure in an operator theory context. From this point of view we transform the operator that embodies the linear dynamics into a multiplication operator (Reed & Simon, 1980), akin to the procedure for diagonalizing matrices.

In Section 2 we describe our class of infinite-dimensional systems and give some examples. In Section 3 we see that the unifying principle of our class is the common Hamiltonian form. Also in this section we describe conservation laws with a particular emphasis on the momentum. Section 4 describes the nontrivial determination of equilibria and the equations for the nearby linear dynamics. The eigenvalue problem is described in Section 5, where the origin of a class of integral transforms, which are generalizations of the Hilbert transform, is also described. In Section 6 the canonization and diagonalization of the class of linear Hamiltonian systems is discussed. Finally in Section 7 we discuss future work.

2. A Class of Infinite-Dimensional Systems

2.1. SCALAR 2 + 1 MEAN FIELD THEORIES

The class of field theories we consider possesses a single independent variable $\zeta(q, p, t)$, which is a density-like variable that depends on the independent variables $z := (q, p)$ as well as time. Associated with the class of field theories are two phase spaces: the field phase space, which is the function space in which the density ζ resides, and the particle phase space of independent variables z. There is cause for confusion here because we introduce action-angle variables associated with both of these phase spaces: the field action-angle variables of (2), our main goal, and en route to this goal, a set of particle action-angle variables that below we denote by (θ, J). We write $\zeta : \mathcal{Z} \times \mathbb{R} \to \mathbb{R}$ where $\mathcal{Z}$ denotes the particle phase space, which we take to be $D_1 \times D_2$, $\Pi \times D_2$, or Π^2, where Π is the one-torus chosen depending on which of p and q is periodic, and $D_{1,2}$ are (not necessarily proper) subsets of $\mathbb{R}$. We will not be specific about the topology of $\mathcal{Z}$.

We suppose the density satisfies an equation of motion of the following form:

$$\frac{\partial \zeta}{\partial t} + [\zeta, \mathcal{E}] = 0\,, \tag{3}$$

where the particle Poisson bracket is defined by the usual expression $[f, g] = f_q g_p - g_q f_p$, where $f_q := \partial f / \partial q$ etc., and the quantity $\mathcal{E}$ is an energy-like quantity that we call the particle energy.

If (3) were a Liouville equation, then $\mathcal{E}$ would be a given function of z, and we would have a linear theory. However, we are concerned about mean field theories which are nonlinear partial integrodifferential equations. Such equations arise, for example, by truncation of BBGKY-like hierarchies, which results in a particular, generally global, functional dependence of the particle energy on the density. We determine our general class of systems by obtaining the particle energy in terms of the field energy given below.

2.2. FIELD ENERGY

The field energy is the integrated energy corresponding to a particular density ζ. We write the field energy in terms of energies corresponding to one-particle, H_1, two-particle, H_2, ... interactions, where

$$\begin{aligned} H_1[\zeta] &= \int_{\mathcal{Z}} h_1(z)\,\zeta(z)\,d^2z\,, \\ H_2[\zeta] &= \frac{1}{2}\int_{\mathcal{Z}}\int_{\mathcal{Z}} \zeta(z)\,h_2(z,z')\,\zeta(z')\,d^2z\,d^2z'\,, \end{aligned} \tag{4}$$

and the generalizations to H_3, H_4, ... are obvious. The quantities h_1 and h_2, the interaction kernels, are left unspecified. But, we suppose the two-particle interaction possesses the symmetry $h_2(z,z') = h_2(z',z)$.

The field energy of our class of systems is then given by $H[\zeta] = H_1 + H_2 + \ldots$. Henceforth, we will only consider field energies with the first two terms, $H[\zeta] = H_1 + H_2$.

The particle energy is obtained from the field energy by functional differentiation

$$\mathcal{E} := \frac{\delta H}{\delta \zeta} = h_1 + \int_{\mathcal{Z}} h_2(z,z')\,\zeta(z')\,d^2z'\,, \tag{5}$$

where the functional derivative is defined as usual by $\delta H = \int_{\mathcal{Z}} \delta\zeta\,\delta H/\delta\zeta\,d^2z$.

2.3. EXAMPLES

2.3.1. *Vlasov-Poisson*

In the case of Vlasov-Poisson we set $z = (x,p)$, which physically corresponds to a one degree-of-freedom particle phase space, and we set $\zeta = f(x,p,t)$, which is the phase space density that is chosen to give zero net charge. The one and two-particle interaction kernels are given respectively by the following:

$$h_1(z) = \frac{p^2}{2m}\,, \qquad h_2(z,z') = c|x - x'|\,, \tag{6}$$

where c is a constant. Thus the field energy is

$$H[\zeta] = \int_{\mathbb{R}^2} \frac{p^2}{2m} f\,dxdp + \frac{1}{8\pi}\int_{\mathbb{R}} E^2\,dx\,. \tag{7}$$

Upon taking the functional derivative, we obtain the usual Vlasov-Poisson particle energy, $\mathcal{E} := \delta H/\delta f = p^2/2m + e\phi[f](x)$.

2.3.2. *2D Euler*

In the case of the two-dimensional Euler fluid equations, we set $z = (x, y)$, which physically corresponds to a two-dimensional configuration space, and we set $\zeta = \zeta(x, y, t)$, which is the scalar vorticity. The one- and two-particle interaction kernels are given respectively by the following

$$h_1(z) \equiv 0\,, \qquad h_2(z, z') = c \ln[(x - x')^2 + (y - y')^2]\,, \tag{8}$$

where c is a constant. Thus the field energy is given by

$$H[\zeta] = \int_{\mathbb{R}^2} \frac{v^2}{2}\, dxdy\,, \tag{9}$$

where the velocity is related to the scalar vorticity by $\zeta = \hat{z} \cdot \nabla \times \mathbf{v}$, and functional differentiation gives $\mathcal{E} := \delta H/\delta\zeta = \psi[\zeta](x)$. For this case the particle energy corresponds to the streamfunction, where $\triangle\psi = \zeta$.

2.3.3. *Other Examples*

Many other examples with physical content exist. For example, Jeans equation for stellar dynamics is obtained by changing the sign of the h_2 of the Vlasov equation and removing the zero net charge condition, and quasi-geostrophy is obtained by changing the relationship between the vorticity and the streamfunction. Interesting examples where the underlying characteristics correspond to integrable n-body problems are the Cologero-Moser system (Moser, 1975; Illner, 2000), and the apparently unstudied cases of Stäckel potential interaction, Smereka's product potential (Smereka, 1998), and the Toda-Vlasov equation.

3. Hamiltonian Form and Conservation Laws

3.1. MEAN FIELD HAMILTONIAN FORM

It is evident from the above that the energy, H, which will turn out to be the Hamiltonian, is quadratic in ζ. Usually quadratic Hamiltonians correspond to linear dynamics, as is the case for the simple oscillations described in Section 1. However, the Vlasov-Poisson equation, the Euler equations, and indeed every equation in the class described in Section 2, are quadratically nonlinear. The discrepancy lies in the fact that the variable ζ does not constitute a set of canonically conjugate field variables. Theories of the kind described in Section 2 possess a description in terms of noncanonical degenerate Poisson brackets, brackets that are sometimes referred to as Lie-Poisson brackets. These brackets are linear in the field variables and thus account for the nonlinearity missing in the Hamiltonians. Much has been written about Lie-Poisson brackets, so we will not dwell, but refer the reader interested in more detail to Morrison (1998) and Marsden & Ratiu (1999). It is this Hamiltonian form that is the unifying theme that defines our class of systems.

The noncanonical Lie-Poisson bracket of our class is given by

$$\{F, G\} = \int_{\mathcal{Z}} \zeta \left[\frac{\delta F}{\delta \zeta}, \frac{\delta G}{\delta \zeta} \right] d^2 z \,. \tag{10}$$

Observe that this bracket depends explicitly upon the variable ζ, unlike usual Poisson brackets that only depend on (functional) derivatives of the canonical variables. Like canonical Poisson brackets, the bracket of (10) is antisymmetric and satisfies the Jacobi identity. Using (10) the equations of motion are obtained in the form

$$\frac{\partial \zeta}{\partial t} = \{\zeta, H\} = -[\zeta, \frac{\delta H}{\delta \zeta}] = -[\zeta, \mathcal{E}] \,, \tag{11}$$

where $H = H_1 + H_2$ is defined by (4). This constitutes the Hamiltonian form.

Associated with this Hamiltonian form are natural constants of motion; viz. the energy or Hamiltonian $H[\zeta] = H_1 + H_2$ and an infinity of constants known as Casimir invariants,

$$C[\zeta] = \int_{\mathcal{Z}} \mathcal{C}(\zeta) \, d^2 z \,, \tag{12}$$

where $\mathcal{C}(\zeta)$ is an arbitray function. The Casimir invariants arise from degeneracies in the Poisson bracket and do not occur in canonical theories. An important additional invariant is the momentum $P[\zeta]$, a quantity that is Hamiltonian dependent. We discuss this invariant further below.

3.2. MOMENTUM

Momentum invariants generally arise from translation symmetries that in the present context might be determined by the form of h_2. This is how the strong version of Newton's third law is built into the n-body problem. We generalize this idea significantly as follows. We state that our system has a momentum invariant if there exists a canonical transformation

$$z = (q, p) \longleftrightarrow \bar{z} := (\chi, \pi)$$

such that in the new particle coordinates $\bar{z} := (\chi, \pi)$, the interactions h_1 and h_2 have the form

$$h_1 \circ z = \bar{h}_1(\pi) \,, \qquad h_2 \circ (z, z') = \bar{h}_2(\pi, \pi', |\chi - \chi'|) \tag{13}$$

upon composition with $z(\bar{z})$. We will refer to the coordinates (χ, π) as *momentum coordinates.*

If such a transformation exists, then the following *momentum* is conserved:

$$P[\zeta] = \int_{\mathcal{Z}} \pi(z)\, \zeta(z)\, d^2 z\,. \tag{14}$$

This can be shown by differentiating (14), changing to momentum coordinates, and using the fact that h_2 depends on $|\chi - \chi'|$.

Building in momentum conservation in the manner above is not the most general way possible. Ultimately, momentum conservation should arise from Nöther's theorem in an action principle, which could then be reduced to obtain the class of systems presented here. However, the definition given is sufficient for our purposes. It includes that for the Vlasov system, where $P = \int pf\, dpdq$ and the two momenta for 2D Euler system, where $P_y = \int x\zeta dxdy$ and $P_x = -\int y\zeta dxdy$. In the case of 2D Euler, there exists a coordinate system in which h_2 possesses translation invariance in both spatial directions, and for this reason we get the two conserved momenta.

4. Equilibria and Linearization

We now effect the usual procedure of expanding about an equilibrium state by setting $\zeta = \zeta_e + \delta\zeta$ and retaining terms of first order in $\delta\zeta$. We will see that our class of Hamiltonian systems has the property that eigenvalue problems resulting from the linearization procedure possess continuous spectra.

4.1. EQUILIBRIA

Equilibria, ζ_e, satisfy

$$\frac{\partial \zeta_e}{\partial t} = 0 = \{\zeta_e, H\}\,, \tag{15}$$

which upon using (10) implies

$$[\zeta_e, \mathcal{E}_e] = 0\,, \tag{16}$$

where we add the subscript 'e' to the particle energy because it depends functionally on ζ_e. Equation (16) implies functional dependence of ζ_e on $\mathcal{E}_e$ or vice versa. More generally it implies the existence of a single variable, say J, such that $\zeta_e(J)$ and $\mathcal{E}_e(J)$.

Given $J(q,p)$ one can obtain a θ such that the pair (θ, J) constitutes a canonically conjugate system of particle coordinates. We call these coordinates *equilibrium coordinates*, which are in general distinct from the momentum coordinates (χ, π) defined above. Thus we have three sets of canonical particle coordinates at our disposal

$$(q,p) \longleftrightarrow (\theta, J) \longleftrightarrow (\chi, \pi)\,.$$

We assume the problem is stated in terms of (q,p) and that we know (χ, π), somehow, possibly because of the physics. We now describe in more detail how one might obtain (θ, J).

If we knew $\mathcal{E}_e(q,p)$ then we could attempt to use the usual procedure for obtaining action-angle variables for a one degree-of-freedom Hamiltonian system. But the question remains, how do we obtain $\mathcal{E}_e(q,p)$ from the equilibrium equation (16)? It is clear that (16) alone cannot determine an equilibrium because this equation allows the choice of a free function, e.g. any function $\zeta_e(\mathcal{E}_e)$ is a solution of (16). In fact there are two routes one can follow in hope of finding a solution: one can assume the function $\zeta_e(\mathcal{E}_e)$ and seek $\mathcal{E}_e = \mathcal{E}_e(z)$ or, conversely, one can assume $\mathcal{E}_e = \mathcal{E}_e(z)$ and seek $\zeta_e(\mathcal{E}_e)$. Once the free function is chosen, we seek to remove the ambiguity by using the expression for the particle energy.

Let us follow the first route and assume a form for the function $\zeta_e(\mathcal{E}_e)$. We then insert this function into the expression for the particle energy (5) to obtain

$$\mathcal{E}_e(z) = h_1(z) + \int_{\mathcal{Z}} h_2(z,z')\, \zeta_e(\mathcal{E}_e(z'))\, d^2z'\,. \tag{17}$$

This integral equation is the generalization of the equilibrium elliptic equations that are obtained in the cases of the Vlasov-Poisson and 2D Euler examples. The point here is that there may be no elliptic equation corresponding to the inverse of the integral operator with the kernel h_2. However, we are fortunate that (17) has the form of a Hammerstein integral equation (Hammerstein, 1930), a nonlinear integral equation about which much is known. In the sequel we will assume we have a solution of this equation, one that is sufficiently well-behaved for our purposes. There are very interesting analysis questions pertaining to (17), but we leave these to a future publication.

We conclude this subsection with a discussion of some special cases. If we have a momentum invariant, then we can rewrite (17) in terms of momentum coordinates as follows:

$$\mathcal{E}_e = \bar{h}_1(\pi) + \int_{\mathcal{Z}} \bar{h}_2(\pi, \pi', |\chi - \chi'|)\, \zeta_e(\mathcal{E}_e)\, d\chi' d\pi'\,.$$

Some advantage is achieved by this form because the difference kernel makes the problem amenable to Fourier transform techniques. Other simplifications occur for the cases below:

4.1.1. *Vlasov-like*

For this case, the momentum and equilibrium particle coordinates coincide. For Vlasov, $\pi = J$, $\bar{h}_2 = \bar{h}_2(\chi - \chi')$, and the convolution form leads to a solution. In addition to the Vlasov equation, defect dynamics (Balmforth *et al.*, 1996) falls into this category. In analogy to Vlasov theory, one can have homogeneous equilibria or the more complicated case of BGK-like equilibria.

4.1.2. *Euler-like*

For this case, $\bar{h}_1 \equiv 0$ and $\bar{h}_2 = \bar{h}_2(\pi - \pi', \chi - \chi')$ and we have two momenta and two difference directions available for Fourier transform techniques.

4.1.3. *Others*

There are many other cases that possess special properties. The integrable systems mentioned above; i.e. the Cologero-Moser systems, systems with Stäckel potential interaction, and the Toda-Vlasov equation would be interesting to analyze.

4.2. LINEARIZATION

Now we suppose that the equilibrium problem has been solved and that we have found the equilibrium coordinates; i.e. it is assumed that the transformation $(q,p) \longleftrightarrow (\theta, J)$ is in hand and the function $\zeta_e(J)$ is known. Moreover, it is assumed known that $\zeta_e(J)$ posseses only a continuous spectrum. Energy-Casimir or Rayleigh-like stability arguments, arguments based on the Green transform (Hille, 1976), or arguments based on the Nyquist method (Balmforth & Morrison, 1998) can be used to rule out stable or unstable discrete spectra.

Setting $\zeta = \zeta_e(J) + \delta\zeta(\theta, J, t)$ and expanding (11) to first order in $\delta\zeta$ gives

$$\frac{\partial \delta\zeta}{\partial t} + [\delta\zeta, \mathcal{E}_e] + [\zeta_e, \delta\mathcal{E}] = 0\,, \tag{18}$$

where $\Omega(J) := d\mathcal{E}_e/dJ$, $\delta\mathcal{E} = \int_{\mathcal{Z}} h_2(z,z')\,\delta\zeta(z')\,d^2z'$ written in terms of (θ, J), and because these equilibrium coordinates are canonical $[f,g] = f_\theta g_J - g_\theta f_J$. Thus, (18) is equivalent to

$$\frac{\partial \delta\zeta}{\partial t} + \Omega(J)\frac{\partial \delta\zeta}{\partial \theta} = \frac{d\zeta_e}{dJ}\frac{\partial \delta\mathcal{E}}{\partial \theta}\,. \tag{19}$$

Equations of the form of (19) generally possess a continuous spectrum, which we now turn to.

5. Spectrum and Integral Transform

5.1. SPECTRUM

To see the origin of the continuous spectrum we insert

$$\delta\zeta = \sum_k \zeta_k(J)\, e^{ik\theta - ik\omega t}$$

into (19) and obtain

$$\Big(\Omega(J) - \omega\Big)\zeta_k = \zeta_e'\, \mathcal{E}_k\,, \tag{20}$$

where $\zeta_e' := d\zeta_e/dJ$ and $\mathcal{E}_k$ is given by the following expression

$$\mathcal{E}_k(J) = \sum_{k'} \int \mathcal{H}_{k,k'}(J,J')\, \zeta_{k'}(J')\, dJ'\,, \tag{21}$$

which is obtained from $\delta\mathcal{E}$ by changing to equilibrium coordinates and Fourier expanding in the angles. (Note that $\mathcal{H}_{k,k'}$ depends upon h_2 and ζ_e.) Equation (20) is an eigenvalue problem for the eigenvalue ω.

The left hand side of (20) vanishes for values of J such that $\omega = \Omega(J)$, and this singularity is recognized in plasma physics and fluid mechanics as the origin of the continuous spectrum. In plasma physics it corresponds to wave-particle resonance, while in the fluid mechanics of shear flow it is called the critical layer (or line) and it corresponds to the matching of a background equilibrium shear velocity to the phase velocity of a perturbation.

Following Van Kampen (1955) we write a solution of (20) in the form:

$$\zeta_k = \lambda_k\, \delta(\Omega - \omega) + \mathcal{P}\, \frac{\zeta_e'\, \mathcal{E}_k}{\Omega - \omega}\,, \tag{22}$$

where δ is the Dirac distribution and $\mathcal{P}$ denotes Cauchy principal value. Equation (22) is of the form of a continuum eigenfunction corresponding to the continuous eigenspectrum labeled by ω. We assume $\Omega(J)$ is monotonic. Thus the eigenfunction labeled by ω can equally well be labeled by J_ω where $\omega = \Omega(J_\omega)$.

Note, the eigenfunction of (22) is indeterminate because the parameter λ_k is unknown and because it is self-referential in that $\mathcal{E}_k$ depends on ζ_k. The parameter λ_k is determined by a normalization condition, e.g. $\int \zeta_k\, dJ = 1$, and the following equation for $\mathcal{E}_k$ is obtained by inserting (22) into (21):

$$\mathcal{E}_k(J, J_\omega) = \sum_{k'} \mathcal{H}_{k,k'}(J, J_\omega) + \sum_{k'} \int \mathcal{E}_{k'}(J', J_\omega)\, \mathcal{F}_{k,k'}(J, J', J_\omega)\, dJ'\,, \tag{23}$$

where the kernel

$$\mathcal{F}_{k,k'}(J,J',J_\omega) := \left[\frac{\mathcal{H}_{k,k'}(J,J') - \mathcal{H}_{k,k'}(J,J_\omega)}{\Omega(J') - \Omega(J_\omega)}\right] \zeta_e'(J')$$

is well-behaved enough to apply the Fredholm theory of integral equations. [See Morrison & Pfirsch (1992), Morrison (2000), and Balmforth & Morrison (2002) for more details in the context of the Vlasov and 2D Euler equations.]

More rigorous statements regarding the spectrum are made in a Banach space, $\mathcal{B}$, setting. In the remainder of this subsection we make a few comments in this regard and leave serious analysis for a possible future publication. To this end we write

$$\mathcal{L}_k\zeta_k := \Omega(J)\zeta_k - \zeta_e' \, \mathcal{E}_k[\zeta_k] = \omega \, \zeta_k \,, \tag{24}$$

where the linear operator $\mathcal{L}_k : \mathcal{B} \to \mathcal{B}$ is our concern. We partition the spectrum of $\mathcal{L}_k$ as follows: $\sigma = \sigma_p \cup \sigma_c \cup \sigma_r$. An eigenvalue ω is in the point spectrum, σ_p, if $\mathcal{L}_k - \omega\mathcal{I}$ is not one-one, where $\mathcal{I}$ is the identity operator. If ω is such that the range of $\mathcal{L}_k - \omega\mathcal{I}$ is not dense in the Banach space of interest, then ω is in the residual spectrum σ_R, and if ω is such that the inverse of $(\mathcal{L}_k - \omega\mathcal{I})$, defined on its range, is unbounded, then ω is in the continuous spectrum σ_c. We find this partition convenient because if σ_r is null, then the approximate or Weyl spectrum corresponds to $\sigma_p \cup \sigma_c$. Note, there are other commonly used decompositions of the spectrum (e.g. Reed & Simon, 1980; Riesz & Nagy, 1955.)

In Section 4.2 we made some comments on how, for a given kernel, an equilibrium can be ensured to have no discrete modes, i.e. ensured to have a null point spectrum. We assume this has been arranged, i.e. we know $\mathcal{L}_k - \omega\mathcal{I}$ is one-one.

The operator $\mathcal{L}_k$ is the sum of a multiplication operator and an integral operator that under mild conditions is bounded. It is well-known that a multiplication operator possesses a continuous spectrum and early theorems by Friedrichs and others (e.g. Kato, 1966) state conditions under which this continuous spectrum survives perturbation by the addition of an integral operator. For example, an operator that is composed of a bounded self-adjoint piece perturbed by the addition of a compact piece retains its continuous spectrum.

One interpretation of the diagonalization procedure we are attempting here is akin to the diagonalization of matrices by coordinate changes. Our goal is to transform the operator $\mathcal{L}_k$ into a pure multiplication operator by a coordinate change. This procedure is described in Reed & Simon (1980) for bounded self-adjoint operators, but it is not confined to such operators. A sum over eigenfunctions of the form of (22) suggests a form for the diagonalizing transform that converts (24) into a pure multiplication operator. We turn to this now.

5.2. INTEGRAL TRANSFORM

The general form of the integral transform that diagonalizes the continuous spectrum of our class of Hamiltonian systems is given by

$$G[g](J) = \epsilon(J)\, g(J) + \mathcal{P}\int \zeta_e'(J)\, \frac{\mathcal{E}(J, J_\omega)\, g(J_\omega)}{\Omega(J) - \Omega(J_\omega)}\, dJ_\omega\,, \tag{25}$$

where g is the function that is being transformed, and the functions ϵ and $\mathcal{E}$ (given below) are determined by the functions h_1 and h_2 that define our system, as well as the equilibrium being studied, ζ_e. In this way the transform is tailored to the problem at hand. Here, for clarity, we have suppressed the dependence on k to display that the transformation is the sum of a multiplication operator, a multiplication of g by a function ϵ, and an integral part that involves the Cauchy principal value, a generalization of the Hilbert transform with a kernal given by $\mathcal{E}\zeta_e'$.

To simplify matters we will assume the momentum and energy coordinates coincide so that $\bar{h} = \bar{h}(J, J', \theta - \theta')$. Consequently (21) and (23) become

$$\mathcal{E}_k(J) = \int h_k(J, J')\, \zeta_k(J')\, dJ'\,, \tag{26}$$

and

$$\mathcal{E}_k(J, J_\omega) = h_k(J, J_\omega) + \int \zeta_e(J')\, \mathcal{E}_k(J', J_\omega)\, \frac{[h_k(J, J') - h_k(J, J_\omega)]}{\Omega(J') - \Omega(J_\omega)}\, dJ'\,, \tag{27}$$

respectively. For this choice of interaction $\mathcal{H}_{k,k'}(J, J') = \delta_{k,k'}\, h_k(J, J')$.

Many properties of transforms of the form of (25) can be proven by techniques similar to those used in Hilbert transform theory (e.g. Stein & Weiss (1971) and other works on Calderón-Zygmund theory). Under mild conditions it can be shown that G is a bounded linear operator on suitable Banach spaces (L_p and $C^{0,\alpha}$). With more difficulty, the inverse can be constructed, and it is of the same general form as (25). In addition there exist identities which can be used to show that the transform diagonalizes the Hamiltonian. The justification of these statements follows closely those for the Vlasov (Morrison, 2000) and shear flow (Balmforth & Morrison, 2002) cases. Below we list the results, but present their justification elsewhere.

5.2.1. *Transform and its Inverse*

Multiplying (24) by an amplitude, $g_k(J_\omega)$, and integrating over J_ω motivates the following form for the transform as a sum over eigenfunctions:

$$G_k[g_k](J, t) := \epsilon_k^{(r)}(J)\, g_k(J, t) + \mathcal{P}\int \frac{\zeta_e'(J)\mathcal{E}_k(J, J_\omega)}{\Omega(J) - \Omega(J_\omega)}\, g_k(J_\omega, t)\, dJ_\omega\,. \tag{28}$$

Observe we have explicitly displayed the t dependence to reinforce the idea that this is a coordinate change. Now, specifically, we define

$$\epsilon_k^{(r)}(J_\omega) := 1 - \mathcal{P}\int \frac{\zeta_e'(J)\mathcal{E}_k(J, J_\omega)}{\Omega(J) - \Omega(J_\omega)}\, dJ \tag{29}$$

and $\mathcal{E}_k(J, J_\omega)$ is the solution to (23).

The inverse of (28), under our assumed conditions of monotonicity and no discrete spectrum, is given by the following:

$$\begin{aligned} \widehat{G}_k[f_k](J_\omega, t) \ :=\ & \frac{1}{|\epsilon_k(J_\omega)|^2}\left[\epsilon_k^{(r)}(J_\omega) f_k(J_\omega, t) + \right. \\ & \left. +\mathcal{P}\int \frac{\zeta_e'(J_\omega)\mathcal{E}_k(J, J_\omega)}{\Omega(J) - \Omega(J_\omega)} f_k(J, t)\, dJ\right], \end{aligned} \tag{30}$$

where $|\epsilon_k(J)|^2 := (\epsilon_k^{(r)})^2 + (\epsilon_k^{(i)})^2$ and

$$\epsilon_k^{(i)}(J_\omega) := \pi \mathcal{E}_k(J_\omega, J_\omega)\frac{\zeta_e'(J_\omega)}{\Omega'(J_\omega)}. \tag{31}$$

That $\widehat{G}$ is the inverse of G is most simply shown by making use of the Poincaré-Bertrand theorem on the interchange of the order of integration for singular integrals.

5.2.2. *Transform Identities*

We record two identities for (30) that will be needed below.

$$\widehat{G}_k[\Omega\, \zeta_k](J_\omega) = \Omega(J_\omega)\widehat{G}_k[\zeta_k](J_\omega) + \frac{\zeta_e'(J_\omega)}{|\epsilon_k|^2(J_\omega)}\mathcal{P}\int \zeta_k(J, t)\, \mathcal{E}_k(J, J_\omega)\, dJ\,, \tag{32}$$

and

$$\widehat{G}_k[\zeta_e'\mathcal{E}_k](J_\omega) = \frac{\zeta_e'(J_\omega)}{|\epsilon_k|^2(J_\omega)}\mathcal{P}\int \zeta_k(J)\, \mathcal{E}_k(J, J_\omega)\, dJ\,. \tag{33}$$

These are shown by techniques similar to those used for verifying the inverse.

6. Linear Canonization and Diagonalization

6.1. LINEAR HAMILTONIAN FORM

In the energy-Casimir method (Holm *et al.*, 1985; Morrison & Eliezer, 1986; Morrison, 1998) one chooses the Casimir invariant C of (12) such that the vanishing of the first variation of the quantity $F := H + C$ gives the equilibrium of interest, and then (modulo some technicalities) one examines the second variation $\delta^2 F$ for positive definiteness in order to prove stability. Physically, $\delta^2 F$ corresponds to the energy content of a perturbation away from equilibrium, and it serves as the Hamiltonian for the linear dynamics. From (4) and (12), it is seen to be given by

$$\delta^2 F = \delta^2 H + \frac{1}{2}\int \mathcal{C}''(\zeta_e)\,(\delta\zeta)^2\, d\theta dJ = \delta^2 H - \frac{1}{2}\int \frac{\mathcal{E}_e'(J)}{\zeta_e'(J)}\,(\delta\zeta)^2\, d\theta dJ\,. \quad (34)$$

Because $\delta^2 F$ is the Hamiltonian for the linear dynamics we rename it H_L. It, together with the linear Poisson bracket,

$$\{F,G\}_L = \int \zeta_e(J)\left[\frac{\delta F}{\delta\delta\zeta}, \frac{\delta G}{\delta\delta\zeta}\right] d\theta dJ\,,$$

generates the linear dynamics as follows:

$$\frac{\partial\delta\zeta}{\partial t} = \{\delta\zeta, H_L\}_L\,.$$

Upon expanding the angle-like dependence, θ, of ζ_k in a Fourier sum, we obtain the following expressions for the linear Hamiltonian and Poisson bracket, respectively:

$$H_L = \frac{1}{2}\sum_{k,k'}\int\!\!\int \zeta_k(J)\,\mathcal{H}_{k,k'}(J,J')\,\zeta_{k'}(J')\,dJdJ' - \frac{1}{2}\sum_k \int \frac{\mathcal{E}_e'(J)}{\zeta_e'(J)}\,\zeta_{-k}\zeta_k\,dJ \quad (35)$$

and

$$\{F,G\}_L = \sum_{k=1}^{\infty} ik \int \zeta_e' \left(\frac{\delta F}{\delta\zeta_k}\frac{\delta G}{\delta\zeta_{-k}} - \frac{\delta G}{\delta\zeta_k}\frac{\delta F}{\delta\zeta_{-k}}\right) dJ\,. \quad (36)$$

Note, because the sum now extends only over positive integers, ζ_k and ζ_{-k} are to be viewed as independent variables. Also, note we have not made any Fourier expansion in the time variable.

6.2. CANONIZATION AND DIAGONALIZATION

If it were not for the presence of $ik\zeta_e'$ in the bracket of (36), ζ_k would be the canonical conjugate of ζ_{-k}. If we define

$$q_k(J,t) := \zeta_k(J,t) \qquad \text{and} \qquad p_k(J,t) = \frac{\zeta_{-k}(J,t)}{ik\zeta_e'}\,, \quad (37)$$

then (36) becomes

$$\{F,G\}_L = \sum_{k=1}^{\infty}\int \left(\frac{\delta F}{\delta q_k}\frac{\delta G}{\delta p_k} - \frac{\delta G}{\delta q_k}\frac{\delta F}{\delta p_k}\right) dJ\,,$$

i.e. we have canonized the bracket.

Diagonalization is a more difficult procedure, but formally we can define the mixed variable generating functional

$$\mathcal{F}[q,P] = \sum_{k=1}^{\infty} \int P_k(J)\, \widehat{G}[q_k](J)\, dJ$$

where $\widehat{G}$ is defined by (30). This type-2 mixed variable generating functional effects the canonical transformation from the old field coordinates (q_k, p_k) to the new field coordinates (Q_k, P_k) according to

$$p_k(J) = \frac{\delta \mathcal{F}[q,P]}{\delta q_k(J)} = \widehat{G}^{\dagger}[P_k](J) \quad \text{and} \quad Q_k(J) = \frac{\delta \mathcal{F}[q,P]}{\delta P_k(J)} = \widehat{G}[q_k](J)\,. \tag{38}$$

Making use of (37) and (26) we write the linear Hamiltonian (35) in the form

$$H_L = \sum_{k=1}^{\infty} ik \int p_k \left[\zeta_e' \mathcal{E}_k - q_k \mathcal{E}_e'\right] dJ\,, \tag{39}$$

into which we insert p_k and q_k from (38), to obtain

$$H_L = \sum_{k=1}^{\infty} ik \int P_k \left(\widehat{G}[\zeta_e' \mathcal{E}_k] - \widehat{G}[\mathcal{E}_k' G[Q_k]]\right) dJ\,, \tag{40}$$

Using (32) and (33) the new Hamiltonian takes the form

$$H_L = -\sum_{k=1}^{\infty} \int ik\, \Omega(J) Q_k(J)\, P_k(J) dJ\,. \tag{41}$$

Demonstrating the above achieves our final goal, because transforming from (41) to (2) is elementary.

7. Conclusions

It is evident that there are many avenues for future work. In conclusion we list some of them.

- Investigate the consequences of the signature of the continuous spectrum; i.e. prove a Krein-Moser theorem in a Banach space setting.
- Investigate the theory of adiabatic invariants in this infinite dimensional Hamiltonian context by adding explicit time dependence to the Hamiltonian. Some results for finite systems in the context of atmospheric models appear in Wirosoetisno & Shepherd (2000).

- Obtain the analog of Birkhoff's nonlinear normal form theory for our class of infinite dimensional Hamiltonian systems with continuous spectra. Some results in this direction appear in Yudichak (2001).
- Obtain our class of infinite dimensional Hamiltonian systems by reduction (Morrison, 1998; Marsden & Ratiu, 1999) of a canonical system with symmetry.
- Investigate the role played by (25) in attempts to understand the function phase space tangent bundle geometry.

Acknowledgements

This research was supported by the US Department of Energy Contract No. DE-FG03-96ER-54346.

References

Arnol'd, V. Conditions for Nonlinear Stability of Stationary Plane Curvilinear Flows of an Ideal Fluid. *Soviet Math. Dokl.*, 6:773–777, 1965.

Balmforth, N. J., D. del-Castillo-Negrete, and W. R. Young. Dynamics of Vortical Defects in Shear. *J. Fluid Mech.*, 333:197–230, 1996.

Balmforth, N. J. and P. J. Morrison. A Necessary and Sufficient Instability Condition for Inviscid Shear Flow. *Studies in Appl. Math.*, 102:309–344, 1998.

Balmforth, N. J. and P. J. Morrison. Hamiltonian Description of Shear Flow. In J. Norbury and I. Roulstone, editors, *Large-Scale Atmosphere-Ocean Dynamics II*, pages 117 – 142. Cambridge, Cambridge, 2002.

Birkhoff, G. D. *Dynamical Systems.* American Mathematical Society Colloquium Publication IX, Providence, Rhode Island, 1927.

Hammerstein, A. Nichtlineare Integralgleichungen nebst Anwendungen. *Acta Math.*, 54:117–176, 1930.

Hille, E. *Ordinary Differential Equations in the Complex Domain.* Wiley, New York, 1976.

Holm, D. D., J. E. Marsden, T. S. Ratiu, and A. Weinstein. Nonlinear Stability of Fluid and Plasma Equilibria. *Phys. Rep.*, 82:1–116, 1985.

Illner, R. Stellar Dynamics and Plasma Physics with Corrected Potentials: Vlasov, Manev, Boltzmann, Smoluchowski. *Fields Inst. Comm.*, 27:98–108, 2000.

Kato, T. *Perturbation Theory for Linear Operators.* Springer-Verlag, Berlin, 1966.

Kreĭn, M. G. A Generalization of Some Investigations on Linear Differential Equations with Periodic Coefficients. *Dokl. Akad. Nauk SSSR A*, 73:445–448, 1950.

Kreĭn, M. G. and V. A. Jakubovič. *Four Papers on Ordinary Differential Equations.* American Mathematical Society, Providence, Rhode Island, 1980.

Kruskal, M. D. and C. Oberman. On the Stability of Plasma in Static Equilibrium. *Phys. Fluids*, 1, 275–280, 1958.

Marsden, J. E. and T. Ratiu. *Introduction to Mechanics and Symmetry.* Texts in Applied Mathematics vol. 17, 2nd edition, Springer-Verlag, Berlin, 1999.

Morrison, P. J. and S. Eliezer. Spontaneous Symmetry Breaking and Neutral Stability in the Noncanonical Hamiltonian Formalism. *Phys. Rev.*, 33A:4205–4214,1986.

Morrison, P. J. Hamiltonian Description of the Ideal Fluid. *Rev. Mod. Phys.*, 70:467–521, 1998.

Morrison, P. J. Hamiltonian Description of Vlasov Dynamics: Action-Angle Variables for the Continuous Spectrum. *Trans. Theory and Stat. Phys.*, 29:397–414, 2000.

Morrison, P. J. and D. Pfirsch. Dielectric Energy Versus Plasma Energy, and Action-Angle Variables for the Vlasov Equation. *Phys. Fluids B*, 4:3038–3057, 1992.

Morrison, P. J. and B. Shadwick. Canonization and Diagonalization of an Infinite Dimensional Noncanonical Hamiltonian System: Linear Vlasov Theory. *Acta Phys. Pol.*, 85:759–769, 1994.

Moser, J. K. New Aspects in the Theory of Stability of Hamiltonian Systems. *Comm. Pure Appl. Math.*, 11:81–114, 1958.

Moser, J. K. Three Integrable Hamiltonian Systems Connected with Isospectral Deformations. *Adv. Math.*, 16:197–220, 1975.

Rayleigh, J. W. S. *Theory of Sound.* Art. 369. Macmillan, London, 1896.

Reed, M. and B. Simon. *Methods of Modern Mathematical Physics I: Functional Analysis.* Academic Press, New York, 1980.

Riesz, F. and B. Sz.-Nagy. *Functional Analysis.* Frederick Ungar Publishing, New York, 1955.

Ripa, P. General Stability Conditions for Zonal Flows in a One-Layer Model on the Beta-Plane or the Sphere. *J. Fluid Mech.*, 126:463–489, 1983.

Shepherd, T. G. Symmetries, Conservation Laws, and Hamiltonian Structure in Geophysical Fluid Dynamics. *Adv. Geophs.*, 32:287–338, 1990.

Smereka, P. Synchronization and Relaxation for a Class of Globally Coupled Hamiltonian Systems. *Physica*, 124D:104–125, 1998.

Stein, E.M. and G. Weiss. *Introduction to Fourier Analysis on Euclidean Spaces.* Princeton University Press, New Jersey, 1971.

Van Kampen, N. G. On the Theory of Stationary Waves in Plasmas. *Physica*, 21:949–963, 1955.

Williamson, J. On the Algebraic Problem Concerning the Normal Forms of Linear Dynamical Systems. *Am. J. Math.*, 58:141–163, 1936.

Wirosoetisno, D. and T. G Shepherd, Averaging, Slaving and Balance Dynamics in a Simple Atmospheric Model. *Physica*, 141D:37–53, 2002.

Yudichak, T. W. *Hamiltonian Methods in Weakly Nonlinear Vlasov-Poisson Dynamics.* PhD thesis, Physics Department, The University of Texas at Austin, 2001.

STABLE VORTICES AS MAXIMUM OR MINIMUM ENERGY FLOWS

JONAS NYCANDER
Department of Meteorology
Stockholm University
S-106 91 Stockholm, Sweden

Abstract. A flow that maximizes or minimizes the energy in a set of isovortical flows is stationary and stable. For example, in ideal, two-dimensional flow, a circular vortex with a monotonic vorticity profile is a maximum energy flow, and therefore stable. An overview is given of more complex flows to which this analysis can be extended. A large class of stable and localised vortex solutions exists when the background is a linear shear flow. The vortex is elongated in the direction of the background flow, and its vorticity anomaly must have the same sign as the background shear. The method is also applied to vortices attached to seamounts. It is found that a large class of stable anticyclones exists. If the seamount is circular there are also stable cyclones, but these are destabilized by noncircularities in the topographic shape, unlike the anticyclones. The results both for vortices in shear flows and vortices attached to seamounts can be extended from two-dimensional barotropic flow to three-dimensional quasigeostrophic flow.

Key words: stable vortices, shear flow, bottom topography

1. Introduction

By introducing the energy-Casimir method, Arnol'd (1965; 1966) extended Rayleigh's and Fjørtoft's classical linear stability theorems to nonlinear stability to finite disturbances. Importantly, he also showed that the stability theorems are rooted in the general invariants of the dynamics (energy, momentum, and the Casimir invariants).

The link between stability and invariants suggests a general way to look for stability theorems, since the invariants are known from general symmetry properties. Perhaps the greatest success of this approach is the proof of "Ripa's theorem" (Ripa, 1983). It gives sufficient stability criteria for parallel shear flows governed by the shallow-water equations. In this case the derivation using Arnol'd's methods preceded the proof of linear stability using traditional Fourier methods.

O.U. Velasco Fuentes et al. (eds.), Nonlinear Processes in Geophysical Fluid Dynamics, 71-86.

All these applications of Arnol'd's method concern the stability of parallel shear flows. However, its application to localised vortices has proved more difficult. To understand why, we consider the simplest case of ideal two-dimensional (2D) flow:

$$\frac{\partial}{\partial t}\nabla^2\phi + J(\phi, \nabla^2\phi) = 0. \tag{1}$$

Here ϕ is the streamfunction and J denotes the Jacobian: $J(f,g) \equiv \partial_x f \partial_y g - \partial_y f \partial_x g$. We assume that there are no boundaries, i.e. the flow is on an infinite plane. Equation (1) describes the Lagrangian conservation of vorticity, i.e. $d\omega/dt = 0$, where the vorticity is given by

$$\omega = \nabla^2\phi. \tag{2}$$

Assume that $\phi_0(\mathbf{r})$ describes a steady flow. Setting $\phi = \phi_0 + \phi_1$ in eq. (1) and linearising, we obtain the equation for the perturbation $\phi_1(\mathbf{r}, t)$ of the streamfunction,

$$\left(\frac{\partial}{\partial t} + \mathbf{U}_0 \cdot \nabla\right)\nabla^2\phi_1 - J(\omega_0, \phi_1) = 0, \tag{3}$$

where $\mathbf{U}_0 = \hat{\mathbf{z}} \times \nabla\phi_0$.

Nonlinear stability is proved by showing that the so-called "pseudo energy" (or "Arnol'd invariant") is either positive definite or negative definite. In the limit of small perturbations it can be written

$$A = \frac{1}{2}\int \left[-\omega_1\phi_1 + \frac{d\phi_0}{d\omega_0}\omega_1^2\right] dS, \tag{4}$$

where $\omega_1 = \nabla^2\phi_1$, and the integral is taken over the xy-plane. It is easily seen that A is conserved by eq. (3). As will be seen below, A is also the second variation of the energy. The first term in eq. (4) can be integrated partially, showing that it is gives a positive contribution, provided that $\int \omega_1 dS = 0$ so that the perturbation energy is finite. If $d\phi_0/d\omega_0 \geq 0$ everywhere, A is therefore positive definite for arbitrary ϕ_1, and the flow is stable. In this case ϕ_0 is a minimum energy flow.

However, localised vortices are usually maximum energy flows rather than minimum energy flows (although this is not necessarily true in the presence of topography, as will be seen below.) We therefore consider the case when $d\phi_0/d\omega_0 \leq 0$ everywhere. To prove Arnol'd stabililty, it is necessary to show that A is negative definite, although the first term of the integrand is positive.

For channel flow it is possible to obtain an upper bound of the perturbation energy $\int(\nabla\phi_1)^2 dS$ in terms of the enstrophy $\int \omega_1^2 dS$. This can then be used to show that if the the channel is narrow enough, A is negative definite, and the flow nonlinearly stable (McIntyre and Shephard, 1987). On an infinite plane, on the other hand, this procedure does not work, since the perturbation energy is not bounded by the enstrophy. Indeed, A cannot be negative definite, as will now be demonstrated. Let ϕ_1 be defined as the solution of the equation

$$\nabla^2\phi_1 = c\frac{d\omega_0}{d\phi_0}\phi_1, \tag{5}$$

where c is a constant coefficient. It is readily seen that if $0 < c < 1$, then A is positive for this particular choice of ϕ_1, and hence cannot be negative definite. It remains to show that eq. (5) has a bounded solution for some c in this interval.

We assume that $d\omega_0/d\phi_0$ is negative and tends to zero rapidly as $r \to \infty$, as is the case if ϕ_0 describes a localised vortex. Equation (5) is then the same as the Schrödinger equation for a particle with zero total energy in a localised 2D potential well (*i.e.* a marginally bound state).

For $c = 1$ eq. (5) has the trivial solutions $\phi_1 = \partial\phi_0/\partial x$ and $\phi_1 = \partial\phi_0/\partial y$. Physically, these solutions correspond to a rigid translation of the equilibrium flow. [For this choice of ϕ_1 we get $A = 0$, which is already sufficient to show that the flow is not Arnol'd stable in the strict sense. This result is known as "Andrews' theorem" (Andrews, 1984; Carnevale and Shephard, 1990). However, it may be argued that this violation of Arnol'd's condition is unimportant, since a translation clearly does not correspond to an instability.] They both have at least one nodal line, which means that they correspond to excited states in the associated quantum mechanical problem. Hence, there also exists a ground state with negative energy (and without nodal lines).

Decreasing the value of c below unity means that the potential well is made more shallow. The marginally bound states then dissappear, but the ground state still exists. As c decreases, the energy of this state increases, until, for some c in $0 < c < 1$, the energy is zero, i.e. the ground state has become marginally bound. It then defines a bounded solution to eq. (5) for which A is positive.

Nevertheless, it is easy to see that a circular vortex is in fact a maximum energy flow if $d\omega_0/d\phi_0$ is negative, provided that the full set of Casimirs is used. (This will be shown in the following section.) Thus, perturbations ϕ_1 for which A is positive cannot be isovortical. We conclude that in order to prove stability of a localised vortex, we must restrict the perturbations to be isovortical, while Arnol'd's original formulation allows them to be arbitrary.

In the case of a circular vortex, these difficulties can be circumvented by transforming the problem to a rotating coordinate system. If this is chosen properly, the vortex is a minimum energy flow in the rotating system, and nonlinear stability follows. Equivalently, one may use the conservation of angular momentum, which adds another term (the "pseudo angular momentum") to the Arnol'd invariant (Carnevale and Shephard, 1990).

In more general, non-circular cases, however, this method cannot be used. The transformation to a rotating system is no longer useful, and the pseudo angular momentum is not conserved. In principle, it should be possible to show that A is negative definite by restricting the perturbations to be isovortical, but the technical difficulties in using this constraint appear to be great. This route was followed by Kloosterziel and Carnevale (1992) in a stability proof for concentric circular vortex patches.

Here we will follow a different route. Instead of working with the invariant A of the linearised equation, we will go back to the fully nonlinear equation and look for a global energy extremum. It will be shown in different situations that for fixed values of all the Casimir invariants (i.e. in a given set of isovortical flows), there exists a maximum energy flow (or, in some cases, a minimum energy flow) corresponding to a localised vortex. Since it is a constrained extremum of a conserved integral, such a flow is stable. Since A is effectively the second variation of the energy, we can also infer that A is sign definite if ϕ_1 is restricted to isovortical perturbations, even though this has not been shown directly.

The method proves the existence of stationary and stable vortices, and gives some of their characteristic properties, but does not give explicit solutions. These would have to be found numerically, for example by using the "modified dynamics" proposed by Vallis *et al.* (1989).

In this paper we will study applications of the method to vortices in shear flows, and to vortices attached to seamounts. This will be done both for 2D ideal flow and for three-dimensional (3D) quasigeostrophic flow. In several of these cases rigorous existence proofs using functional analysis have been found, but these will not be given here. Instead, an overview of the heuristic arguments will be given.

2. Ideal two-dimensional flow

2.1. CIRCULAR VORTEX

In this subsection it will be shown that a circular vortex with monotonic vorticity profile is a maximum energy flow. This does not in itself yield any new results; as mentioned above, Arnol'd stability can be proved by transformation to a rotating coordinate system. The purpose here is rather to introduce the method and explain the basic concepts.

We assume that the flow is governed by eq. (1). This equation conserves the total energy, which can be written as

$$E = -\frac{1}{2}\int \omega\phi \, dS = -\frac{1}{4\pi}\iint \omega(\mathbf{r})\omega(\mathbf{r}') \ln|\mathbf{r}-\mathbf{r}'| \, dS dS', \tag{6}$$

where ω is defined in eq. (2). Equation (1) also conserves the angular momentum,

$$M = \int \omega r^2 \, dS. \tag{7}$$

and the infinite set of Casimir integrals,

$$C_F = \int F[\omega(\mathbf{r})] \, dS, \tag{8}$$

where F is an arbitrary function. Flows that have the same value of all the Casimir integrals are called *isovortical* flows, and perturbations of the vorticity field that keep the flow in the same set of isovortical flows are said to be isovortical perturbations. Also, fields $\omega(\mathbf{r})$ in the same set of isovortical flows are said to be *rearrangements* of each other. A rearrangement can be thought of as an incompressible deformation.

A general first order isovortical perturbation of a given field $\omega(\mathbf{r})$ (i.e. one that satisfies $\delta C_F = 0$ for any F) is given by $\delta\omega = J(\xi, \omega)$, where $\xi(\mathbf{r})$ is arbitrary. To higher orders we have

$$\Delta\omega = \delta\omega + \frac{1}{2}\delta^2\omega + \ldots = J(\xi,\omega) + \frac{1}{2}J(\xi, J(\xi,\omega)) + \ldots \tag{9}$$

The energy variation caused by such a perturbation can be calculated from eq. (6). To first order we obtain

$$\delta E = -\int \phi\delta\omega \, dS = \int \xi J(\phi,\omega) \, dS.$$

Thus, if $\delta E = 0$ for arbitrary isovortical perturbations, then $J(\phi, \omega) \equiv 0$, which is the equation for a stationary flow. Thus, as is well known, stationary points of the energy integral, using the Casimir integrals as constraints, correspond to stationary flows.

To second order we obtain

$$\begin{aligned} \delta^2 E &= -\frac{1}{4\pi}\iint [\delta\omega(\mathbf{r})\delta\omega(\mathbf{r}') + \delta^2\omega(\mathbf{r})\omega(\mathbf{r}')] \ln|\mathbf{r}-\mathbf{r}'| \, dS dS' \\ &= -\frac{1}{2}\int (\delta\phi\delta\omega + \phi\delta^2\omega) \, dS \\ &= \frac{1}{2}\int [-\delta\phi\delta\omega + \frac{d\phi}{d\omega}(\delta\omega)^2] \, dS, \end{aligned} \tag{10}$$

where we assumed that the flow is stationary, so that ϕ and ω are functionally related, and performed a partial integration. We see that $\delta^2 E$ is the same as the pseudo energy A, if we identify $\delta\phi$ with ϕ_1 and $\delta\omega$ with ω_1. For a flow that maximises the energy in a set of isovortical flows, $\delta^2 E$ is negative. Therefore A must be negative definite if we restrict the perturbations to be isovortical. Clearly such a flow is linearly stable. It is also nonlinearly stable, at least in a practical sense, as argued by Benjamin (1976), analogously to Lyapunov stability for a system with a finite number of degrees of freedom. However, this cannot be formalised to stability in some norm, as in "Arnol'd stability".

We now think of ω as the density of some incompressible matter. If we change the sign of eq. (6), it has exactly the same form as the potential energy due to two-dimensional attractive gravitational forces. If ω has only one sign, it is intuitively obvious that the minimum potential energy (i.e. the maximum of E) is attained by a circular distribution with the densest matter at the center. This can also be rigorously proved with the help of theorems about symmetrisations (Sobolev, 1963).

Hence, a circular vortex with monotonic vorticity profile is a maximum energy flow, and therefore stable.

A similar conclusion can be reached by using the conservation of angular momentum. Equation (7) can be interpreted as the potential energy of the density distribution $\omega(\mathbf{r})$ in an external potential well proportional to r^2. In this case the minimum of M (or the maximum, if ω is negative) is attained by a circular vortex with monotonic vorticity profile. Again, the conclusion is that such a vortex is stable.

2.2. VORTEX IN SHEAR FLOW

In this subsection we assume that there is a bakground shear flow given by $\mathbf{U} = -sy\hat{\mathbf{x}}$, where $s > 0$ is the shear strength. We can then rewrite eq. (1) as

$$\frac{\partial \omega}{\partial t} + J(\Psi, \omega) = 0, \tag{11}$$

where $\Psi \equiv \frac{1}{2}sy^2 + \phi$ is the streamfunction of the total flow $\mathbf{U} + \mathbf{u}$, where $\mathbf{u} = \hat{\mathbf{z}} \times \nabla\phi$ is the velocity anomaly, and $\omega = \nabla^2\phi$ is the vorticity anomaly (i.e. the vortex contribution to the vorticity). The Casimir invariants are still defined as in eq. (8), while the energy can now be written

$$E = -\frac{1}{4\pi}\iint \omega(\mathbf{r})\omega(\mathbf{r}') \ln|\mathbf{r} - \mathbf{r}'| \, dSdS' - \int \frac{sy^2}{2}\omega \, dS. \tag{12}$$

In the gravitational analogy from subsection 2.1, the first, quadratic term still describes the mutual gravitational attraction between the incompressible matter elements. The linear term proportional to s is new, and if $\omega \geq 0$

it can be regarded as an external potential well with the shape $sy^2/2$. It is intuitively clear that the minimum potential energy (i.e. the maximum of E) is attained when the "lump of matter" lies on the bottom of the well. It will be squeezed together by the external gravitational field, so that it is no longer circular, but the density still decreses monotonically outward from its center.

A rigorous proof supporting this intuitive argument exists. What has been shown can be formulated as follows. A maximum energy flow (and hence a stationary and stable flow) exists in any set of isovortical flows that satisfies the following conditions: $s \geq 0$ and $\omega \geq 0$, and ω has compact support. [The support of a function $f(\mathbf{r})$ is defined as the region where $f \neq 0$.] This flow describes a vortex with ω decreasing monotonically outward, and elongated in the direction of the external flow.

This theorem was first proved by Nycander (1995), who needed the additional technical condition that ω satisfies $0 < a \leq \omega \leq b < \infty$ on its support, for some finite constants a and b. However, this technical condition was removed by Emamizadeh (2000).

The case when both s and ω are negative is of course equivalent to both being positive. If, on the other hand, s and ω have different signs, we can neither prove existence nor stability of any stationary solution. If a stationary solution exists, however (as is the case for an elliptic vortex patch in a uniform shear flow), the arguments above show that it corresponds to a saddle point of the energy, and we can expect it to be unstable.

2.3. VORTEX ATTACHED TO SEAMOUNT

In this subsection we assume that there is an isolated topographic feature described by the function $h(\mathbf{r})$, the height of the bottom topography relative to a constant background value. The barotropic vorticity equation is

$$\frac{\partial \omega}{\partial t} + J(\phi, \omega + h) = 0, \tag{13}$$

where the relative vorticity ω is still defined by eq. (2). This equation describes the Lagrangian conservation of potential vorticity (PV), i.e. $dq/dt = 0$, with the PV defined by

$$q \equiv \omega + h. \tag{14}$$

Isovortical flows have PV-fields that are rearrangements of each other. Eq. (13) conserves the energy E, which can be written as

$$\begin{aligned} E &= -\frac{1}{4\pi} \iint [q(\mathbf{r}) - h(\mathbf{r})]\,[q(\mathbf{r}') - h(\mathbf{r}')] \ln|\mathbf{r} - \mathbf{r}'|\, dS dS' \\ &= -\frac{1}{4\pi} \iint q(\mathbf{r}) q(\mathbf{r}') \ln|\mathbf{r} - \mathbf{r}'|\, dS dS' + \int q\eta\, dS - \frac{1}{2} \int h\eta\, dS, \end{aligned} \tag{15}$$

where we defined

$$\eta(\mathbf{r}) = \frac{1}{2\pi} \int h(\mathbf{r}') \ln |\mathbf{r} - \mathbf{r}'| \, dS', \tag{16}$$

so that $\nabla^2 \eta = h$. Note that the last term of eq. (15) is independent of q.

The energy is conserved regardless of the shape of the seamount. If, however, the seamount is circularly symmetric, i.e. $h = h(r)$, eq. (13) in addition conserves the angular momentum M:

$$M = \int q r^2 \, dS. \tag{17}$$

We first assume that the seamount is circular. The conservation of angular momentum can then be used exactly as in subsection 2.1 to show that a circular vortex with positive $q(r)$ decreasing monotonically to zero outward is stable, since it minimizes M. Likewise, a circular vortex with negative $q(r)$ increasing monotonically to zero outward maximizes M. Hence, any circular vortex (cyclone or anticyclone) which has a monotonic PV-profile and is centered on a circular seamount is stable.

We then allow the seamount to have an arbitrary, noncircular shape. Assume h to have compact support, and $h \geq 0$. (Note that a cyclone over a seamount is equivalent to an anticyclone over a depression.) The function $\eta(\mathbf{r})$ defined in eq. (16) then has a minimun at the seamount, and increases outward; η is proportional to $\ln(r)$ as $r \to \infty$. Also assume the support of q to be compact, and q to have the same sign (either positive or negative) everywhere on its support.

An important solution in this case is given by $q(\mathbf{r}) \equiv 0$. This is the anticyclonic Taylor column, which is obviously stable. Indeed, no isovortical perturbations are possible: the set of isovortical flows (or rearrangements) contains only one flow.

We then assume that $q \leq 0$. As in subsection 2.1, the first, quadratic term of the second version of eq. (15) is maximised by placing fluid elements with large negative values of q as close to each other as possible. The second, linear term is maximised by placing fluid elements with large negative values of q where η is small. Thus, there is almost no conflict between the requirements for maximising the two terms, and the energy maximum is clearly attained by a localised vortex. A rigorous proof of this can be found along the lines of Emamizadeh (2000); the mathematical details will be published elsewhere (Nycander and Emamizadeh, 2003).

This proves the existence of a large class of stationary and stable anticyclonic vortices attached to a seamount. The radial profile of q is monotonic, with the minimum value at the seamount, but otherwise arbitrary. The shape of the vortex is noncircular, unless the seamount is circularly symmetric. It rotates in the same direction as the Taylor column, but faster.

A vortex of this kind exists in every set of isovortical flows for which $q \leq 0$ everywhere.

We then assume that $q \geq 0$. Clearly, the energy maximum is not attained by a vortex attached to the seamount in this case, since the linear term of the second version of eq. (15) is maximised by placing fluid elements with large values of q as far away from the seamount as possible. Instead, we look for a minimum energy state. To minimise the quadratic term one should spread out the fluid elements with non-zero q as much as possible, but to minimise the linear term they should be placed at the seamount. Since these requirements are in conflict, the result depends on the relative strength of q and h.

A necessary condition for the existence of a minimum energy flow is

$$\int_0^R q^* r\, dr \leq \int_0^R h^* r\, dr \tag{18}$$

for any $R > 0$. Here q^* and h^* are the symmetrisations of q and h, respectively. (A symmetrisation is defined as the unique rearrangement which is circularly symmetric and monotonic decreasing outward from the origin.) To show this, we note that a minimum energy flow has $d\phi/dq \geq 0$ everywhere, as can be seen from the second variation $\delta^2 E$ in eq. (10). Since ∇q is directed inward in a localised vortex with $q \geq 0$, this means that $\nabla \phi$ is also directed inward. Using Gauss' theorem, we conclude that a minimum energy vortex must satisfy $\int_S \omega dS < 0$, where S is any region bounded by a streamline. Since $\omega = q - h$, this in turn implies the condition (18), which is therefore a necessary condition for the existence of a minimum energy flow.

It has not yet been proved rigorously that the condition (18) is also sufficient. However, it is possible to use an analogy with electrostatic forces to make this very plausible. In this case the first form of eq. (15) is the most appropriate one, and q is analogous to the density of free charges, and $-h$ to the density of bound charges. The details of the argument will be published elsewhere (Nycander and LaCasce, 2003).

Notice that the condition $\int_S \omega dS < 0$ implies that the circulation along any streamline is anticyclonic. Thus, the minimum energy vortices are antycyclones, but unlike the maximum energy vortices they rotate more slowly than the Taylor column. A stable vortex of this kind exists in every set of isovortical flows in which $q \geq 0$ everywhere and q satisfies (18).

We conclude that there exists a large set of stable anticyclonic vortex solutions over an arbitrarily shaped seamount, and they can be either maximum or minimum energy flows. Stable cyclones, on the other hand, only exist if the seamount is circular.

3. Three-dimensional quasi-geostrophic flow

3.1. VORTEX IN SHEAR FLOW

In this subsection the results of subsection 2.2 will be extended to 3D quasi-geostrophic flow. We assume that there is a background zonal flow with linear horizontal and vertical shear: $\mathbf{U} = -2y(s_0 + s_1 z)\hat{\mathbf{x}}$, where s_0 and s_1 are constant coefficients. The corresponding background potential vorticity $2(s_0 + s_1 z)$ is layerwise uniform. We will show that a large family of stationary and stable vortex solutions exists in this background flow. The flow is governed by the equation

$$\frac{\partial q}{\partial t} + J(\Psi, q) = 0, \tag{19}$$

where $\Psi \equiv (s_0 + s_1 z)y^2 + \phi$ is the streamfunction of the total flow $\mathbf{U} + \mathbf{u}$, and the PV-anomaly (i.e. the vortex contribution to the PV) is defined by

$$q = \nabla^2 \phi. \tag{20}$$

In this section ∇^2 denotes the 3D Laplacian, in contrast to in section 2. We assume that q has compact support, and that $q \geq 0$ everywhere. We also assume the domain of the flow to be unbounded. The Casimir invariants can now be written

$$C_F = \int F[z, q(\mathbf{r})]\, dV, \tag{21}$$

where F is an arbitrary function of both arguments. Again, flows with the same value of all the Casimir integrals are called isovortical flows. Fields $q(\mathbf{r})$ in the same set of isovortical flows are said to be *stratified rearrangements* of each other; they can be transformed to each other by rearranging the PV-elements separately on levels with constant z.

Equation (19) conserves the energy, which can be written

$$E = \frac{1}{8\pi} \iint \frac{q(\mathbf{r})q(\mathbf{r}')}{|\mathbf{r} - \mathbf{r}'|}\, dV\, dV' - \int (s_0 + s_1 z) y^2 q\, dV. \tag{22}$$

We want to find the stratified rearrangement that maximizes the energy. As in the case of ideal 2D flow, an analogy with gravitation is useful. Changing the sign of E, the first, quadratic term above describes the potential energy due to 3D attractive gravitational forces, while the second, linear term corresponds to an external potential well. If $s_0 + s_1 z$ is positive this well has the minumim at $y = 0$. The incompressible matter elements are constrained to move along levels $z = const$. It is intuitively clear that the minimum potiential energy (i.e. the maximum of E) is attained with the densest matter on each level near $y = 0$. This corresponds to a localised vortex with $q(\mathbf{r})$ decreasing monotonically outward from its center, and elongated in the direction of the background flow.

A rigourous proof of this has been given by Burton and Nycander (1999). More precisely, they proved that a maximum energy flow exists in any set of isovortical flows that satisfies the following conditons: q must have compact support, it must have the same sign everywhere, and the sign of the background shear $2(s_0+s_1z)$ must be the same as the sign of q over the interval in z to which the support of q is confined. The maximum energy flow is a localised vortex, with q symmetric decreasing in x and y for every fixed z (symmetric increasing if $q \leq 0$).

The proof is valid for flow in an unbounded domain, but 3D quasi-geostrophic flow is usually considered with vertical boundaries. A similar theorem is probably valid also for this case, although this has not been proved. However, the heuristic argument given above is valid also in the vertically bounded case.

3.2. VORTEX ATTACHED TO SEAMOUNT

In this subsection the results of subsection 2.3 will be extended to 3D quasi-geostrophic flow. In contrast to the previous subsections, the results shown here are new, and have not yet been proved rigorously.

We assume that $h(\mathbf{r})$ (the height of the bottom relative to a constant background value) is non-negative and has compact support. For simplicity, we also assume that the fluid is unbounded from above. The governing equation is

$$\frac{\partial q}{\partial t} + J(\phi, q) = 0, \quad z > 0, \tag{23}$$

where q is still given by eq. (20). The boundary condition at the bottom is defined by the equations

$$\frac{\partial T}{\partial t} + J(\phi, T + h) = 0, \quad z = 0, \tag{24}$$

and

$$\frac{\partial \phi}{\partial z} = T, \quad z = 0. \tag{25}$$

Here T is the deviation from the stably stratified background temperature.

We define the potential temperature Θ by

$$\Theta = T + h. \tag{26}$$

Then Θ is a Lagrangian invariant of the flow along the bottom. Physically, this describes the effect that a fluid particle which is elevated by the bottom topography is colder than the surrounding particles at the elevated level.

We first assume that the seamount is circular. Equations (23)-(25) then conserve the angular momentum, which can be written

$$M = \int_{z>0} q r^2 dV + \int_{z=0} \Theta r^2 dS. \tag{27}$$

We see that a potential temperature anomaly at the lower boundary is equivalent to a singular sheet of PV. Defining the "generalized PV" by

$$Q = q + \Theta\delta(z), \tag{28}$$

eq. (27) takes the form

$$M = \int_{z\geq 0} Q r^2 dV, \tag{29}$$

where it is understood that the integration domain includes the singular sheet at the bottom. The conservation of angular momentum can be used exactly as in subsection 2.3 to show that a circular vortex with positive $Q(r)$ decreasing monotonically to zero outward is stable, since it minimizes M. Likewise, a circular vortex with negative $Q(r)$ increasing monotonically to zero outward maximizes M. (Note that the potential temperature Θ at the bottom must have the same sign as q.) Hence, any circular vortex (cyclone or anticyclone) which has a monotonic Q-profile, and is centered on a circular seamount, is stable.

If the seamount is noncircular, the angular monentum is not conserved, and we must use the energy to show existence of stationary and stable flows. The energy is defined by

$$E = \frac{1}{2}\int_{z>0} (\nabla\phi)^2 dV.$$

To express it in terms of the Lagrangian invariants we invert eq. (20) with the boundary condition (25):

$$\phi(\mathbf{r}) = \int_{z'>0} q(\mathbf{r}')G(\mathbf{r},\mathbf{r}')dV' + \int_{z'=0} T(\mathbf{r}')G(\mathbf{r},\mathbf{r}')dS'. \tag{30}$$

The Green's function is given by

$$G(\mathbf{r},\mathbf{r}') = -\frac{1}{4\pi}\left(\frac{1}{\left[a_\perp^2 + (z-z')^2\right]^{1/2}} + \frac{1}{\left[a_\perp^2 + (z+z')^2\right]^{1/2}}\right), \tag{31}$$

where $a_\perp^2 = (x-x')^2 + (y-y')^2$. In terms of the generalized PV, eq. (30) can be rewritten as

$$\phi(\mathbf{r}) = \int_{z'\geq 0} \left[Q(\mathbf{r}') - h(\mathbf{r}')\delta(z')\right] G(\mathbf{r},\mathbf{r}')dV'. \tag{32}$$

The energy can then be written

$$
\begin{aligned}
E &= -\frac{1}{2}\iint_{z,z'\geq 0} [Q(\mathbf{r}) - h(\mathbf{r})\delta(z)]\,[Q(\mathbf{r}') - h(\mathbf{r}')\delta(z')]\,G(\mathbf{r},\mathbf{r}')\,dV\,dV' \\
&= -\frac{1}{2}\iint_{z,z'\geq 0} Q(\mathbf{r})Q(\mathbf{r}')G(\mathbf{r},\mathbf{r}')\,dV + \int_{z\geq 0} q\eta\,dV - \frac{1}{2}\int_{z=0} h\eta\,dS, \quad (33)
\end{aligned}
$$

where we defined

$$\eta(\mathbf{r}) = \int h(\mathbf{r}')G(\mathbf{r},\mathbf{r}')\,dS'. \quad (34)$$

Since $h \geq 0$, η has a minimum at the seamount and increases toward zero away from it.

Notice the similarity between eq. (33) and eq. (15). In the present case the anticyclonic Taylor column is given by $Q(\mathbf{r}) \equiv 0$, which implies that $\nabla^2\phi \equiv 0$ and $\partial\phi/\partial z = -h$ at $z = 0$. The streamfunction is given by $\phi = -\eta$. Again, the Taylor column is obviously stable, since it is the only flow that belongs to its own set of isovortical flows.

Then assume that $Q \leq 0$. As in section 2.3, the first, quadratic term of the second version of eq. (33) is maximized by placing fluid elements with large negative values of Q (i.e. of q and Θ) as close to each other as possible, while the second, linear term is maximized by placing them as close to the minimum of η as possible. Again, it is intuitively clear that there exists a maximum energy flow, consisting of a localised anticyclonic vortex that rotates faster than the Taylor column. Both q and Θ increase toward zero outward. The shape is non-circular, unless the seamount is circular.

Finally assume that $Q \geq 0$. In this case no maximum energy flow exists, and we instead look for a minimum energy flow. Again, there is a conflict between minimizing the linear and the quadratic terms in the second version of eq. (33). A stationary flow satisfies $\phi = F[q(\mathbf{r}), z]$, and a minimum energy flow must satisfy $(\partial\phi/\partial q)_z \geq 0$ everywhere (i.e. on every streamline on every vertical level). Since q decreases outward in a localised vortex, ϕ must therefore also decrease outward. This means that only an anticyclone can be a minimum energy vortex.

However, it is more difficult than in the corresponding 2D case to characterise the PV-distributions that can be deformed into a minimum energy flow. A necessary conditon for the existence of a minimum energy flow is

$$\int_0^R \Theta^* r\,dr \leq \int_0^R h^* r\,dr \quad (35)$$

for any $R > 0$. Here Θ^* and h^* are the symmetrisations of Θ and h, respectively. To see that this condition is necessary, we calculate the integral $\int_{S_1} \nabla\phi \cdot \hat{\mathbf{n}} dS$, where $\hat{\mathbf{n}}$ is the unit normal to the cylindrical surface S_1, which is parallel to the z-axis and intersects the xy-plane along an arbitrary streamline C. Since ϕ decreases outward this integral is negative. Using Gauss' theorem this implies that the integral $\int_{S_2} T dS$ is negative, where S_2 is the surface in the xy-plane bounded by the streamline C. Since $T = \Theta - h$, this implies (35).

A particularly simple case is when $q \equiv 0$, so that the dynamics is entirely driven by the temperature anomaly at the lower boundary. This is the "surface quasi-geostrophic flow" discussed by Held *et al.* (1995). In this case the condition (35) is probably also sufficient for the existence of a minimum energy flow, as can be argued by the same electrostatic analogy as used in the 2D case. However, if $q \neq 0$, the condition (35) is certainly not sufficient; indeed, it appears difficult to formulate a sharp criterion for the existence of a minimum energy flow.

Qualitatively, the results of the present section are similar as for ideal 2D flow: there exists a large set of stable anticyclonic vortex solutions over an arbitrarily shaped seamount, and they can be either maximum or minimum energy flows. Stable cyclones, on the other hand, only exist if the seamount is circular.

4. Conclusion

According to a fundamental variational principle, a flow that maximizes or minimizes the energy in a set of isovortical flows is stationary and stable. Thus, if it can be shown that an energy extremum is attained by some localised distribution of vorticity, it follows that a stationary and stable vortex solution exists.

An overview has been given of cases where this variational approach gives useful results. The first concerns vortices in a background shear flow, and the second vortices attached to seamounts. They have been analysed both with the 2D barotropic vorticity equation and with the 3D quasigeostrophic equation, with qualitatively similar results.

It has been shown that a large family of stable vortex solutions exists in a linear shear flow. The vorticity anomaly of the vortex must have the same sign as the background shear, and the vortex is elongated in the direction of the external flow. The radial vorticity profile is monotonic, but otherwise arbitrary.

These results agree very well with numerous observations of shear flows in nature, laboratory experiments (Sommeria *et al.*, 1988) and numerical simulations (Marcus, 1990; Toh *et al.*, 1991; Bracco *et al.*, 1999). Vortices often appear spontanously, and most of them rotate in the same direction as the background flow. These are very robust, while those that rotate in the opposite direction are quickly destroyed by the external flow.

It has also been shown that a large family of stable vortex solutions attached to a seamount exists. If the seamount is circular, they can be both cyclones and anticyclones, with a monotonic radial PV-profile. If the seamount has a noncircular shape, it is possible to show the existence of stable anticyclones, but not cyclones. The anticyclones can be both minimim energy flows and maximum energy flows. The former have positive PV and rotate more slowly than the Taylor column (which has zero PV), while the latter have negative PV and rotate faster. The radial PV profile must in both cases be monotonic.

These results have been confirmed by numerical simulations of ideal 2D flow (Nycander and LaCasce, 2003). In particular, it is seen that attached cyclones are destabilised by a noncircularity in the shape of the seamount, while anticyclones adjust and remain stable.

Trapped anticyclones also occur in the ocean. One example is the Zapiola Drift, an isolated topographic high occuring in an abyssal plain of the South Atlantic, where strong anticyclonic flow has been observed (Saunders and King, 1995). Another example is the anticyclonic flow observed over the Fieberling Seamount (Kunze and Toole, 1997). Interestingly, the PV anomaly of this flow was observed to be strongly negative, hence it appears to be a maximum energy flow. The present results help explaining its dynamic stability.

The present analysis does not consider how the stable vortices might be created. This question is addressed by statistical mechanics theories, which predict that the final result of the flow evolution is a state of maximum entropy (Robert and Sommeria, 1991). Since such states are also maximum energy flows, there is a close relation between these theories and the present variational analysis.

Most of the vortices studied here are maximum energy flows, and as pointed out in the Introduction, they are not Arnol'd stable. Yet, it is clear that they are nonlinearly stable in a practical sense. Thus, if one wants to study the nonlinear stability of localised vortices, it is necessary to go beyond the concept of Arnol'd stability, by relaxing the stability definition. (Arnol'd stability represent a strong form of normed stability.) The present work is an attempt to take a step in this direction.

References

Andrews, D. G. On the existence of nonzonal flows satisfying sufficient conditions for stability. *Geophys. Astrophys. Fluid Dyn.*, 28:243–256, 1984.

Arnol'd, V. I. Conditions for nonlinear stability of stationary plane curvilinear flows of an ideal fluid. *Dokl. Akad. Nauk SSSR*, 162:975-978, 1965.

Arnol'd, V. I. On an a priori estimate in the theory of hydrodynamical stability. *Izv. Vyssh. Uchebn. Zaved. Matematika*, 54(5):3-5, 1966.

Benjamin, T. B. The alliance of practical and analytical insights into the nonlinear problems of fluid mechanics. In *Lecture Notes in Mathematics*, vol 503:8–29, Springer, 1976.

Bracco, A., P. H. Chavanis, A. Provenzale and E. A. Spiegel. Particle aggregation in a turbulent Keplerian flow. *Phys. Fluids*, 11:2280–2287, 1999.

Burton, G. R. and J. Nycander. Stationary vortices in three-dimensional quasigeostrophic shear flow. *J. Fluid Mech.*, 389:255–274, 1999.

Carnevale, G. F. and T. G. Shephard. On the interpretation of Andrews' theorem. *Geophys. Astrophys. Fluid Dyn.*, 51:1–17, 1990.

Emamizadeh, B. Steady vortex in a uniform shear flow. *Proc. Roy. Soc. Edin.*, 130A:801–812, 2000.

Held, I. M., R. T. Pierrehumbert, S. T. Garner and K. L. Swanson. Surface quasi-geostrophic dynamics. *J. Fluid Mech.*, 282:1–20, 1995.

Kloosterziel, R. C. and G. F. Carnevale. Formal stability of circular vortices. *J. Fluid Mech.*, 242:249–278, 1992.

Kunze, E. and J. M. Toole. Tidally driven vorticity, diurnal shear, and turbulence atop Fieberling Seamount. *J. Phys. Oceanogr.*, 27:2663–2693, 1997.

Marcus, P. S. Vortex dynamics in a shearing zonal flow. *J. Fluid Mech.*, 215:393–430, 1990.

McIntyre, M. I. and T. G. Shephard. An exact local conservation theorem for finite-amplitude disturbances to non-parallel shear flows, with remarks on Hamiltonian structure and on Arnol'd's stability theorems. *J. Fluid Mech.*, 181:527–565, 1987.

Nycander, J. Existence and stability of stationary vortices in a uniform shear flow. *J. Fluid Mech.*, 287:119–132, 1995.

Nycander, J. and B. Emamizadeh. Variational problem for vortices attached to seamounts, Submitted to *Nonlin. Analysis.*

Nycander, J. and J. H. LaCasce. Stable and unstable vortices attached to seamounts, Submitted to *J. Fluid Mech.*.

Ripa, P. General stability conditions for zonal flows in a one-layer model on the β-plane or the sphere. *J. Fluid Mech.*, 126:463–489, 1983.

Robert, R. and J. Sommeria. Statistical equilibrium states for two-dimensional flows. *J. Fluid Mech.*, 229:291–310, 1991.

Saunders, P. M. and B. A. King. Bottom currents derived from a shipborne ADCP in WOCE cruice A11 in the South Atlantic. J. Phys. Oceanogr., 25(3):329–347, 1995.

Sobolev, S. L. On a theorem of functional analysis. *Am. Math. Soc. Transl.*, 34(2):39–68, 1963.

Sommeria, J., S. D. Meyers and H. L. Swinney. Laboratory simulation of Jupiter's great red spot. *Nature*, 331:689–693, 1988.

Toh, S., K. Ohkitani and M. Yamada. Enstrophy and momentum fluxes in two-dimensional shear flow turbulence. *Physica D*, 51:569–578, 1991.

Vallis, G. K., G. F. Carnevale and W. Y. Young. Extremal energy properties and construction of stable solutions of the Euler equations. *J. Fluid Mech.*, 207:133–152, 1989.

NONLINEAR OUTFLOWS ON A β PLANE

DORON NOF
Department of Oceanography and
The Geophysical Fluid Dynamics Institute
Florida State University
Tallahassee, FL 32306-4320, U.S.A.

Abstract. Nonlinear analytical solutions and numerical simulations show that the behavior of inviscid (light) outflows on a β plane strongly depend on the orientation of the boundary along which they are situated. Both outflows associated with eastern and southern boundaries split into two branches. One branch contains a chain of eddies and the other corresponds to a steady boundary current. Outflows situated along a western boundary, on the other hand, produce a steady gyre and outflows situated along a northern boundary produce an exceptionally broad steady current.

Key words: outflows, β plane

The manner in which water of anomalous density empties into an ocean has been of theoretical interest to oceanographers for decades. In particular, various attempts have been made to understand how the anomalous water is distributed once it debouches into the ocean (e.g. Takano, 1955; Defant, 1961; Nof, 1978a,b; Garvine, 1987, 1996, 2001; Chao and Boicort, 1986; O'Donnell, 1990; Oey and Mellor, Oey and Mellor; Kourafalou *et al.*, 1996; Yankosky and Chapman, 1997). In everyday life, a source of anomalous water emptying into a large container tends to spread evenly in all directions. In the ocean, however, the earth's rotation tends to confine the outflow to the coast (in the Kelvin wave sense) forming a longshore current. The complications added by rotation do not end here, and recent analytical and numerical studies have shown that such an outflowing current on an f plane can *never be steady* (see Figure 1, and Pichevin and Nof, 1997; Nof and Pichevin, 2001). Furthermore, numerical simulations and analytical work demonstrate that the outflow balloons in the sense that a forever-growing eddy is generated near the coast (Pichevin and Nof, 1997; Fong, 1998; Nof and Pichevin, 2001); i.e. a part of the outflow goes into a gyre and the

O.U. Velasco Fuentes et al. (eds.), Nonlinear Processes in Geophysical Fluid Dynamics, 87-90.

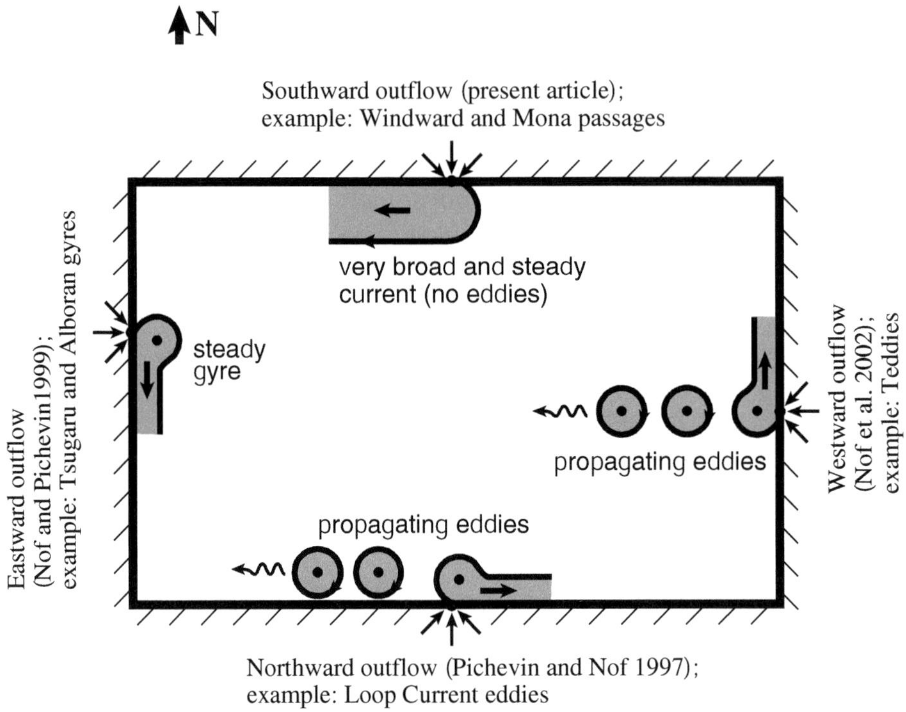

Figure 1. The four scenarios of the β outflow problem. A northward outflow (Pichevin and Nof 1997) is situated next to a southern boundary; a westward outflow is situated along an eastern boundary (Nof *et al.*, 2002); a southward outflow is situated next to a northern boundary (this paper); and an eastward outflow is situated next to a western boundary (Nof and Pichevin, 1999). The southward outflow, which is the focus of this study, is the only case where neither a steady gyre nor eddies are formed.

remaining part goes into a boundary current.

On a β plane the situation is still more complicated. There are four possible scenarios – a northward outflow, a westward outflow, a southward outflow and an eastward outflow (Figure 1) – all of which have different solutions. Of these four, three outflows (northward, westward and eastward) involve eddies or gyres. These three have been addressed previously in Pichevin and Nof (1997), Nof and Pichevin (1999), and in Nof *et al.* (2002). The last unsolved case, which is the focus of this study, is the southward outflow which, as we shall see, does not involve eddies. Ironically, it is the simplest solution of the four and yet was the most difficult for us to obtain (probably because we erroneously assumed *a priori* that it would also involve eddies). We show analytically that the formation of a very broad current is a fundamental property of any nonlinear *southward outflow* regardless of the fluid's vorticity. It results from the impossibility of balancing the flow-force (associated with the long-shore current) without the establishment of such a broad current. The results of our theory can be summarized as follows:

1. The nonlinear inviscid southern outflow involves a very broad westward flow consisting of a nearly stagnant region near the wall and a jet on the ocean side. The nearly stagnant region width is of the order of the equatorial Rossby radius, R_{de} $[(g'H)^{1/4}/\beta^{1/2}]$. It is $1.228R_{de}$ for a zero PV outflow, $0.816R_{de}$ for a uniform PV whose potential vorticity depth matches the undisturbed depth. The jet's width is the familiar mid-latitude Rossby radius, R_d.

2. The general analytical solution corresponds to a balance between two longshore forces, the rocket-like "jet" force resulting from the dynamic momentum flux of the intense narrow flow near the edge and the westward β -induced force associated with the nearly stagnant region. Such a balance does not hold in the northward outflow case (examined by Pichevin and Nof 1997) because, in this case, the jet and β forces are pointing in the same direction. As a result, an entirely different, time-dependent balance takes place and eddies are generated on the west side.

3. The above results are in very good agreement with the numerics. The main difference between the analytical and the numerical simulations is the neglect of the jet's width and friction in the analytics.

4. Friction enters the problem in two ways. First, it brings the region into which the steady outflow does not penetrate (i.e. the broad region next to the wall) to rest by changing its potential vorticity. Second, it spreads the jet causing it to broaden downstream.

The above calculation should have numerous applications (e.g. the surface waters flowing from the Atlantic to the Caribbean via the Windward and Mona Passages) but it is one of these rare cases where the theory is ahead of the observations simply because there are presently no measurements to support or refute it. Although measurements in and around those passages have been made (e.g. Johns *et al.*, 1998), it is not presently possible to say what the downstream width is. One would hope that general numerical simulations would help to bridge this gap but such models do not usually resolve the passages' dynamics so that they cannot be used for this purpose. It is hoped that this study will encourage future observational programs to look at the outflow length scale issue.

Acknowledgements

This study was supported by the National Aeronautics and Space Administration under grant $NGT5-30164$; National Science Foundation contract OCE 9911324; and Office of Naval Research grant $N00014-01-0291$.

References

Chao, S.-Y., and B. Boicourt. Onset of Estuarine Plumes. *J. Phys. Ocean.*, 16: 2137–2149, 1986.

Defant, A. *Physical Oceanography*, Vol. I. Pergamon Press, 729 pp. (see Chap. 16), 1961.

Fong, D.A. *Dynamics of Freshwater Plumes: Observations and Numerical Modelling of the Wind-forced Reponse and Alongshore Freshwater Transport.* PhD thesis, Woods Hole Oceanographic Institution, Massachusetts Institute of Technology, 172 pp., 1998.

Garvine, R. W. Estuary Plumes and Fronts in Shelf Waters, A Layer Model. *J. Phys. Ocean.*, 17: 1877–1896, 1987.

Garvine, R. W. Buoyant Discharge on the Inner Continental Shelf: A Frontal Model. *J. Mar. Res.*, 54: 1–33, 1996.

Garvine, R. W. The Impact of Model Configuration in Studies of Buoyant Coastal Discharge. *J. Mar. Res.*, 59: 193–225, 2001.

Johns, W. E., T. N. Lee, R. C. Beardsley, J. Candela, R. Limeburner, and B. Castro. Annual Cycle and Variability of the North Brazil Current. *J. Phys. Ocean.*, 28: 103–128, 1998.

Kourafalou, V. K., L. Y. Oey, J. D. Wang and T. N. Lee. The Fate of River Discharge on the Continental Shelf. 1. Modeling the River Plume and the Inner Shelf Coastal Current. *J. Geophys. Res.*, 101: 3415–3434, 1996.

Nof, D. and T. Pichevin. The Establishment of the Tsugaru and the Alboran Gyres. *J. Phys. Ocean.*, 29: 39–54, 1999.

Nof, D. and T. Pichevin. The Ballooning of Outflows. *J. Phys. Ocean.*, 31: 3045–3058, 2001.

Nof, D. On Geostrophic Adjustment in Sea Straits and Wide Estuaries: Theory and Laboratory Experiments. Part I: One-Layer System. *J. Phys. Ocean.*, 8: 690–702, 1978a.

Nof, D. On Geostrophic Adjustment in Sea Straits and Wide Estuaries: Theory and Laboratory Experiments. Part II: Two-Layer System. *J. Phys. Ocean.*, 8: 861–872, 1978b.

Nof, D., T. Pichevin and J. Sprintall. "Teddies" and the Origin of the Leeuwin Current. *J. Phys. Ocean.*, 32: 2571–2588, 2002.

O'Donnell, J. The Formation and Fate of a River Plume: A Numerical Model. *J. Phys. Ocean.*, 20: 551–569, 1990.

Oey, L.-Y. and G. L. Mellor. Subtidal Variability of Estuarine Outflow, Plume, and Coastal Current: A Model Study. *J. Phys. Ocean.*, 23: 164–171, 1993.

Pichevin, T. and D. Nof. The Momentum Imbalance Paradox. *Tellus*, 49: 298–319, 1997.

Takano, K. On the Velocity Distribution off the Mouth of a River. *J. Ocean. Soc. Japan*, 10: 60–64, 1954.

Takano, K. A Complementary Note on the Diffusion of the Seaward River Flowing off the Mouth. *J. Ocean. Soc. Japan*, 11: 147–149, 1955.

Yankovsky, A. E. and D. C. Chapman. A Simple Theory for the Fate of Buoyant Coastal Discharges. *J. Phys. Ocean.*, 27: 1386–1401, 1997.

GENERATION OF INTERNAL GRAVITY WAVES BY UNSTABLE OVERFLOWS

GORDON E. SWATERS
Department of Mathematical and Statistical Sciences, and Institute of Geophysical Research
University of Alberta
Edmonton, Alberta, T6G 2G1 Canada

Abstract. A model is introduced which describes the baroclinic generation of internal gravity waves in the water column overlying an abyssal overflow in which bottom friction, rotation and down slope gravitational acceleration are all present. A brief description is given of the internal waves associated with unstable supercritical abyssal overflows.

Key words: internal gravity waves, overflows, instability

1. Introduction

The deep western undercurrent (DWUC) in the Atlantic ocean has, as one of its sources, the Denmark Strait Overflow (DSO). The DSO is an example of a sill-controlled abyssal gravity current. Greatly simplified, overflows of this kind initially exhibit pronounced down slope motion which subsequently evolves into more or less along slope motion, which is banked against sloping topography. This picture is, of course, far from complete. Baroclinic interactions with the overlying water column and non-conservative processes such as entrainment and friction are present. In addition, there is considerable spatial and temporal variability associated with these flows in both the near-sill and down stream regions.

Bruce (1995), examining satellite imagery, and Krause (1996), examining buoy trajectories, showed the development of down stream cyclonic eddies associated with the DSO. Recently analyzed observations and numerical simulations (Krauss and Käse, 1998; Käse and Oschlies, 2000; Girton and Sanford, 2001, 2002; Käse *et al.*, 2002) suggest that key aspects of the down stream mesoscale variability can be understood in the context of the (non-quasigeostrophic) baroclinic instability mechanism described by Swaters (1991) for abyssal currents, interpreted, of course, in the context of

O.U. Velasco Fuentes et al. (eds.), Nonlinear Processes in Geophysical Fluid Dynamics, 91-102.

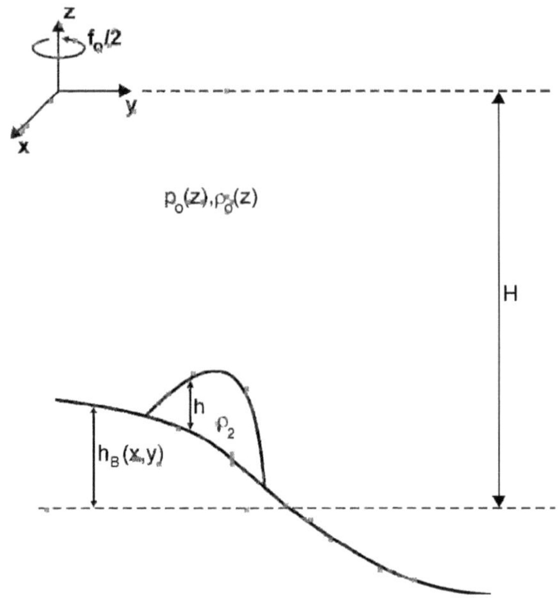

Figure 1. Geometry of the model used in this paper.

realistic physical oceanographic properties (e.g. Jiang and Garwood, 1996; Jungclaus *et al.*, 2001).

Käse *et al.* (2002), analyzing oceanographic data for the DSO region, from four different cruises over a three year period, and examining high resolution numerical simulations, describe the differing dynamical regimes between the near-sill and down stream overflow. In contrast to the down stream flow, the near-sill overflow is predominately down slope, strongly influenced by bottom friction and is near, and even possibly super, critical (with respect to long internal gravity waves). Girton and Sanford (2002), using estimates derived from the aforementioned cruise data, argue, perhaps not surprisingly, that the near-sill momentum balance for the overflow is principally between rotation, down slope gravitational acceleration and bottom friction.

These dynamical balances suggest another source for overflow variability, particularly in the near-sill region, and one which has not been explored before in this context. Frictional down slope flows which are super critical can be unstable. In the absence of rotation and baroclinicity, the instabilities are classical roll waves (Jeffreys, 1925; Whitham, 1974). For oceanographically relevant scales, the instabilities will, as we show, manifest themselves in the overlying ocean as amplifying long internal gravity waves. Within the overflow itself, the instabilities take the form of down slope propagating, growing periodic bores or pulses.

A summary is presented here of a simple theory for the frictional destabilization of abyssal overflows, with rotation and baroclinicity present, and of the characteristics of the internal gravity field, in the overlying ocean, associated with the instability. Full details can be found in Swaters (2003).

2. Model Equations

The underlying geometry is sketched in Figure 1. We assume $f-$plane dynamics for a stably and continuously stratified fluid of finite depth overlying a well mixed abyssal current layer with variable bottom topography. The upper, i.e. the continuously stratified, layer is denoted as layer one. The abyssal current, i.e. the lower layer, is denoted as layer two. The upper and abyssal layer dynamical quantities will be denoted, unless otherwise specified, with a 1 and 2 subscript, respectively.

The theoretical model is based on the incompressible adiabatic equations under a Boussinesq approximation for a continuously stratified fluid for the upper layer and the shallow water equations for the abyssal layer. We assume a rigid ocean surface which will filter out the external gravity wave modes in the model and focus attention on the baroclinic aspects of the dynamics.

Assuming that $O(h/H)$ is small, the *nondimensional* model can be written in the form (see Swaters, 2003 for full details)

$$\left(\partial_{tt} + f^2\right)\left(B^{-1}\varphi_{zt}\right)_z + \Delta\varphi_t = 0, \tag{1}$$

subject to

$$\varphi_{zt} = 0 \text{ on } z = 0, \text{ and } \varphi_{zt} = -B(-1)\, h_t \text{ on } z = -1, \tag{2}$$

with the auxiliary upper layer relations

$$\left(\partial_{tt} + f^2\right)\mathbf{u}_1 = f\mathbf{e}_3 \times \nabla\varphi - \nabla\varphi_t,\; \rho = -\varphi_z,\; w = -B^{-1}\varphi_{zt}, \tag{3}$$

where $h(x, y, t)$ is determined from the abyssal layer equations

$$(\partial_t + \mathbf{u}_2 \cdot \nabla)\,\mathbf{u}_2 + f\mathbf{e}_3 \times \mathbf{u}_2 = -\nabla(h + h_B) + \frac{1}{R_e}\frac{\nabla \cdot (h\nabla \mathbf{u}_2)}{h} - \frac{c_D\,|\mathbf{u}_2|\,\mathbf{u}_2}{h}, \tag{4}$$

$$h_t + \nabla \cdot (\mathbf{u}_2 h) = 0, \tag{5}$$

where the abyssal layer's Reynolds number, R_e, nondimensional Coriolis parameter (or, equivalently, the reciprocal of the temporal Rossby number),

f, scaled bottom drag coefficient (or, equivalently, the reciprocal of the nonrotating Froude number), c_D, and the Burger number, $B(z)$, are given by, respectively,

$$f = \frac{f^* \sqrt{h_*/g'}}{s^*}, \quad B(z) = \frac{s^* N^2 (Hz) H^2}{g' h_*}, \quad R_e = \frac{h_* \sqrt{g' h_*}}{A_H s^*}, \quad c_D = \frac{c_D^*}{s^*},$$

where c_D^*, f^*, h_*, A_H, s^* are the bottom friction coefficient, local Coriolis parameter, scale abyssal layer thickness, horizontal eddy coefficient and scale bottom slope parameter, respectively. The Brunt-Väisälä frequency, $N(z)$, and the reduced gravity, g', are given by

$$N^2(z) = -\frac{g}{\rho_2} \frac{d\rho_0(z)}{dz} > 0, \quad g' = \frac{g(\rho_2 - \rho_0(-H))}{\rho_2} > 0,$$

respectively where $\rho_0(z)$ and ρ_2 are the upper layer hydrostatic background and abyssal layer density, respectively. These equations remain valid in both the nonrotating $f = 0$ and inviscid $c_D = A_H = 0$ ($R_e \to \infty$) limits.

3. Normal Mode Stability Equations

We examine steady "slab" solutions (see e.g. Jeffreys, 1925; Whitham, 1974; Baines, 1995) given by

$$\mathbf{u}_2 = (U, V), \quad h = 1, \tag{6}$$

for the linearly sloping bottom $h_B = -y$. These uniform flows are equivalent to the solutions found for "stream tube" models, without along-stream variation, which have been used to examine aspects of the dynamics of rotating turbidity and abyssal currents (e.g. Smith, 1975; Killworth, 1977; Price and Baringer, 1994; Emms, 1998).

Substitution of (6) into (4) yields (continuity is trivially satisfied)

$$fV = c_D \left(U^2 + V^2\right)^{\frac{1}{2}} U, \tag{7}$$

$$fU = 1 - c_D \left(U^2 + V^2\right)^{\frac{1}{2}} V, \tag{8}$$

which can be solved to give

$$(U, V) = \left(f\gamma^2, c_D\gamma^3\right); \; \gamma \equiv \sqrt{\frac{2}{f^2 + \sqrt{f^4 + 4c_D^2}}} > 0. \tag{9}$$

Figure 2 is a stick plot of the steady uniform velocity (U, V), determined by (7) and (8), as a function of the bottom friction coefficient c_D and the

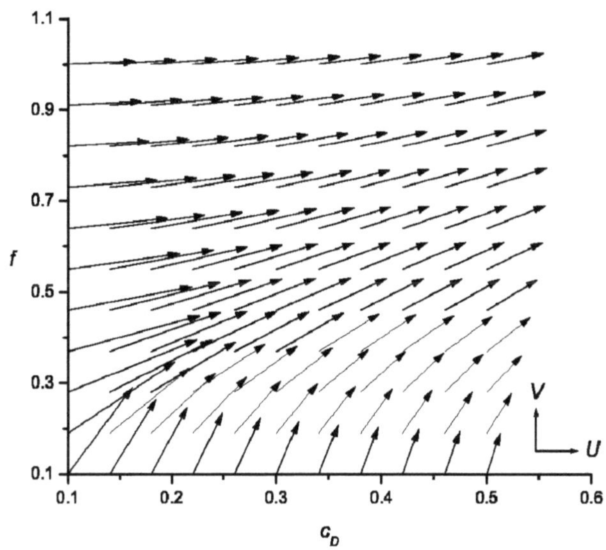

Figure 2. $(U,\ V)$ in the $(c_D,\ f)$ – plane.

nondimensional Coriolis parameter f for the range $0.1 \leq c_D \leq 0.5$ and $0.1 \leq f \leq 1.0$. In order to ensure that the vectors remain within the plot boundaries, the velocity vectors have been scaled so that the maximum speed, which occurs for the velocity vector located at $c_D = f = 0.1$ in Figure 2 (the velocity vector located at the lower left hand corner), has length 0.2. In addition, the vectors are oriented so that down slope motion is indicated by the vector pointing in the direction of increasing f. Rightward deflected (which occurs for positive f) along slope motion is indicated by the vector pointing in the direction of increasing c_D. The orientation is shown in the lower right corner in Fig. 2 by a (U, V) coordinate axes. We see the general trend from down (along) slope motion to along (down) slope motion as f (c_D) increases for a given c_D (f), within the context that the speed monotonically decreases as c_D and f individually increase.

The general normal mode stability problem is determined by substituting

$$(u, v, h) = (U, V, 1) + \left(\widetilde{u}, \widetilde{v}, \widetilde{h}\right) \exp\left[ikx + ily + (\sigma - ikU - ilV)\, t\right] + c.c., \tag{10}$$

$$\varphi = \psi\,(z) \exp\left[ikx + ily + (\sigma - ikU - ilV)\, t\right] + c.c., \tag{11}$$

into (1), (2), (4) and (5), giving

$$\psi_{zz} - \lambda^2 \psi = 0;\ \lambda^2 \equiv \frac{(k^2 + l^2)\, B}{(\sigma - ikU - ilV)^2 + f^2}, \tag{12}$$

subject to

$$\psi_z = 0 \text{ on } z = 0, \text{ and } \psi_z = -Bh \text{ on } z = -1, \tag{13}$$

with

$$\mathcal{M}\, [u, v, h]^{\top} = \mathbf{0}, \tag{14}$$

where we have assumed, for convenience, a constant Burger number, dropped the tildes, and where $\mathcal{M}$ is the 3×3 matrix

$$\begin{bmatrix} \sigma + \frac{k^2+l^2}{R_e} + c_D\gamma\,(1 + \gamma^2 f^2) & f\,(c_D^2\gamma^4 - 1) & ik - c_D f \gamma^3 \\ f\,(1 + c_D^2\gamma^4) & \sigma + \frac{k^2+l^2}{R_e} + c_D\gamma\,(1 + c_D^2\gamma^4) & il - c_D^2\gamma^4 \\ ik & il & \sigma \end{bmatrix}. \tag{15}$$

The vertical structure of the normal modes in the upper layer is determined by the solution to (12) and (13), and is given by

$$\psi\,(z) = \frac{Bh \cosh\,(\lambda z)}{\lambda \sinh\,(\lambda)}, \tag{16}$$

which implies that the vertical structure in the (normal mode) vertical velocity will be described by

$$\widetilde{w}\,(z) = \frac{(ikU + ilV - \sigma)}{B}\psi_z = \frac{(ikU + ilV - \sigma)\, h \sinh\,(\lambda z)}{\sinh\,(\lambda)}. \tag{17}$$

Equation (14) has nontrivial solutions if and only if

$$\det \mathcal{M} = 0, \tag{18}$$

which gives rise to a cubic polynomial in σ which can be solved to give solutions of the form

$$\sigma = \sigma\,(c_D,\ f,\ R_e,\ k,\ l)\,. \tag{19}$$

Instability occurs if the growth rate $Re\,(\sigma) > 0$.

In the non-rotating limit it can be shown (Swaters, 2003) that (18) reduces, for the non-trivial modes, to

$$\sigma = -\left(\sqrt{c_D} + \frac{l^2}{2R_e}\right) \pm \sqrt{\left(\sqrt{c_D} + \frac{l^2}{2R_e}\right)^2 - (il + l^2)}, \tag{20}$$

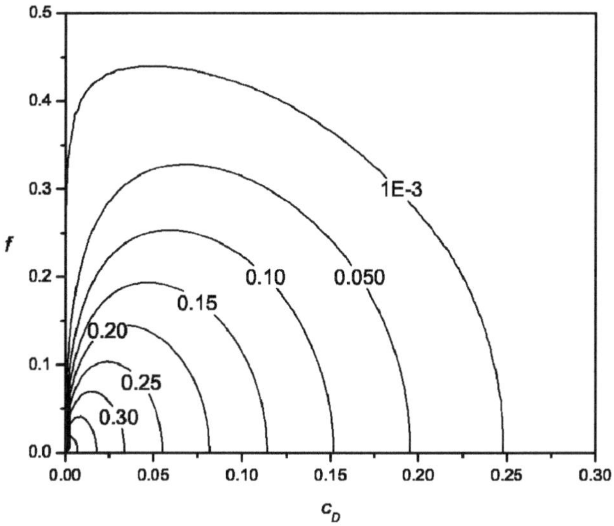

Figure 3. $Re\,(\sigma_{\max})$ in the $(c_D,\ f)$ – plane.

where we have assumed $k = 0$. A mode for a given l will be stable provided

$$Re\left\{\sqrt{\left(\sqrt{c_D}+\frac{l^2}{2R_e}\right)^2-(il+l^2)}\right\}\leq\sqrt{c_D}+\frac{l^2}{2R_e},$$

which is satisfied *if and only if* $\sqrt{c_D} + l^2/\,(2R_e) \geq 1/2$. Thus, in the non-rotating limit, $c_D \geq \frac{1}{4} \Longleftrightarrow$ stability which is just the classical roll wave stability result (see e.g. Jeffreys, 1925; Whitham, 1974; Baines, 1995).

4. Stability Characteristics

Figure 3 is a contour plot of the growth rate of the most unstable mode (denoted as $Re\,(\sigma_{\max})$) in the $(c_D,\ f)$ – plane assuming $R_e = 400.0$. The most unstable mode, denoted as $\sigma_{\max}$, is that normal mode with the largest value of $Re\,(\sigma)$ considered as a function of the wave numbers (k, l) for a given value of the parameters $(c_D,\ f,\ R_e)$. We denote the wave number of the most unstable mode by $(k_{\max},\ l_{\max})$.

In Figure 3 we can see that the sharp cutoff value for instability associated with the bottom friction coefficient continues to exist (but decreases) as f increases from zero. Instability continues to occur for $f > 0$, but does so for a smaller range of c_D values. In the non-rotating limit, the bottom friction coefficient cutoff value is given exactly by $c_D = 0.25$.

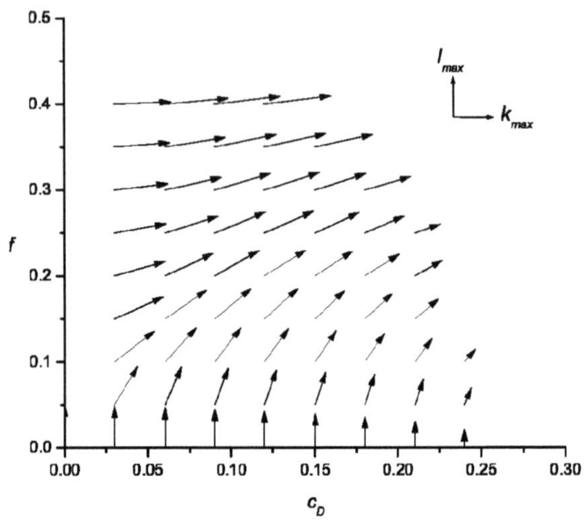

Figure 4. $(k_{\max},\ l_{\max})$ in the $(c_D,\ f)$ – plane.

Although it is difficult to discern clearly in Fig. 3, there is a sharp boundary with respect to f, i.e. there is a distinct marginal stability curve, between the region of instability (where the growth rate is positive) and the region of stability (where the most unstable mode has zero growth rate, i.e. the abyssal flow is neutrally stable). The contour labelled 0.001 is very close to this boundary. (When we tried to contour the 0–growth rate isoline exactly the contour package introduced a highly irregular multiply connected pattern.)

As f increases, Fig. 3 shows that the growth rate of the most unstable mode decreases monotonically. Nevertheless, Fig. 3 suggests that the frictional destabilization of abyssal overflows is possible for physically realizable values of f or, equivalently, the inverse Rossby number. For example, for $c_D = 0.1$ and $f = 0.25$ (i.e. a Rossby number of about 4.0), the most unstable mode has a (nondimensional) growth rate of about 0.09 which corresponds to a (dimensional) $e-$ folding time of about 24 hours (the time scale is about 2.2 hours). For $c_D = 0.1$ and $f = 0.25$, $U = 1.84$ and $V = 1.99$ which imply a *dimensional* along and cross slope velocity for the mean overflow of about 110 cm/s and 120 cm/s, respectively (the abyssal velocity scale is about 60 cm/s).

Figure 4 is a stick plot of the wave number vector $(k_{\max},\ l_{\max})$ as a function of the bottom friction coefficient c_D and the nondimensional Coriolis parameter f for the range $0 \leq c_D \leq 0.3$ and $0 \leq f \leq 0.5$. In order to ensure that the vectors remain within the plot boundaries, the velocity vectors have been scaled so that the length of the wave number vector located at $c_D = f = 0$ in Fig. 4 (located at the lower left hand corner), has length 0.05 (its actual length is about 2.58).

The wave number vectors are oriented so that positive $l_{\max}$ is indicated by the vector pointing in the direction of increasing f. Positive $k_{\max}$ is indicated by the vector pointing in the direction of increasing c_D. The orientation is shown in the upper right corner in Fig. 4 by a $(k_{\max},\ l_{\max})$ coordinate axes. The region of stability in the $(c_D,\ f)-$ plane has no wave number vector shown because $(k_{\max},\ l_{\max}) = \mathbf{0}$ there. The reason that *both* $l_{\max}$ and $k_{\max}$ will be zero in the region of stability, away from the marginal stability boundary, is a consequence of the presence of horizontal friction in the linear stability problem. The *most unstable mode* in the region of stability, when horizontal friction is present (which is proportional to the Laplacian operator), occurs when the magnitude of the wave number vector is zero since any other wave number pair will necessarily result in a more negative growth rate, when all other parameters are held constant. One can also see the general trend for the increasing orientation toward along slope propagation as f increases for a given c_D. In addition, one can see the general trend of diminishing $k_{\max}$ as c_D increases for a given f as well as an overall decrease in the magnitude of the wave number vector (i.e. a trend to longer waves).

It is useful to give a dimensional estimate of $(k_{\max}, l_{\max})$. If we consider $c_D = 0.1$ and $f = 0.25$, we find that $k_{\max} \approx 1.55$ and $l_{\max} \approx 1.29$, which implies a *dimensional* along slope wave length of about 20 km and a cross slope wave length of about 24 km for a total wave length of about 31 km for the most unstable mode associated with $c_D = 0.1$ and $f = 0.25$ (the length scale is about 5 km).

Figure 5 is a contour plot of the *geostationary* frequency of the most unstable mode, given by

$$\omega_{\max}^{geo} = Uk_{\max} + Vl_{\max} - Im\left(\sigma_{\max}\right),$$

in the $(c_D,\ f)-$ plane for $R_e = 400.0$. The geostationary frequency is the frequency one would measure using bottom moored instruments. As f increases, away from $c_D = 0$ (the $f-$axis), we see the general trend to lower, yet not sub-inertial, frequencies. If we consider $c_D = 0.1$ and $f = 0.25$, we find that $\omega_{\max}^{geo} \approx 7.45$, which implies a *dimensional geostationary* period of about 2 *hours*.

Figure 6 is a contour plot of a horizontal section of the *total* vertical velocity in the overlying ocean, given by

$$w\left(x,y,z,t\right) = Re\left\{\widetilde{w}\left(z\right)\exp\left[ikx + ily + \left(\sigma - ikU - ilV\right)t\right]\right\},$$

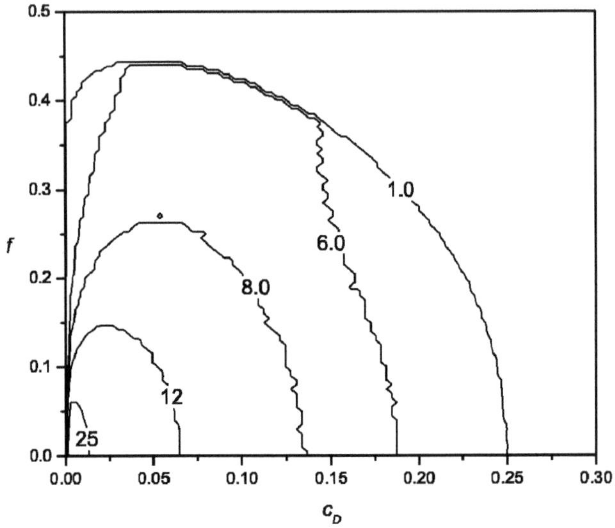

Figure 5. $\omega_{\max}^{geo}$ in the $(c_D,\ f)$ – plane.

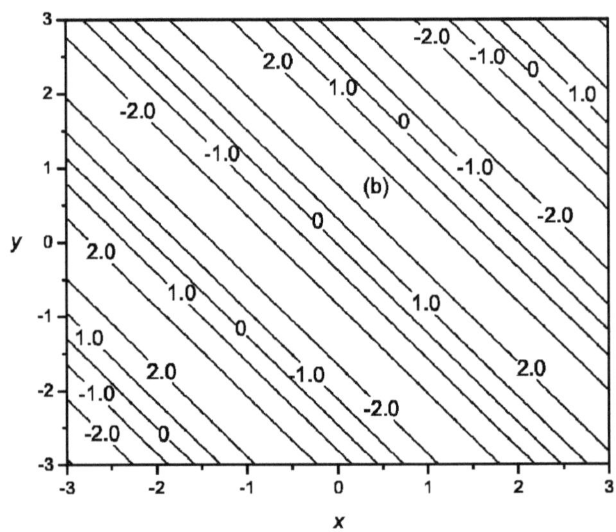

Figure 6. $w\,(x,\ y,\ -1,\ 0)$ in the $(x,\ y)$ – plane.

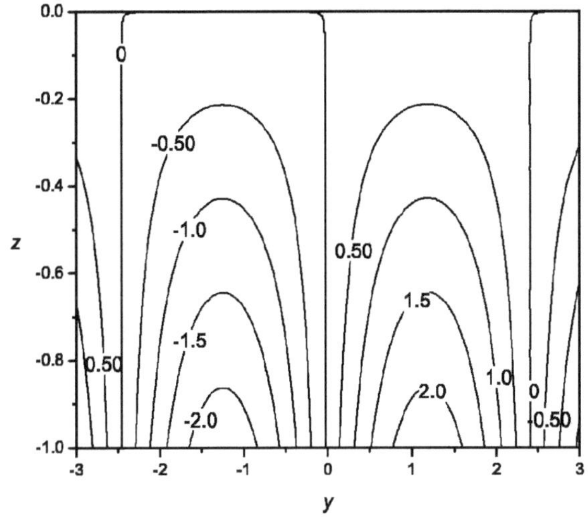

Figure 7. $w(0,\ y,\ z,\ 0)$ in the $(y,\ z)$ – plane.

for $z = -1$ (immediately above the abyssal overflow), $t = 0$, $B = 1$ and $h = 1$ (for convenience) for the most unstable mode

$$\sigma_{\max} \approx 0.09 - 2.04i,\ k_{\max} \approx 1.55,\ l_{\max} \approx 1.29,$$

$$U \approx 1.84,\ V \approx 1.99,$$

for $c_D = 0.1$ and $f = 0.25$. We recall that the bottom topography is given by $h_B = -y$ so that the depth increases with increasing y. The wave field propagates from the lower left corner toward the upper right corner.

The upper layer vertical velocity scale is about 1.2 cm/s. Thus, assuming a *nondimensional* perturbation thickness in the abyssal current of about 0.1 (corresponding to about a *dimensional* abyssal current height anomaly of about 10 m), Fig. 6 suggests a *dimensional* vertical velocity in the overlying water column, immediately above the abyssal current, associated with the generated internal gravity wave field of about 0.25 cm/s. We note again that Fig. 6 assumes that $h = 1$, so that $h = 0.1$ would reduce the w values by a factor of 10.

Figure 7 is a contour plot of a vertical section of the *total* vertical velocity in the overlying ocean along $y = 0$ with $t = 0$ and $h = 1$ (again, for convenience) for the most unstable mode for $c_D = 0.1$ and $f = 0.25$. One can see the bottom intensification in the internal wave field and, for $B = 1$, the approximate linear decrease in the magnitude of w with decreasing depth. For larger values of B, the internal gravity wave field is increasingly bottom intensified.

Acknowledgements

Preparation of this manuscript was supported in part by Research Grants awarded by the Natural Sciences and Engineering Research Council (NSERC) of Canada. Email: gordon.swaters@ualberta.ca.
URL: pacific.math.ualberta.ca/gordon.

References

Baines, P. G. Topographic Effects in Stratified Flows. Cambridge University Press, 482 pp., 1995.

Bruce, J. G. Eddies southwest of Denmark Strait. *Deep-Sea Res.,* 42:13-29, 1995.

Emms, P. W. A streamtube model of rotating turbidity currents. *J. Mar. Res.*,56:41-74, 1998.

Girton, J. B., and T. B. Sanford. Synoptic sections of the Denmark Strait overflow. *Geophys. Res. Lett.*, 28:1619-1622, 2001.

Girton, J. B., and T. B. Sanford. Descent and modification of the Denmark Strait overflow. Submitted to *J. Phys. Oceanogr.*, 2002

Jeffreys, H. The flow of water in an inclined channel of rectangular bottom. *Phil. Mag.*, 49: 793-807, 1925.

Jiang, L., and R. W. Garwood. Three-dimensional simulations of overflows on continental slopes. *J. Phys. Oceanogr.,* 26:1214-1233, 1996.

Jungclaus, J. H., J. Hauser, and R. H. Käse. Cyclogenesis in the Denmark Strait Overflow plume. *J. Phys. Oceanogr.,* 31:3214-3228, 2001.

Käse, R. H., Girton, J. B., and T. B. Sanford. Structure and variability of the Denmark Strait overflow: Model and observations. Submitted to *J. Geophys. Res.*, 2002.

Käse, R. H., and A. Oschlies. Flow through Denmark Strait. *J. Geophys. Res.*, 105:28527-28546, 2000.

Krauss, W. A note on overflow eddies. *Deep-Sea Res.,* 43:1661-1667, 1996.

Killworth, P. D. Mixing on the Weddell Sea continental slope. *Deep-Sea Res.,* 24:427-448, 1977.

Krauss, W., and R. H. Käse. Eddy formation in the Denmark Strait overflow. *J. Geophys. Res., 103:15523-15538, 1998.*

Price, J. F., and M. O. Baringer. Outflows and deep water production by marginal seas. *Progr. Oceanogr.*, 33:161-200, 1994.

Smith, P. C. A streamtube model for bottom boundary currents in the ocean. *Deep-Sea Res.,* 22:853-873, 1975.

Swaters, G. E. On the baroclinic instability of cold-core coupled density fronts on a sloping continental shelf. *J. Fluid Mech.*, 224:361-382, 1991.

Swaters, G. E. Baroclinic characteristics of frictionally destabilized abyssal overflows. *J. Fluid Mech.*, 2003, in press.

Whitham, G. B. *Linear and Nonlinear Waves.* Wiley, 636 pp., 1974.

ON THE EFFECT OF HEAT AND FRESH WATER FLUXES ACROSS THE OCEAN SURFACE, IN VOLUME-CONSERVING AND MASS-CONSERVING MODELS

PEDRO RIPA
Departamento de Oceanografía Física, CICESE
Ensenada, Baja California, México

Abstract. A simple two-layer model is used to study changes in sea level forced by fresh water and heat fluxes across the ocean surface, including the dynamical effects related to induced pressure gradients. These solutions (η_E & η_Q) are compared with the expected expansion of the water column (η_S & η_T) related to the salinity and temperature changes produced by the *local* fresh water and heat fluxes which are restricted to a surface layer. Deeper salinity and temperature variations produced by the induced pressure gradients, are excluded from the definition of η_S and η_T.

Two conditions must be met to attribute sea level variations to an expansion or contraction of the water column. First, the sea level produced by the surface fluxes, η_E & η_Q, must be close to η_S & η_T. Second, this result must be obtained with a mass-conserving model that allows for the thermohaline expansion of seawater, but not with the usual volume-conserving model.

The information on the horizontal wavenumber and frequency of the forcing is contained in a single variable with two very different normalizations: κ_0 & κ_1, where $\kappa_0 = 1$ and $\kappa_1 = 1$ represent the dispersion relation for barotropic and baroclinic Poincaré waves respectively.

For very long horizontal scales, $\kappa_0 \ll 1$, the total response to heat forcing η_Q coincides with η_T. This effect is not predicted by a volume-conserving model. On the other hand, the effect of precipitation and evaporation is to raise and lower the surface by adding or subtracting water, its impact on water density is much less important, *i.e.* $\eta_E \gg \eta_S$ and can be safely modelled with a volume-conserving model. At long scales, $\kappa_0 \approx O(1)$, the solution is related to the forcing of the barotropic mode. Mass-conserving equations are still crucial to obtain η_Q correctly, but are not necessary for η_E. A single (thermohaline active) layer model with a rigid bottom, behaves like the two-layer model at these scales.

Finally, at short scales, $\kappa_1 \approx O(1)$, the response is controlled by the forcing of the baroclinic mode and mass-conserving equations are not needed. If the top layer is relatively shallow, the behavior at short and intermediate ($\kappa_1 \sim \kappa_0^{-1} \ll 1$) scales is similar to that of a reduced gravity model with a single (thermohaline active) layer. In this case, sea level equals the dynamic height relative to the lower layer. This effect is totally unrelated to the expansion or contraction of the water column.

Key words: Sea-level changes, surface fluxes, mass-conserving and volume-conserving models

O.U. Velasco Fuentes et al. (eds.), Nonlinear Processes in Geophysical Fluid Dynamics, 103-126.

1. Introduction

Suppose that sea level variations are found to be correlated with subsurface changes in temperature and salinity in the form

$$\Delta\eta = \int_{-H_*}^{0} (\alpha_T \Delta T - \alpha_S \Delta S)\, \mathrm{d}z, \tag{1}$$

where α_T and α_S are respectively the thermal expansion and haline contraction coefficients, and $z = -H_*$ represents the ocean bottom or a very deep level (Pattullo *et al.*, 1955). A possible interpretation of this relationship is that the change $\Delta\eta$ in the water column height is produced by the expansion (or contraction) related to the changes of its temperature or salinity. Moreover, there might be a temptation to believe that the latter are caused by the *local* heat and fresh water fluxes across the ocean surface (henceforth, HFW). For instance Gill and Niiler (1973) conclude that "steric changes in sea-level [are] produced by expansion and contraction of the water column above the seasonal thermocline due to changing [HFW]." This explanation is at odds with the usual assumption of a non-divergent total velocity field, namely, the equation of *volume conservation*, which is generally taken to be a very good approximation to that of *mass conservation* (Greatbatch, 1994).

However, there is quite a different interpretation of (1). The hydrostatic balance –another common approximation, totally unrelated to HFW– renders the deep pressure as

$$\frac{1}{g\rho_0}\, \Delta p|_{z=-H_*} = \Delta\eta - \int_{-H_*}^{0} (\alpha_T \Delta T - \alpha_S \Delta S)\, \mathrm{d}z. \tag{2}$$

On the right hand side one easily recognizes the difference between both sides of equation (1); interpreting Δ as a horizontal variation or gradient condition, (1) is no more than that of no-motion at level $z = -H_*$. A deep level of no motion may be a good approximation for baroclinic signals but it is certainly not correct for a barotropic one. For instance, if it is assumed that density is conserved following a fluid element, then making a vertical normal modes expansion of the dynamical fields it follows that equation (1) is satisfied mode by mode, with an extra factor $[1 - G_n(-H_*)/G_n(0)]$ in the left hand side, where $G_n(z)$ is the $n^{\underline{\text{th}}}$ mode structure function for horizontal velocity and pressure fields (Ripa, 1997). This factor is approximately equal to unity for the baroclinic modes (if H_* is deep enough) and to zero for the barotropic mode.

There are, then, two different interpretations of condition (1):

1. It expresses the *expansion or contraction* of the water column due to the local HFW.
2. It shows evidence of surface intensified *baroclinic* motion in a deep enough ocean.

Of course, HFW may *indirectly* contribute to baroclinic motions through the set up of pressure gradients. Needless to say, wind forced signals and free waves may have an important baroclinic contribution to sea level, totally independent of HFW.

The validity of both explanations is explored here in wavenumber/frequency space, with the simplest useful model that the author can imagine: HFW are allowed assuming that salinity, temperature, and horizontal velocity are depth independent within an "active" top layer. This type of model is not new. It was first developed, to the best knowledge of the author, by Dronkers (1969) in a one-layer rigid bottom set up, with the purpose of studying the tides in a coastal area. Afterwards it was used in a one-layer reduced gravity setting by Lavoie (1972) and Schopf and Cane (1983), for an atmospheric and oceanic problem, respectively. It has been employed by many other authors ever since, but always with volume-conserving equations. In Ripa (1999) it is shown that neglecting vertical variations within the heterogeneous layer is a good approximation, even with finite horizontal variations of the density and velocity fields, as long as the perturbation horizontal scale is not much smaller than the resolved deformation radii. Here, the field variations will be considered infinitesimal –not finite– and therefore we ought to be on safe ground (within the same wavelength-scale restriction).

Probably the most often used model with only baroclinic (barotropic) dynamics is one with an homogeneous layer in a reduced-gravity (rigid-bottom) setting. Heterogeneous one- and two-layer models are developed here in §2 and §3, respectively, including the correct forcing terms for the HFW (Beron-Vera, Ochoa and Ripa, 1999). In particular, it is analyzed to what extent the one-layer reduced gravity and rigid bottom results are equivalent to the baroclinic and barotropic mode contributions in a two-layer model. Since the first interpretation of (1) is at odds with the usual assumption of a divergenceless velocity field, volume conserving and mass conserving equations are compared. An expansion in continuous vertical modes is usually based on the assumption that the surface temperature and salinity are laterally uniform; here it will be necessary to go beyond this assumption, in order to accommodate the surface buoyancy flux. The response to these forcings is discussed in §4 and final conclusions are given in §5.

TABLE I. Reduced gravity model variables

Layer:	Active	Passive
Salinity	$S(\mathbf{x},t) = S_1 + S'(\mathbf{x},t)$	S_2
Temperature	$T(\mathbf{x},t) = T_1 + T'(\mathbf{x},t)$	T_2
Density	$\rho(\mathbf{x},t) = \rho_0(1 - \alpha_T T' + \alpha_S S')$	$\rho_0(1+\varepsilon)$
Velocity	$\mathbf{u}(\mathbf{x},t)$	$\mathbf{0}$
Depth	$h(\mathbf{x},t)$	∞

2. Volume- or mass-conserving one-layer models

In the reduced gravity case, the active layer is assumed to be on top of a passive one and the dynamical variables have the form shown in Table I where S_1, S_2, T_1, T_2 and ρ_0 are constants, and

$$\varepsilon = \alpha_T(T_1 - T_2) - \alpha_S(S_1 - S_2) \tag{3}$$

is the relative density excess of the lower layer (in the reference state). In the rigid bottom case, on the other hand, the active layer is the only existing one. (In the following sections the lower layer will be allowed to have a finite depth).

Since S, T, and $\mathbf{u}$ are assumed to be depth independent within an active layer of depth h, the first three equations of motion are (Beron-Vera, Ochoa and Ripa, 1999)

$$\begin{aligned} h(\partial_t S + \mathbf{u}\cdot\nabla S) &= SE_e \\ \rho_0 C_p h(\partial_t T + \mathbf{u}\cdot\nabla T) &= Q \\ \partial_t h + \nabla\cdot(\mathbf{u}h) &= -E_e \end{aligned}$$

where Q and $-E_e$ are the local heat and fresh water inputs across the ocean surface; $E_e = E - P$ (the difference between evaporation and precipitation rates) is called the "effective evaporation." If the model conserves the mass of each fluid element, rather than its volume, then these equations must be modified into

$$h(\partial_t S + \mathbf{u}\cdot\nabla S) = SE_e \tag{4a}$$

$$\rho C_p h(\partial_t T + \mathbf{u}\cdot\nabla T) = Q \tag{4b}$$

$$\partial_t(h\rho) + \nabla\cdot(\mathbf{u}h\rho) = -\rho E_e \tag{4c}$$

where $\rho(\mathbf{x}, t)$ is given in Table I. These equations are equivalent to the transport balances for salt, fresh water, and heat content:

$$\partial_t (h\rho s) + \nabla \cdot (\mathbf{u}h\rho s) = 0 \tag{5a}$$
$$\partial_t [h\rho (1 - s)] + \nabla \cdot [\mathbf{u}h\rho (1 - s)] = -\rho E_e \tag{5b}$$
$$C_p [\partial_t (h\rho T) + \nabla \cdot (\mathbf{u}h\rho T)] = Q - \rho C_p T E_e \tag{5c}$$

where $s \approx 10^{-3} \times S$ is the mass salt fraction. Notice that there is no salt flux across the ocean surface and that the last term in the heat balance renders that equation independent of the temperature origin, as it must be (therefore T could be replaced by T').

The pressure fields in the upper and lower layer are calculated from the hydrostatic balance and the condition of uniform pressure at the surface (no atmospheric pressure forcing is considered here, for simplicity), which give

$$p = \begin{cases} p_u = g\rho(\eta - z) & : -h + \eta < z < \eta \\ p_l = g\rho_2(\eta - z) - g(\rho_2 - \rho)h & : z < -h + \eta \end{cases}$$

Taking the horizontal gradient and *then* vertically averaging the first one it is found

$$h^{-1}\int_{-h+\eta}^{\eta} \nabla p_u \, \mathrm{d}z = g\rho\nabla\eta + \tfrac{1}{2}gh\nabla\rho,$$
$$\nabla p_l = g\nabla(\rho_2\eta - (\rho_2 - \rho)h). \tag{6}$$

The next step is different for each type of model: In the reduced gravity case, η is diagnosed from $\nabla p_l = 0$ and then used in the vertical average of ∇p_u, whereas in the rigid bottom case ∇p_l clearly does no exist (there is only one layer) and in ∇p_u it is used $\nabla\eta = \nabla h$. Finally, making the Boussinesq approximation, for simplicity, the driving force per unit mass is found to be equal to $-\frac{1}{2}h^{-1}\nabla(\vartheta h^2)$, where

$$\vartheta(\mathbf{x}, t) = \begin{cases} g(\varepsilon + \alpha_T T' - \alpha_S S') & : \text{reduced gravity} \\ g(1 - \alpha_T T' + \alpha_S S') & : \text{rigid bottom} \end{cases} ; \tag{7}$$

notice the difference in signs (Ripa, 1993).

The dynamical system is then completed with the evolution equation for the (depth independent) horizontal velocity field,

$$\partial_t\mathbf{u} + \mathbf{u}\cdot\nabla\mathbf{u} + f\,\hat{\mathbf{z}}\times\mathbf{u} + \tfrac{1}{2}h^{-1}\nabla\left(\vartheta h^2\right) = \mathbf{0}; \tag{8}$$

for simplicity, f is assumed constant and wind stress forcing is not included. The equations of the volume conserving model are those of (4) –or (5)– and (8), with ρ replaced by ρ_0.

The relationship between the active layer depth $h(\mathbf{x},t)$ and the sea level $\eta(\mathbf{x},t)$ is quite different in both cases, namely

$$\begin{array}{ll} h = H - \zeta; \quad \eta = (\alpha_T T' - \alpha_S S')\, h - \varepsilon\zeta & : \text{reduced gravity} \\ h = H + \eta & : \text{rigid bottom} \end{array} \tag{9}$$

where $\zeta(\mathbf{x},t)$ represents the interface elevation field. (Even though the same symbol, H, is used for the reference h in both the reduced gravity and rigid bottom cases, their actual values are usually quite different: H represents an upper layer mean depth, in the first case, or the total ocean depth, in the second one.) In the rigid bottom model $\eta = h - H$ is a prognostic variable. For the reduced gravity model, on the other hand, η is a diagnosed variable, obtained from the no motion condition, $\nabla p = 0$, inside the passive layer, where p given by (2) at $z = -H_* < -h$.

Instead of T' and S' we define equivalent variables, with units of length, by

$$(\eta_T,\ \eta_S) = (\alpha_T T',\ -\alpha_S S')\, H. \tag{10}$$

Linearizing (4a) and (4b) it follows that S' and T' (and, therefore, η_S & η_T) are solely determined by the HFW, because the basic values S_1 and T_1 are but constants. More precisely,

$$\begin{aligned} \partial_t \eta_T &= \tfrac{\alpha_T}{\rho_0 C_p} Q =: W_T, \\ \partial_t \eta_S &= -\varepsilon_1 E_e =: W_S, \end{aligned} \tag{11}$$

where

$$\varepsilon_1 - \alpha_S S_1, \tag{12}$$

controls the changes in salinity induced by the fresh water output E_e. Recall that typical values of H are quite different in the reduced gravity and rigid bottom cases, but so are T' and S', since in both cases (11) shows that η_T & η_S, defined in (10), are independent of the value of H or, in fact, of any assumption on how is heat and fresh water mixed down the water column. The same definition (10) will be used with the two-layer model, discussed next.

According to hypothesis 1 in the Introduction, the freshening and heating of the active layer produced by the surface fluxes E_e and Q should produce a sea level change given by $\eta \approx \eta_T + \eta_S$. This is also the dynamic height (1) relative to $H_* = h \approx H$. In the reduced gravity model, however, definition (9) shows that η is the dynamic height (1) relative to $H_* > h$, and therefore has two contributions: $\eta_T + \eta_S$ and $-\varepsilon\zeta$. Hypothesis 2 from the Introduction is always satisfied in the reduced gravity model, by construction. Hypothesis 1, on the other hand, not only requires $\eta \approx \eta_T + \eta_S$ but,

TABLE II. Two layer model variables

Layer:	**Top**	**Lower**
Salinity	$S(\mathbf{x},t) = S_1 + S'(\mathbf{x},t)$	S_2
Temperature	$T(\mathbf{x},t) = T_1 + T'(\mathbf{x},t)$	T_2
Density	$\rho(\mathbf{x},t) = \rho_0(1 - \alpha_T T' + \alpha_S S')$	$\rho_2 = \rho_0(1+\varepsilon)$
Velocity	$\mathbf{u}(\mathbf{x},t)$	$\mathbf{u}_2(\mathbf{x},t)$
Depth	$h(\mathbf{x},t) = H + \eta(\mathbf{x},t) - \zeta(\mathbf{x},t)$	$h_2(\mathbf{x},t) = H_2 + \zeta(\mathbf{x},t)$

in addition, this result has to be obtained only with the mass-conserving equations. In the rigid bottom model η_T and η_S are not explicit in (9), but might show up as a result of the model equations. In §4 the response to HFW is obtained and both hypotheses from the Introduction are tested.

3. Volume- or mass-conserving two-layer models

If the lower layer in the model described in Table I is now assumed to have a finite mean depth H_2, then its actual depth $h_2(\mathbf{x},t)$ and velocity $\mathbf{u}_2(\mathbf{x},t)$ must be prognostic. Keeping the temperature and salinity –and, therefore the density– of the lower layer as constants, the variables are redefined as shown in Table II. Now, both the sea surface and interface elevations, $\eta(\mathbf{x},t)$ and $\zeta(\mathbf{x},t)$, are prognostic.

The evolution equations are the whole set (4) plus

$$\begin{aligned} \partial_t h_2 + \nabla\cdot(\mathbf{u}_2 h_2) &= 0 \\ \partial_t \mathbf{u} + \mathbf{u}\cdot\nabla\mathbf{u} + f\,\hat{\mathbf{z}}\times\mathbf{u} + g\nabla\eta - \tfrac{1}{2}h\nabla\vartheta &= \mathbf{0} \\ \partial_t \mathbf{u}_2 + \mathbf{u}_2\cdot\nabla\mathbf{u}_2 + f\,\hat{\mathbf{z}}\times\mathbf{u}_2 + \nabla(g\eta - \vartheta h) &= \mathbf{0} \end{aligned} \qquad (13)$$

where $\vartheta = g(\varepsilon + \alpha_T T' - \alpha_S S') = (\rho_2 - \rho)/\rho_0$, i.e. the definition (7) for the reduced gravity case. In the last two equations, the terms $g\nabla\eta - \frac{1}{2}h\nabla\vartheta$ and $\nabla(g\eta - \vartheta h)$ are the Boussinesq approximation of the right hand sides of (6) (Ripa, 1993). The model set up is determined by two independent non-dimensional numbers: ε ($\ll 1$), defined in (3), and

$$\gamma = \frac{H_2}{H_1 + H_2}, \qquad (14)$$

where $H_1 = H$ is the mean depth of the top layer.

Unlike system (13), the linearized pressure forces are equal to minus the gradient of two scalars:

$$\begin{aligned} p_1 &= g\eta - \tfrac{1}{2} g\left(\eta_T + \eta_S\right), \\ p_2 &= g\eta + \varepsilon g\zeta - g\left(\eta_T + \eta_S\right). \end{aligned} \tag{15}$$

These definitions include the contributions of both the variations of layer thickness, η & ζ, and the heterogeneities T' and S', through η_T & η_S. The linearized evolution of the layer fields is driven by

$$\partial_t \begin{pmatrix} p_1 \\ p_2 \end{pmatrix} + \mathcal{L}\nabla\cdot\begin{pmatrix} \mathbf{u} \\ \mathbf{u}_2 \end{pmatrix} = g\begin{pmatrix} \varepsilon_1^{-1} W_S + \left(\lambda - \frac{1}{2}\right)\left(W_S + W_T\right) \\ \varepsilon_1^{-1} W_S + \left(\lambda - 1\right)\left(W_S + W_T\right) \end{pmatrix}, \tag{16a}$$

$$\left(\partial_t + f\,\hat{\mathbf{z}}\times\right)\begin{pmatrix} \mathbf{u} \\ \mathbf{u}_2 \end{pmatrix} + \nabla\begin{pmatrix} p_1 \\ p_2 \end{pmatrix} = 0, \tag{16b}$$

where $\lambda = 0$ ($\lambda = 1$) corresponds to the volume (mass) conserving case and the coupling of layer variables evolution is controlled by the non-diagonal matrix

$$\mathcal{L} = g\begin{pmatrix} H_1 & H_2 \\ H_1 & (1+\varepsilon)\,H_2 \end{pmatrix} \tag{17}$$

[with $H_1 = H$ and $H_2 = H\gamma/(1-\gamma)$ it follows that $\mathcal{L}$ is equal to gH times a non-dimensional matrix, function of γ and ε]. The term $\varepsilon_1^{-1} W_S = -E_e$ in (16a) is just the contribution to $\partial_t \eta$ due to the addition or subtraction of water by the effective evaporation. The term $\lambda\left(W_S + W_T\right)$, with $\lambda = 1$, models the contribution of T' and S' to $\rho^{-1}\partial_t(\rho\eta)$. Finally, the last terms proportional to $\left(W_S + W_T\right)$ come from using (11) in the time derivative of the terms $\left(\eta_T + \eta_S\right)$ of (15). The results are obtained more easily using the normal modes, discussed next.

The matrices

$$\mathcal{S}_1 = \frac{1}{1+\varepsilon\gamma_1^2}\begin{pmatrix} \gamma_1 \\ \gamma_1 - 1 \end{pmatrix}\left(\,1+\varepsilon\gamma_1 \;\; -1\,\right), \tag{18a}$$

$$\mathcal{S}_0 = \frac{1}{1+\varepsilon\gamma_1^2}\begin{pmatrix} 1 \\ 1+\varepsilon\gamma_1 \end{pmatrix}\left(\,1-\gamma_1 \;\; \gamma_1\,\right), \tag{18b}$$

are constructed with the right and left eigenvectors of $\mathcal{L}$, which satisfy $\mathcal{L}\mathcal{S}_j = \mathcal{S}_j\mathcal{L} = c_j^2\mathcal{S}_j$, with the eigenvalues

$$c^2 = \begin{cases} c_1^2 = \dfrac{\varepsilon g H_1 H_2}{H_1 + H_2} \dfrac{\gamma_1/\gamma}{1 + \varepsilon\gamma_1} & : \text{baroclinic} \\ \\ c_0^2 = g\left(H_1 + H_2\right) \dfrac{1 + \varepsilon\gamma_1}{\gamma_1/\gamma} & : \text{barotropic} \end{cases} \tag{19}$$

where

$$\gamma_1 = \frac{2\gamma}{1 - \varepsilon\gamma + \sqrt{\left(1 + \varepsilon\gamma\right)^2 - 4\varepsilon\gamma\left(1 - \gamma\right)}}$$

$= \gamma + O\left(\varepsilon\right)$. The matrices $\mathcal{S}_1$ & $\mathcal{S}_0$ are also found to satisfy

$$\begin{aligned} \mathcal{S}_0\mathcal{S}_1 &= \mathcal{S}_1\mathcal{S}_0 = 0, \\ \mathcal{S}_j^2 &= \mathcal{S}_j, \quad \mathcal{S}_1 + \mathcal{S}_0 = 1. \end{aligned}$$

The first line are the orthogonality conditions between right and left eigenvectors (*i.e.* the direct and dual basis), whilst the second line shows that the normalization has been chosen so that $\mathcal{S}_1$ and $\mathcal{S}_0$ are *projection operators* into the baroclinic and barotropic components of the pressure or velocity layer variables. The uncoupled evolution equations for the contribution of either mode to these variables are obtained applying $\mathcal{S}_1$ or $\mathcal{S}_0$ to both equations in (16) and taking the first row. The contribution of each mode to the lower layer fields, $g\eta + \varepsilon g\zeta - g\left(\eta_T + \eta_S\right)$ and $\mathbf{u}_2$, are obtained from the proportionality implied in the column vectors in $\mathcal{S}_1$ & $\mathcal{S}_0$.

Since $0 < \gamma < 1$, in the final results γ_1 can be replaced by γ and $\varepsilon\gamma_1$ can be neglected compared to 1. For instance, making these approximation in (19) the well known expressions $c_1^2 \approx \varepsilon g H_1 H_2 / \left(H_1 + H_2\right)$ and $c_0^2 \approx g\left(H_1 + H_2\right)$ are easily obtained.

4. Results

In order to assess the importance of the different forcing and response variables, we consider an infinitesimal plane wave,

$$\left(E_e,\ Q;\ S',\ T',\ \eta,\ \mathbf{u},\ \cdots\right) \propto a\, \mathrm{e}^{i(kx+ly-\omega t)} + O\left(a^2\right)$$

which renders the linearized version of the evolution equations –(4) and either (8) or (13), with ρ given by Table I or replaced by ρ_0– a very simple algebraic problem.

The surface inputs of fresh water $(P - E)$ and heat Q produce directly the density changes represented by η_S & η_T, and indirectly, through the set

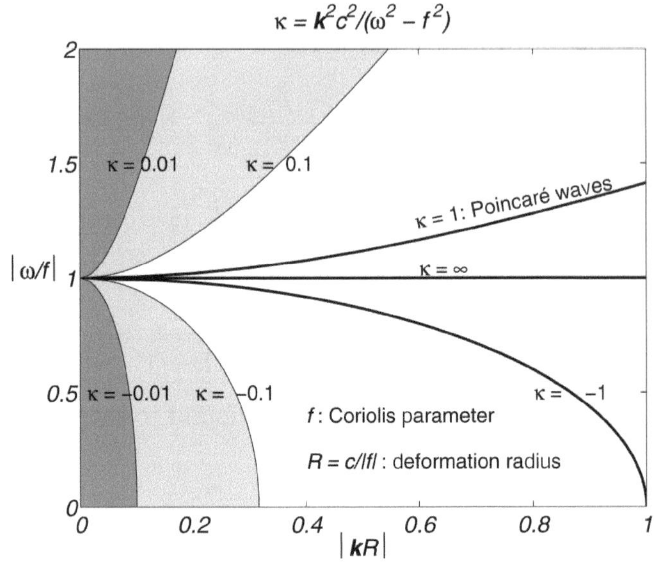

Figure 1. The heat input (output) Q through the ocean surface, with wavenumber $\mathbf{k}$ and frequency ω, produces a sea level η_Q, part of which, η_T, corresponds to the direct warming (cooling) of an upper layer, but another part of η_Q is due to the pressure gradient generated by Q. In the shaded region ($\kappa \ll 1$), though, it is $\eta_Q \simeq \eta_T$. In order to obtain this result, a model must conserve mass instead of volume, *i.e.* the three-dimensional velocity field must have the possibility of a non-vanishing divergence. The variable κ defined here is also used in the following figures, with different normalizations, defined by the choice of c^2.

up of pressure gradients, those represented in ζ. By linearizing the evolution equations, the total sea level forced by each of both fluxes is calculated separately, say

$$\left.\begin{array}{r} E_e \mapsto (\eta_E,\, \mathbf{u}_E,\, \ldots) \\ Q \mapsto (\eta_Q,\, \mathbf{u}_Q,\, \ldots) \end{array}\right\} : (\eta = \eta_E + \eta_Q,\; \mathbf{u} = \mathbf{u}_E + \mathbf{u}_Q,\; \ldots)\,.$$

An obvious advantage of using linearized model equations is that the general solution is easily obtained, allowing for a study of the whole parameters space. Another advantage, for the purpose of this paper, is that η_T and η_S in (10) are solely produced by HFW, as shown in (11). Clearly η_T and η_S are totally independent of the atmospheric pressure- and wind-forced signals and of the free wave solutions; as long as these have an important surface intensified baroclinic contribution, they will satisfy (1) for $H_* > h$ independently of HFW. The terms η_T and η_S are only part of η_Q and η_E, as the horizontal gradients of temperature and salinity imply a force proportional to $h\nabla\vartheta$ in each active layer, which has a further effect on η.

In order to test both hypotheses, it is necessary to 1) evaluate the ratio between η_E [η_Q] –the total sea level due to E_e [Q]– and η_S [η_T] –the local effect of fresh water [heat] flux– and 2) check whether the latter is due to the expansion or contraction of the water column, that is, the effect of a variable ρ in (4c), or to the effect of a variable ρ in the hydrostatic balance $\partial_z p = -g\rho$, whose integral is (2).

The evaluation of the ratios η_E/η_S and η_Q/η_T for the one-layer models of §2 needs of only three non-dimensional parameters: ε and ε_1, defined in (3) and (12), and

$$\kappa = \frac{k^2 + l^2}{\omega^2 - f^2} c^2, \tag{20}$$

which is the ratio of the squared forcing wavenumber with that of a free Poincaré wave with the same frequency as the forcing one, where

$$c^2 = \begin{cases} \varepsilon g H & : \text{reduced gravity} \\ g H & : \text{rigid bottom} \end{cases}$$

see Figure 1. In the two-layer model, the eigenvalues (19) are similarly used to define κ_1 and κ_0 with equation (20). Clearly κ_1 and κ_0 are not independent; in fact

$$\frac{\kappa_1}{\kappa_0} = \frac{c_1^2}{c_0^2} = \frac{\varepsilon\gamma\,(1-\gamma)}{(1+\varepsilon\gamma_1)^2} \tag{21}$$

$\approx \varepsilon\gamma\,(1-\gamma)$. The evaluation of the ratios η_E/η_S and η_Q/η_T for the two-layer models of §3 requires of four non-dimensional parameters: ε, ε_1, γ and either κ_0 or κ_1.

As explained next, the linearized equations of the reduced gravity and rigid bottom one-layer models, as well as those of both modal equations for the two-layer model, can be cast in the form

$$\begin{aligned} \partial_t \eta' + g^{-1} c^2 \nabla \cdot \mathbf{u}' &= (M\lambda - N)\,(W_S + W_T) + (M/\varepsilon_1)\,W_S, \\ \partial_t \mathbf{u}' + f\,\hat{\mathbf{z}} \times \mathbf{u}' + g\nabla\eta' &= \mathbf{0}, \end{aligned} \tag{22}$$

where η' is chosen so that there is no term on the right hand side of the $\mathbf{u}'$ equation. [The meaning of every forcing term is explained right after (17).] Thus, for the one-layer reduced gravity model it is $\eta' = -\varepsilon\zeta + \frac{1}{2}(\eta_T + \eta_S) = \eta - \frac{1}{2}(\eta_T + \eta_S)$. For the one-layer rigid bottom model, on the other hand, it is $\eta' = \eta - \frac{1}{2}(\eta_T + \eta_S)$. Finally, for the two-layer model it is $p_1/g = \eta - \frac{1}{2}(\eta_T + \eta_S) = \eta^{(0)} + \eta^{(1)}$ and $\mathbf{u} = \mathbf{u}^{(0)} + \mathbf{u}^{(1)}$: the uncoupled evolution equations for either $\left(\eta^{(1)}, \mathbf{u}^{(1)}\right)$ or $\left(\eta^{(0)}, \mathbf{u}^{(0)}\right)$ are obtained applying the projection operators $\mathcal{S}_1$ or $\mathcal{S}_0$ from (18) to both equations in (16) and taking

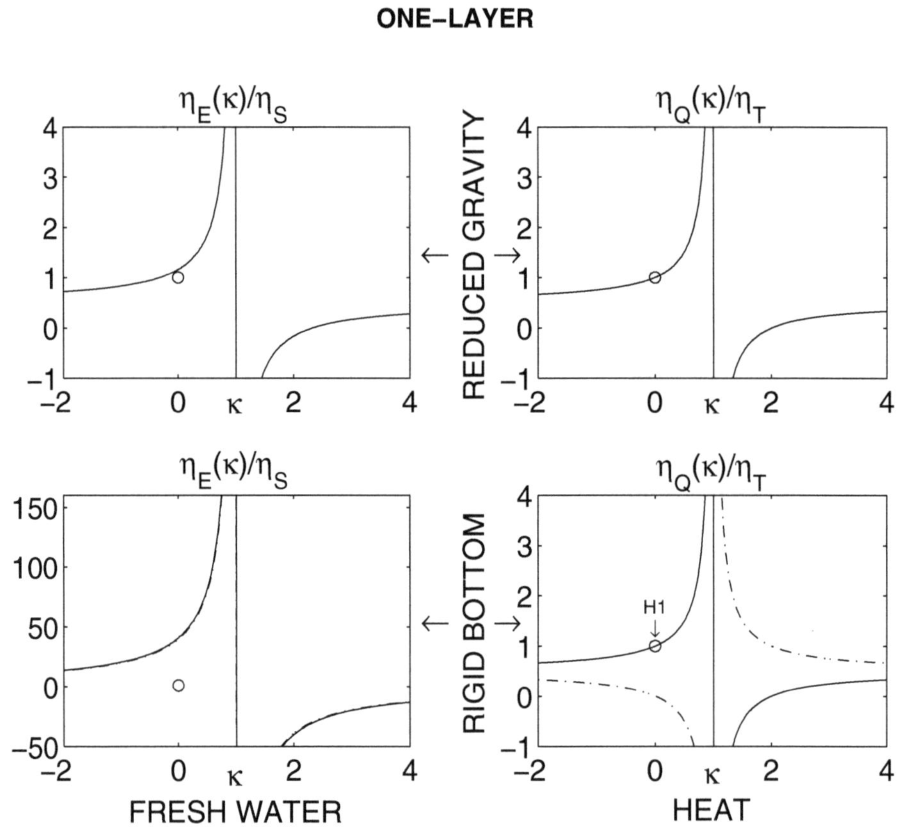

Figure 2. Ratio of the total sea level forced by the fresh water flux (left panels) or the heat flux (right panels) across the ocean surface, to that part of sea level directly produced by the same forcing. The solutions of the volume (mass) conserving models are shown with broken (solid) lines. All the relevant information on the wavenumber and frequency of the forcing, as well as the vertical structure of the response, is contained in the variable κ, defined in (20) and in Figure 1. The small circles, at $\kappa = 0$ and unit ratio, indicate the dynamic height due to the heating or freshening of the active layer. The formulae are presented in equation (25). The symbol H1 shows that $\eta_Q \approx \eta_T$ due to the thermal expansion of the water column, *i.e.* fulfillment of hypothesis 1 from the Introduction. Notice the different scale in the lower left panel: η_E is of $O\left(\varepsilon_1^{-1}\eta_S\right)$ rather than $O\left(\eta_S\right)$.

the first row (the second row equations are just proportional to this one, by construction). The coefficients M & N come from projecting the forcing into that mode. Consequently, all four cases take the form (22),

	η'	$\mathbf{u}'$	M	N
reduced gravity	$\eta - \frac{1}{2}(\eta_T + \eta_S)$	$\mathbf{u}$	ε	$-\frac{1}{2}$
rigid bottom	$\eta - \frac{1}{2}(\eta_T + \eta_S)$	$\mathbf{u}$	1	$\frac{1}{2}$
baroclinic	$\eta^{(1)}$	$\mathbf{u}^{(1)}$	$\frac{\varepsilon\gamma_1^2}{1+\varepsilon\gamma_1^2}$	$-\frac{\gamma_1(1-\gamma_1\varepsilon)}{2(1+\varepsilon\gamma_1^2)}$
barotropic	$\eta^{(0)}$	$\mathbf{u}^{(0)}$	$\frac{1}{1+\varepsilon\gamma_1^2}$	$\frac{1+\gamma_1}{2(1+\varepsilon\gamma_1^2)}$

Using $\nabla = i\mathbf{k}$, $\partial_t = -i\omega$, and (20) the forced solution is easily found to be

$$\begin{pmatrix} \eta' \\ \mathbf{u}' \end{pmatrix} = \begin{pmatrix} 1 \\ \mathbf{U}\,|\kappa|^{\frac{1}{2}}\,g/c \end{pmatrix} \frac{A\eta_S + B\eta_T}{1-\kappa}, \tag{23}$$

where $A = M(\lambda + 1/\varepsilon_1) - N$, $B = M\lambda - N$ and, denoting with $\hat{\mathbf{k}}$ a unit vector in the direction of the wavenumber $\mathbf{k}$,

$$\mathbf{U} = \frac{\omega\hat{\mathbf{k}} - if\,\hat{\mathbf{z}} \times \hat{\mathbf{k}}}{\sqrt{|\omega^2 - f^2|}}\mathrm{sgn}(\kappa)$$

$\equiv (\omega\mathbf{k} - if\,\hat{\mathbf{z}} \times \mathbf{k})\left(\omega^2 - f^2\right)^{-1} c/|\kappa|^{\frac{1}{2}}$ is defined so that $|\mathbf{U}| = 1$ for either $\omega \gg f$ or $\omega \ll f$.

For the one-layer models, (23) has the coefficients

$$\begin{aligned} &A = \tfrac{1}{2} + \varepsilon/\varepsilon_1 + \varepsilon\lambda, \;\; B = \tfrac{1}{2} + \varepsilon\lambda \;\; : \text{reduced gravity} \\ &A = 1/\varepsilon_1 + \lambda - \tfrac{1}{2}, \;\;\; B = \lambda - \tfrac{1}{2} \;\;\; : \text{rigid bottom} \end{aligned} \tag{24}$$

which are used in the forced sea level and velocity solutions

$$\begin{aligned} \eta &= \frac{A\eta_S + B\eta_T}{1-\kappa} + \tfrac{1}{2}(\eta_S + \eta_T), \\ \mathbf{u} &= \mathbf{U}\,\frac{|\kappa|^{\frac{1}{2}}}{1-\kappa} g\,(A\eta_S + B\eta_T)/c. \end{aligned} \tag{25}$$

In the solution for η, the whole term proportional to η_S (η_T) is η_E (η_Q). This solution is composed of two parts: the first part, proportional to $1/(1-\kappa)$, is a forced wave that varies continuously, as the frequency decreases from suprainertial to subinercial, from Poincaré-like to geostrophic-like (Ripa and Zavala-Garay, 1999). The factor $1-\kappa$ in the denominator is proportional to $(k - k_+(\omega))(k - k_-(\omega))$, where for suprainercial frequencies, $k_\pm(\omega)$ are the inverse of the dispersion relation of Poincaré waves. In the forced problem there are two eigenvalues k for each ω, even though in the free waves problem there are three eigenvalues ω for each k (Philander, 1978; Ripa and Zavala-Garay, 1999). For subinertial frequencies, $k_\pm$ are complex. However, if we had included effects of Earth's curvature, there would be another resonance, with $k_\pm$ real, corresponding to the dispersion relation of the Rossby waves.

The second part of the solution, $\eta = \frac{1}{2}(\eta_S + \eta_T)$ & $\mathbf{u} = \mathbf{0}$, is what in Ripa (1996) is called the "force compensating mode": the balance between the gradients of h and ϑ in the $\mathbf{u}$ equation (8); this is obviously the dominant term for very large $|\mathbf{k}|$. In Ripa (1999) it is argued that the force compensating mode is related to the geostrophic solution in a higher vertical mode, which has $\omega = 0$ even with the β effect, as if having a "vanishing deformation radius," due to the approximation made when assuming that the dynamical fields are depth independent in the active layer. Consequently, the model fails at very short scales, of the order of the deformation radius of the (ill resolved) higher vertical mode. In the notation of this paper, it is equivalent to making $\kappa' = 0$, where κ' has the normalization corresponding to the value of c^2 of the ill resolved (higher) vertical mode.

The one-layer solutions are illustrated in Figure 2. Typical oceanic values are $\varepsilon \approx 1 - 4 \times 10^{-3}$ and $\varepsilon_1 \approx 2.5 \times 10^{-2}$, and therefore volume conserving ($\lambda = 0$) and mass conserving ($\lambda = 1$) models give essentially the same result for all cases, except for the heat flux forced solution in the rigid bottom model, $\eta_Q/\eta_T = \left(\lambda - \frac{1}{2}\kappa\right)/(1-\kappa)$, for which the volume conserving model ($\lambda = 0$) gives the incorrect result $\eta_Q/\eta_T = -\frac{1}{2}\kappa/(1-\kappa)$, which, in particular, vanishes for $\kappa \to 0$, instead of tending to 1. Hypothesis 1 in the Introduction is verified for κ ($= \kappa_{\text{rigid bottom}}$) $\ll 1$. This condition means wavenumbers smaller than that of a Poincaré wave with a frequency equal to ω, if $\omega^2 > f^2$, or wavenumbers smaller than the inverse reduced gravity deformation radius R^{-1}, if $\omega^2 \ll f^2$. Figure 1 shows the region in the forcing wavenumber/frequency space where κ is small enough for the direct heat forcing to have a significant effect on sea level. As $\kappa \to 1$, η_T greatly underestimates η_Q because free waves are resonantly excited by the forcing. As $\kappa \to 2$, on the other hand, η_Q is greatly overestimated by η_T because there is a cancellation between this (directly forced) part of the solution and the part produced through the set up of pressure gradients. For $\kappa \gg 1$ the model is very likely not valid (Ripa, 1999).

For the two-layer model, the solutions (23) for both normal modes, say

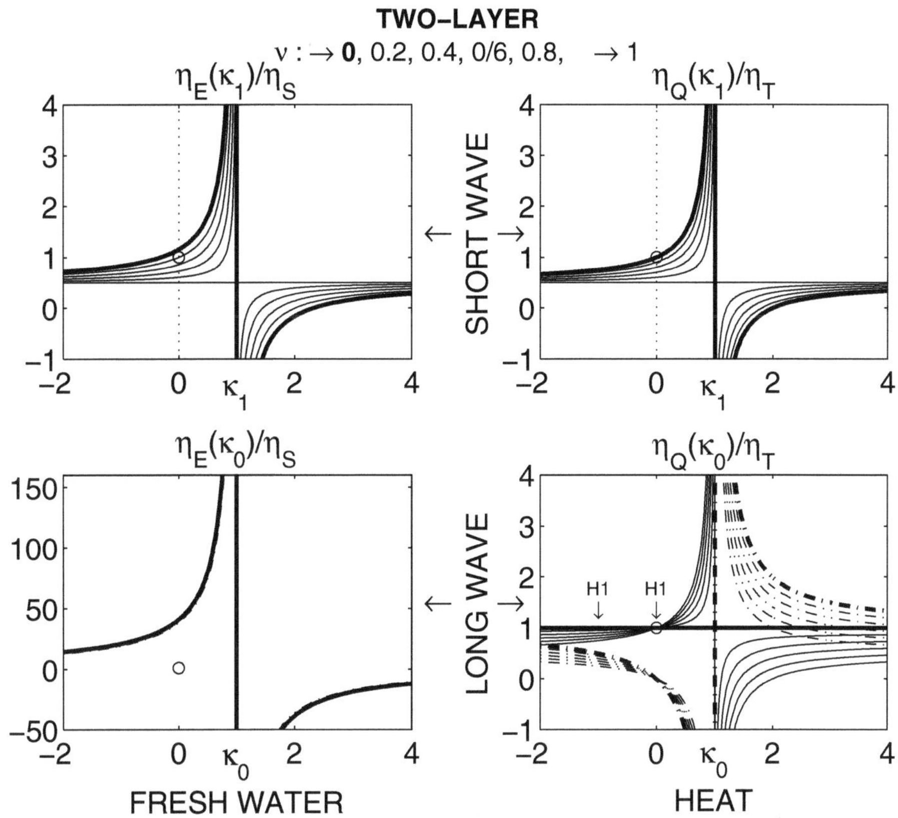

Figure 3. As in Figure 2, for the two-layer model, where γ is the ratio of the lower layer thickness to the total ocean depth. Thick lines correspond to $\gamma \to 1$ (very shallow upper layer). Since $\kappa_1 \approx \kappa_0\,\varepsilon\gamma\,(1-\gamma) \ll \kappa_0$, the lower graphs are like a boundary layer at the origin of the top ones (where the dotted line is).

$\eta^{(\alpha)}\,(\kappa_\alpha) = (A_\alpha \eta_S + B_\alpha \eta_T)\,/(1-\kappa_\alpha)$ for $\alpha = 0$ & 1, have the parameters

$$\begin{aligned}
A_0\left(1+\varepsilon\gamma_1^2\right) &= 1/\varepsilon_1 + \lambda - \tfrac{1}{2} - \tfrac{1}{2}\gamma_1,\\
B_0\left(1+\varepsilon\gamma_1^2\right) &= \lambda - \tfrac{1}{2} - \tfrac{1}{2}\gamma_1,\\
A_1\left(1+\varepsilon\gamma_1^2\right) &= \tfrac{1}{2}\gamma_1 + \varepsilon\left(1/\varepsilon_1 + \lambda - \tfrac{1}{2}\right)\gamma_1^2,\\
B_1\left(1+\varepsilon\gamma_1^2\right) &= \tfrac{1}{2}\gamma_1 + \varepsilon\left(\lambda - \tfrac{1}{2}\right)\gamma_1^2,
\end{aligned} \tag{26}$$

where the terms on the right hand side have been ordered by importance, according to $\varepsilon \ll \varepsilon_1 \ll 1$. The effect of λ is negligible in all parameters, except for B_0 and for any value of γ. In other words, the usual volume-conserving model ($\lambda = 0$) performs as well as the mass-conserving one

($\lambda = 1$) for the fresh-water forcing at all scales, and for the heat forcing at short enough scales so that the term $B_0/(1-\kappa_0)$ can be neglected. For each vertical mode, the upper, $g\eta - \frac{1}{2}g(\eta_T + \eta_S)$ & $\mathbf{u}$, and lower layer fields, $g\eta + \varepsilon g\zeta - g(\eta_T + \eta_S)$ & $\mathbf{u}_2$, are in the ratio of the components in the column vectors in the definition of $\mathcal{S}_1$ & $\mathcal{S}_0$. Consequently, the whole solution is found to be given by

$$\begin{aligned}
\eta &= \eta^{(0)}(\kappa_0) + \eta^{(1)}(\kappa_1) + \tfrac{1}{2}(\eta_S + \eta_T), \\
\zeta &= \gamma_1\eta^{(0)}(\kappa_0) - \varepsilon^{-1}\gamma_1^{-1}\eta^{(1)}(\kappa_1) + \varepsilon^{-1}\tfrac{1}{2}(\eta_S + \eta_T), \\
\mathbf{u} &= \mathbf{U}\left[|\kappa_0|^{\frac{1}{2}}\, g\eta^{(0)}(\kappa_0)/c_0 + |\kappa_1|^{\frac{1}{2}}\, g\eta^{(1)}(\kappa_1)/c_1\right], \\
\mathbf{u}_2 &= \mathbf{U}\left[(1+\varepsilon\gamma_1)|\kappa_0|^{\frac{1}{2}}\, g\eta^{(0)}(\kappa_0)/c_0 - \left(1-\gamma_1^{-1}\right)|\kappa_1|^{\frac{1}{2}}\, g\eta^{(1)}(\kappa_1)/c_1\right].
\end{aligned} \tag{27}$$

In the solution for η, the whole term proportional to η_S (η_T) is η_E (η_Q). These results are illustrated in Figure 3. Since $\kappa_1 \ll \kappa_0$, the whole structure of η_E/η_S or η_Q/η_T must be shown in two parts: The short wave response [upper graphs: $\kappa_1 = O(1)$ & $\kappa_0 = O(\varepsilon^{-1})$] is made of the force compensating and baroclinic modes. The long wave response [lower graphs: $\kappa_1 = O(\varepsilon)$ & $\kappa_0 = O(1)$], on the other hand, is made of the force compensating and barotropic modes plus the $\kappa_1 \to 0$ limit of the baroclinic one.

The magnitudes of κ_0 and κ_1, related by (21), are used to distinguish five different regions: the scales of the upper and lower graphs in Figure 3 and their asymptotes. (These are regions in wave number/frequency space; their names reflect the size of $\mathbf{k}$ for fixed ω.) The properties of the solution of the two-layer model in each region are discussed next. Only the leading orders terms are usually shown, except that in some cases are also included terms with a relative $O(\varepsilon)$, in order to show explicitly the correct matching between the different regions.

4.1. VERY LONG SCALES: $\kappa_0 \ll 1$

The solution at these scales corresponds to (27) with $\kappa_0 = \kappa_1 = 0$, which gives

$$\begin{aligned}
&\eta \sim (1/\varepsilon_1)\,\eta_S + \lambda\eta_T, \\
&\zeta \sim 0, \ \mathbf{u} \sim \mathbf{u}_2 \sim \mathbf{0}.
\end{aligned}$$

Note that, at these scales, a mass conserving model ($\lambda = 1$) is essential to get the correct η_Q, and no perturbation is transmitted to the interface elevation and the lower layer.

The main consequence of the effective evaporation is to raise or lower the surface by adding or subtracting water; its effect on water density is

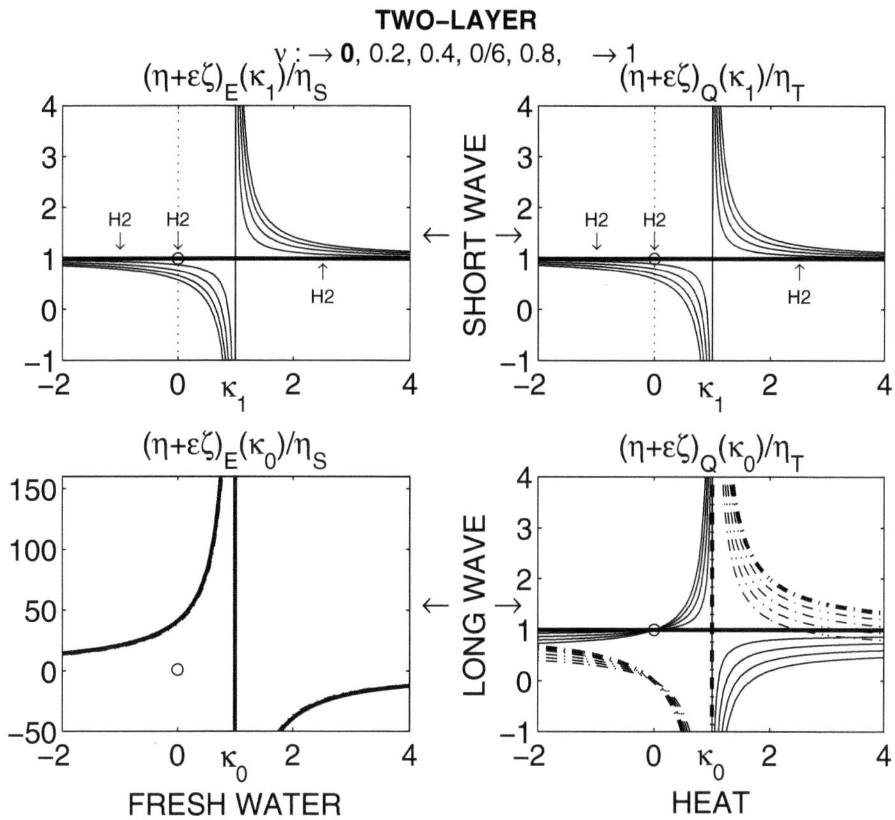

Figure 4. As in Figure 3, but subtracting from η the dynamic height due to deep ocean density variations $-\varepsilon\zeta$, where ε is the relative density excess of the lower layer (in the reference state) and ζ is the interface elevation field. The symbol H2 indicates where η is equal to the total dynamic height, *i.e.* the effect of upper layer temperature and salinity variations is included. Since this equality happens in both volume and mass conserving equations, this is a confirmation of hypothesis 2 from the Introduction.

an $O(\varepsilon_1)$ less important: $\partial_t \eta_E = -E_e = (1/\varepsilon_1)\, \partial_t \eta_S$. Consequently, this response can be safely modelled with a volume-conserving model.

Heat flux, on the other hand, raises or lowers the sea surface by means of a true expansion or contraction of the upper layer. This effect can only be modelled with a mass-conserving model ($\lambda = 1$); a volume conserving model ($\lambda = 0$) gives an incorrect result: Hypothesis 1 from the Introduction is satisfied at these scales and for this forcing.

4.2. LONG SCALES: $\kappa_0 = O(1)$

The solution at these scales corresponds to (27) with $\kappa_0 \neq 0$ and $\kappa_1 = 0$, which gives

$$\begin{aligned} \eta &\sim \frac{(1/\varepsilon_1)\left(1-\varepsilon\gamma^2\kappa_0\right)}{1-\kappa_0}\eta_S + \frac{\lambda - \left(\frac{1}{2}+\frac{1}{2}\gamma\right)\kappa_0}{1-\kappa_0}\left(\eta_S+\eta_T\right), \\ \zeta &\sim \frac{(\gamma/\varepsilon_1)\,\kappa_0}{1-\kappa_0}\eta_S, \\ \mathbf{u} \sim \mathbf{u}_2 &\sim \mathbf{U}\frac{|\kappa_0|^{\frac{1}{2}}}{1-\kappa_0} g\left[(1/\varepsilon_1)\,\eta_S + \left(\lambda - \tfrac{1}{2} - \tfrac{1}{2}\gamma\right)\eta_T\right]/c_0. \end{aligned}$$

The structure of the response in wavenumber/frequency space is dominated by the forcing of the barotropic mode. See lower graphs in Figure 3. As before, a mass-conserving model ($\lambda = 1$) is essential for the correct evaluation of η_Q and the corresponding velocity field. Note that

$$\eta_Q - \eta_T \sim \tfrac{1}{2}\frac{1-\gamma}{1-\kappa_0}\kappa_0\eta_T + O(\varepsilon),$$

and therefore Hypothesis 1 is satisfied for $\gamma \to 1$ as long as $1-\kappa_0 \ll 1-\gamma$; see thick line in the bottom right graph of Figure 3.

The leading order contribution to the interface elevation produced by the heat flux, $\zeta \sim \gamma\left(\lambda - \frac{1}{2} - \frac{1}{2}\gamma\right)\frac{\kappa_0}{1-\kappa_0}\eta_T$, is sensitive to the difference between mass-conserving ($\lambda = 1$) and volume conserving ($\lambda = 0$) equations. However, this term is very small for an interface elevation and therefore is included neither here nor in the following subsection.

4.3. INTERMEDIATE SCALES: $\kappa_0^{-1} \sim \kappa_1 \sim \sqrt{\varepsilon\gamma(1-\gamma)} \ll 1$

The solution at these scales corresponds to (27) with $\kappa_0 \to \infty$ and $\kappa_1 = 0$, which leaves just the baroclinic and force compensating modes

$$\begin{aligned} \eta &\sim \left(\tfrac{1}{2}+\tfrac{1}{2}\gamma\right)\left(\eta_S+\eta_T\right) + \gamma^2\left(\varepsilon/\varepsilon_1\right)\eta_S, \\ \zeta &\sim -\left(\gamma/\varepsilon_1\right)\eta_S, \\ \mathbf{u} &\sim \mathbf{u}_2 \sim \mathbf{0}. \end{aligned}$$

If $\gamma \to 1$ (*i.e.* an infinitesimally thin upper layer) it is $\eta \approx \eta_S + \eta_T$. Note that this effect is not due to the expansion or contraction of the upper layer because the same result is obtained with a volume conserving ($\lambda = 0$) or a mass conserving ($\lambda = 1$) model. Rather is a confirmation of hypothesis 2, namely

$$\eta + \varepsilon\zeta = \eta_S + \eta_T + O(1-\gamma, \varepsilon)$$

see thick line in the bottom right graph of Figure 4.

4.4. SHORT SCALES: $\kappa_1 = O(1)$

The solution at these scales corresponds to (27) with $\kappa_0 \to \infty$ and $\kappa_1 \neq 0$, which gives

$$\begin{aligned}
\eta &\sim \tfrac{1}{2}\frac{1+\gamma-\kappa_1}{1-\kappa_1}\left(\eta_S+\eta_T\right)+\frac{\gamma^2\varepsilon/\varepsilon_1}{1-\kappa_1}\eta_S,\\
\zeta &\sim -\tfrac{1}{2}\frac{\kappa_1/\varepsilon}{1-\kappa_1}\left(\eta_S+\eta_T\right)-\frac{\gamma/\varepsilon_1}{1-\kappa_1}\eta_S,\\
\mathbf{u} &\sim \tfrac{1}{2}\gamma\mathbf{U}\frac{|\kappa_1|^{\frac{1}{2}}}{1-\kappa_1}g\left(\eta_S+\eta_T\right)/c_1,\\
\mathbf{u}_2 &\sim -\tfrac{1}{2}\left(1-\gamma\right)\mathbf{U}\frac{|\kappa_1|^{\frac{1}{2}}}{1-\kappa_1}g\left(\eta_S+\eta_T\right)/c_1.
\end{aligned}$$

The structure of the response in wavenumber/frequency space is dominated by the forcing of the baroclinic mode. See upper graphs in Figure 3. Note that

$$\eta+\varepsilon\zeta=\left(1-\tfrac{1}{2}\frac{1-\gamma}{1-\kappa_1}\right)\left(\eta_S+\eta_T\right)+O\left(\varepsilon\right)$$

and therefore hypothesis 2 is satisfied for $\gamma \to 1$ as long as $1-\kappa_1 \ll 1-\gamma$; see thick line in the bottom right graph of Figure 4.

4.5. VERY SHORT SCALES: $\kappa_1 \gg 1$

The solution at these scales corresponds to (27) with $\kappa_0 \to \infty$ and $\kappa_1 \to \infty$, which leaves just the contribution of the force compensating mode

$$\begin{aligned}
\eta &\sim \tfrac{1}{2}\left(\eta_S+\eta_T\right),\\
\zeta &\sim \tfrac{1}{2}\varepsilon^{-1}\left(\eta_S+\eta_T\right),\\
\mathbf{u} &\sim \mathbf{0},\ \mathbf{u}_2 \sim \mathbf{0}.
\end{aligned}$$

This means that hypothesis 2 is verified:

$$\eta+\varepsilon\zeta \sim \eta_S+\eta_T,$$

for any value of γ. However, as mentioned above, the validity of keeping the upper layer dynamical fields depth independent, probably breaks down at these short scales. With more vertical structure (e.g. one more layer) in the model setup, there would be terms proportional to $1/\left(1-\kappa_2\right)$ due to the second baroclinic mode; typically, κ_2/κ_1 is equal to 4 or thereabouts.

We finish this section addressing a question raised in the Introduction, on the relationship between the one-layer solutions (25), for both the reduced gravity and rigid bottom cases, and the two-layer ones (27). All these models share the force compensating mode, $\eta = \frac{1}{2}(\eta_S + \eta_T)$ & $\mathbf{u} = \mathbf{0}$.

The forced wave solution, that is, the part proportional to $(1-\kappa_j)^{-1}$, of the rigid bottom and reduced gravity models, coincide in some limit with the forced wave solutions of the barotropic and baroclinic modes of the two-layer model respectively. To see that, the eigenvalues c_j^2 and coefficients in (26) are analyzed, and shown to yield the coefficients in (24) in the appropriate limit.

Firstly, in the limit of an infinitesimally thin lower layer, $\gamma \to 0$, both $A_1 \to 0$ and $B_1 \to 0$: the contribution of the baroclinic mode vanishes in (26). Moreover, $A_0 \to 1/\varepsilon_1 + \lambda - \frac{1}{2}$ and $B_0 \to \lambda - \frac{1}{2}$, which gives precisely the results for the one-layer rigid bottom case in (24), as expected. This corresponds to the regions discussed in §§4.1 through 4.3, i.e. before the baroclinic resonance shows up.

Secondly, the limit $\gamma \to 1$ of an infinitesimally thin upper layer, implies $A_1 = \frac{1}{2} + \varepsilon/\varepsilon_1 + O(1-\gamma, \varepsilon)$ and $B_1 = \frac{1}{2} + O(1-\gamma, \varepsilon)$ in (26), which coincide with the A and B from the one-layer reduced gravity case (24). Notice, however, that A_0 and B_0 do not tend to zero as $\gamma \to 1$; in fact it is $A_0 \to 1/\varepsilon_1 + \lambda - 1$ and $B_0 \to \lambda - 1$. Consequently, the two-layer solution tends to that of the reduced gravity case for $\gamma \to 1$ *as long as* $|\kappa_1| \gg \varepsilon$, because then (21) makes $\kappa_0 \gg 1$ and therefore the terms proportional to the barotropic pole $(1-\kappa_0)^{-1}$ can be neglected. This corresponds to the regions discussed in §§4.3 through 4.5.

5. Conclusions

The coincidence between actual sea level and the steric integral given by (1) may or not be an *observational* fact. Its explanation, in particular, one of the two hypotheses stated in the Introduction, is model dependent. The first hypothesis (H1) attributes a sea surface elevation η to the expansion of the water column produced by the local surface heat Q and/or fresh water $P - E = -E_e$ input. The second one (H2) sees η as the dynamic height relative to the abyss, associated to a surface intensified baroclinic signal. H1 needs mass-conserving (rather than volume-conserving) equations and presumably of very long horizontal scales, say, a forcing wavenumber "$\mathbf{k} \to 0$" (Greatbatch, 1994). H2 works fine with the volume-conserving approximation and needs a pressure gradient "$\mathbf{k} \neq 0$". (Baroclinic signals can, of course, be in the form of free waves as well as wind and atmospheric pressure forced motions, but this paper focuses on the Q and E_e forced signals).

The complete response to Q and E_e forcing in a f-plane two-layer model is obtained and used, in particular to test in which regions of parameter space hypotheses H1, H2 are valid or not. Q and E_e are directly responsible for local changes of temperature and salinity in the upper layer, scaled as the variables η_T and η_S defined in (10). Q and E_e may also produce deeper changes in temperature and salinity (through the set up of pressure gradients) represented here by the interface elevation field ζ, (see table II). Both hypotheses in the Introduction can then be cast, for the simple models used here, in the form

$$\mathrm{H1} : \eta \approx \eta_S + \eta_T \qquad \text{only for } \lambda = 1$$

$$\mathrm{H2} : \eta + \varepsilon\zeta \approx \eta_S + \eta_T \quad \text{for any } \lambda$$

where λ is a flag used to distinguish the usual volume-conserving models ($\lambda = 0$) from the mass-conserving models ($\lambda = 1$). In order to consider the order of magnitude of the signals studied here, assume that Q has an amplitude of $200\,\mathrm{W\,m^{-2}}$ at $\omega = 2\times10^{-7}\,\mathrm{s}$ (one cycle per year); then η_T will have an amplitude of about 6 cm. Similarly, if $P-E$ has an amplitude of $3\times 10^{-8}\,\mathrm{m\,s^{-1}}$ (one meter/year) at the same frequency, then the amplitude of η_S will be about 4 mm, but that of $\varepsilon_1^{-1}\eta_S$ will be about 16 cm. The velocity scale depends also on the eigenvalues c_j^2 of the barotropic and baroclinic modes. Thus, for $\eta = 10$ cm, it is $g\eta/c_0 \approx 5\,\mathrm{mm\,s^{-1}}$ for a barotropic signal but $g\eta/c_1 \approx 50\,\mathrm{cm\,s^{-1}}$ for a baroclinic one.

The exact response to Q and E_e forcing is discussed in §§4.1 through 4.5, illustrated in Figures 3 and 4, and summarized in Table III. For η and ζ, the whole information on the forcing wavenumber and frequency is contained in a single variable (see Figure 1) with two quite different normalizations, κ_0 & κ_1, where $\kappa_0 = 1$ ($\kappa_1 = 1$) is the dispersion relation of barotropic (baroclinic) Poincaré waves and $\kappa_0 \gg \kappa_1$. More precisely

$$\kappa_0 = \frac{k^2 + l^2}{\omega^2 - f^2} g\,(H_1 + H_2)\,,$$

$$\frac{\kappa_1}{\kappa_0} = \frac{\varepsilon H_1 H_2}{(H_1 + H_2)^2} \leq \tfrac{1}{4}\varepsilon.$$

The disparity of κ_0 and κ_1 calls for a description of the solution in quite different regions in parameter space. (For instance, in Figures 3 and 4, the lower graphs are like a very thin boundary layer in the upper ones.)

The first hypothesis is realized only for the heat flux forcing and for very long scales ($\kappa_0 \ll 1$). It is also realized for long scales ($\kappa_0 \sim 1$) if the upper layer, where the heat is absorbed, is very shallow ($H_1 \ll H_2$). Mass conserving equations are required for these two regions, regardless of the relative values of H_1 & H_2.

TABLE III. Validity of both hypotheses in the Introduction (H1, only for the heat forcing) and domain of the one-layer models from §2 and of the modes of the two-layer model from §3. The five columns indicate the regions on the κ axis described in §§4.1 through 4.5.

<table>
<tr><th></th><th>$\kappa_0 \ll 1$</th><th>$\kappa_0 = O(1)$</th><th>$\kappa_0 \sim \kappa_1^{-1} \gg 1$</th><th>$\kappa_1 = O(1)$</th><th>$\kappa_1 \gg 1$</th></tr>
<tr><td>Hypothesis 2</td><td>no</td><td>no</td><td>$H_1 \ll H_2$</td><td>$H_1 \ll H_2$</td><td>yes</td></tr>
<tr><td>One-layer</td><td>no</td><td>no</td><td colspan="3">reduced gravity</td></tr>
<tr><td>Two-layer</td><td colspan="5">force compensating mode</td></tr>
<tr><td>Two-layer</td><td colspan="4">baroclinic mode</td><td>no</td></tr>
<tr><td>Two-layer</td><td colspan="2">barotropic mode</td><td>no</td><td>no</td><td>no</td></tr>
<tr><td>One-layer</td><td colspan="3">rigid bottom</td><td>no</td><td>no</td></tr>
<tr><td>Hypothesis 1</td><td>yes</td><td>$H_1 \ll H_2$</td><td>no</td><td>no</td><td>no</td></tr>
</table>

Gill and Niiler (1973) calculated the response to heat and fresh water forcing at the annual frequency, $\omega = 1\,\mathrm{cpy}$, excluding the band $\pm 15°$ from the equator, and restricting the zonal and meridional scales to $k^{-1} \geq 3\,\mathrm{Mm}$ & $l^{-1} \geq 1\,\mathrm{Mm}$. This roughly corresponds to $-30 < \kappa_0 < 0$ and $\omega \ll f$; therefore their conclusion that "steric changes in sea-level [are] produced by expansion and contraction of the water column above the seasonal thermocline due to changing fluxes of heat and water across the surface" may depend upon the shallowness of the heat mixing layer ($H_1 \ll H_2$ in the notation of this paper) and is only valid for the heat flux. The main effect of evaporation (precipitation) is to lower (raise) sea level, with $\eta = O\left(\varepsilon_1^{-1}\eta_S\right)$, simply by the flux of water across the surface. Haline contraction of the water column has much smaller effect, $\eta = O(\eta_S)$ [the statement by Gill and Niiler (1973) not withstanding for the fresh water flux], and therefore volume-conserving equations can be safely used for this forcing.

Greatbatch (1994) proposed to evaluate sea level in rigid-lid volume-conserving numerical models by

$$\partial_t \eta + \nabla \cdot \left(\int_{bottom}^{surface} \mathbf{u}\, \mathrm{d}z \right) = \frac{\alpha_T}{\rho_0 C_p} \bar{Q}(t)$$

where $\bar{Q}$ represents a global average. In the notation of this paper, this recipe may be stated as $\eta = \eta|_{\lambda=0} + \bar{\eta}_T$, or adding $+1$ to the broken curves in the lower right graph of Figure 3 at $\kappa_0 = 0$. The Figure indicates that Greatbatch (1994) solution works very well for the very long scales $\kappa_0 \ll 1$.

However, heat forcing at long scales ($\kappa_0 = O(1)$), needs to include the forced pressure gradient part and use mass conserving equations.

One simple way to improve upon Greatbatch's recipe is to add the mass-conserving barotropic solution of this paper (see §4.2), to the solution from a volume-conserving model (indicated below by $\cdots$):

$$\partial_t \eta = \cdots + \frac{\alpha_T}{\rho_0 C_p} \iint \left[1 + \frac{1-\gamma}{2} \frac{\kappa_0}{1-\kappa_0} \right] \hat{Q}(\mathbf{k}, \omega) \, \mathrm{e}^{i\mathbf{k}\cdot\mathbf{x} - i\omega t} \mathrm{d}\mathbf{k} \mathrm{d}\omega,$$

where $Q(\mathbf{x}, t) = \iint \hat{Q}(\mathbf{k}, \omega) \, \mathrm{e}^{i\mathbf{k}\cdot\mathbf{x} - i\omega t} \mathrm{d}\mathbf{k} \mathrm{d}\omega$, and $1-\gamma$ is a parameterization of the depth of the layer where the incoming heat is mixed up relative to the total ocean depth. The factor $\kappa_0/(1-\kappa_0)$ might be important in shallow areas, see Dronkers (1969) with a one-layer rigid bottom model (which corresponds to making $\gamma = 0$ in the above formula).

Hypothesis 2 is related to the dominance of surface intensified baroclinic signals, which may be free waves or wind-forced motions, totally independent of the fresh water and heat surface fluxes see Ripa (1997). As far as these buoyancy forcings are concerned, and for the present model, Hypothesis 2 is realized for very short scales ($\kappa_1 \gg 1$). In addition, it works for short ($\kappa_1 \sim 1$) and medium ($\kappa_0^{-1} \sim \kappa_1 \ll 1$) scales if the top layer is sufficiently shallow (so that the baroclinic signal is surface intensified). In any case, these effects can be safely modelled with the usual volume-conserving equations.

References

Beron-Vera F. J., J. Ochoa and P. Ripa. A note on boundary conditions for salt and fresh- water balances. *Ocean Modelling*, 1:111–118, 1999.

Beron-Vera F. J. and P. Ripa. Three- dimensional aspects of the seasonal heat balance in the Gulf of California. *J. Geophys. Res.*, 105:11441–11457, 2000.

Beron-Vera F. J. and P. Ripa. Seasonal salinity balance in the Gulf of California. *J. Geophys. Res.*, 107(C8):10.1029/2000JC000769, 2002.

Dronkers, A. Tidal computations in rivers, coastal areas and seas. *J. of Hydraulics Division* ASCE, 95:44–77, 1969.

Gill, A. E. and P. Niiler. The theory of the seasonal variability in the ocean. *Deep-Sea Res.*, 20:141–177, 1973.

Greatbatch, R. A note on the representation of steric sea level in models that conserve volume rather than mass. *J. Geophys. Res.*, 99:12767–12771, 1994.

Lavoie, R. A mesoscale numerical model of lake-effect storms. *J. Atmos. Sci.*, 29:1025–1040, 1972.

Pattullo, J., W. Munk, R. Revelle and E. Strong. The seasonal oscillation in sea level. *J. Mar. Res.*, 14:88–155, 1955.

Philander, G. Forced oceanic waves. *Rev. Geophys.*, 16:15–46, 1978.

Ripa, P. Seasonal circulation in the Gulf of California,. *Annales Geophysicae*, 8:559–564, 1990.

Ripa, P. Conservation laws for primitive equations models with inhomogeneous layers. *Geophys. Astrophys. Fluid Dyn.*, 70:85–111, 1993.

Ripa, P. Linear waves in a one-layer ocean model with thermodynamics. *J. Geophys. Res.*, C101:1233–1245, 1996.

Ripa, P. Towards a physical explanation of the seasonal dynamics and thermodynamics of the Gulf of California. *J. Phys. Oceanogr.*, 27:597–614, 1997.

Ripa, P. On the validity of layered models of ocean dynamics and thermodynamics with re- duced vertical resolution. *Dyn. Atmos. Oceans*, 29:1–40, 1999.

Ripa P. and J. Zavala-Garay. Ocean channel modes. *J. Geophys. Res.*, 104:15479–15494, 1999.

Schopf, G. and M. Cane, On equatorial dynamics, mixed layer physics and sea surface temperature. *J. Phys. Oceanogr.*, 13:917–935, 1983.

BAROCLINIC WAVES IN CLIMATES OF THE EARTH'S PAST

A.B.G. BUSH
Department of Earth and Atmospheric Sciences
University of Alberta
Edmonton, Alberta, T6G 2E3 Canada

Abstract. Our understanding of the climates that have existed on Earth through its history has increased tremendously through a combination of geophysical fluid dynamics, geological evidence, and numerical modelling. Evolution of our Earth's orbit redistributes incoming solar radiation latitudinally and temporally and is believed to have been responsible for changing the strength of the south Asian monsoon, expanding and contracting desert regions, and perhaps even initiating the (geologically) recent cycles of glaciation. Evolution of the atmosphere itself, in terms of the amount of atmospheric greenhouse gases in it, affects the amount of water vapour in the atmosphere, global temperatures, and meridional temperature gradients. Topographic forcing by the massive continental ice sheets that have existed in the past is believed to have significantly altered the jet stream circulation.

All of these factors–the distribution of incoming solar radiation, atmospheric greenhouse gases, and topograhic forcing–affect the mean baroclinic structure of our atmosphere, the amount of baroclinic wave activity present, and the eddy heat and momentum fluxes associated with these eddies. The equatorward flux of easterly momentum during the barotropic decay phase of these waves, in particular, plays a key role in determining the strength of upper level convergence and subsidence in the subtropics and, hence, the low-level meridional pressure gradient that helps to maintain the tropical trade winds. Through a series of numerical experiments with a coupled atmosphere-ocean general circulation model, it is shown that all of the factors listed above have played a role in determining the amount of baroclinic wave activity in climates of the Earth's past and that changes in tropical circulations are consistent with the notion that the baroclinic eddy field plays an important role in determining the mean state in the tropics.

Key words: baroclinic waves, past climates, snowball earth

1. Introduction

Maintenance of the atmospheric and oceanic general circulations involves a complex balance between boundary condition forcing (e.g. bottom topography and incoming solar radiation) and the nonlinear dynamical processes that occur within each medium (e.g. baroclinic eddies and convection, to name only two). The relative roles of the dynamical processes in producing

O.U. Velasco Fuentes et al. (eds.), Nonlinear Processes in Geophysical Fluid Dynamics, 127-139.

the mean atmospheric general circulation that we observe today has received much attention (e.g. Held and Hou, 1980; Pfeffer, 1981; Lindzen and Hou, 1988; Haynes and Shepherd, 1989; Chang, 1996; Becker *et al.*, 1997; Kim and Lee, 2001a,b). While the role of midlatitude baroclinic eddies in fully generating and sustaining the Ferrel cell has been established (e.g. Held and Hou, 1980), their role in governing observed mean tropical circulations such as the Hadley cell is more debatable. It was at first believed that their contribution is small and that diabatic heating from tropical convection was the primary driving force for the Hadley circulation (e.g. Pfeffer, 1981). More recently, however, the relative role of eddies in driving the Hadley circulation has been shown to be much larger (approximately 75%) when the feedback between eddy momentum flux, surface friction, and tropical moisture convergence is taken into account in the computation of direct diabatic forcing (Kim and Lee, 2001a,b).

Changes in mean wind strength in the tropics are extremely important in a coupled atmosphere-ocean system because of positive dynamical feedbacks (e.g. Philander, 1985; Xie, 1998). The tropical oceans are quite sensitive to changes in Hadley cell strength because, through angular momentum conservation, an increase in Hadley cell strength implies stronger surface easterlies which, in turn, produce stronger oceanic upwelling and colder sea surface temperatures (SSTs) through the feedbacks that are responsible for the El Niño Southern Oscillation (e.g. Philander, 1990). Therefore, midlatitude baroclinic eddies are likely to contribute, at least in part, to the generation of the mean state of the tropical oceans.

Some aspects of our climate's past may provide insight into this hypothesis because our atmosphere-ocean system has evolved through many states quite different than the one in which it is presently found and, moreover, geological proxy data exist and give clues as to what climatic conditions existed in the past.

For example, repeated glaciations over the past 900,000 years have dramatically altered surface topography by imposing massive Northern Hemisphere ice sheets up to 3 kilometers thick (e.g. Peltier, 1994) that reflect much of the incoming solar radiation and alter the path of the Northern Hemisphere jet stream (Bush and Philander, 1999; Hall *et al.*, 1996). Geological evidence, as inferred from aeolian deposits (e.g. Sarnthein *et al.*, 1981; Farrell *et al.*, 1995), productivity estimates (Pedersen, 1983; Lyle *et al.*, 1992), grain size analysis (Rutter, 1992), and planktonic foraminifera (Andreasen and Ravelo, 1997), suggest that the mean atmospheric circulation was stronger during these glacial periods.

Conversely, during equable climates when atmospheric carbon dioxide was higher than it is today, such as the Cretaceous (Berner, 1991), high latitude surface temperatures were much warmer and hence, through thermal wind balance, weaker westerlies and a weaker general circulation have been proposed (e.g. Barron and Washington, 1982; Sloan and Barron, 1990).

During periods such as the early-mid Holocene [~10,000-5,000 years before present (B.P. hereafter)] when the seasonal and latitudinal distribution of incoming solar radiation favoured a much more seasonal climate in the Northern Hemisphere, atmospheric winds were different, particulary those of the south Asian monsoon (e.g. Wright *et al.*, 1993; Prell, 1984; Clemens and Prell, 1990; Prell and Kutzbach, 1992).

Baroclinic instability of the atmosphere's jet streams is highly dependent on the shears present in the climatological mean state. The factors listed above (i.e. solar forcing, topographic forcing, and the amount of carbon dioxide present in the atmosphere) all have the potential to alter the mean state of the atmosphere and hence the strength and frequency of baroclinic eddies. Are the changes that would be produced in the tropics by changing these factors consistent with the notion that baroclinic eddies play an important role in governing tropical circulations?

This question is addressed through analysis of a series of numerical experiments performed with the coupled atmosphere-ocean general circulation model developed at the Geophysical Fluid Dynamics Laboratory in Princeton, New Jersey. The results of Bush (2001) are extended here to include analyses of four experiments that explore a broader range of boundary conditions that have occured in Earth's past. First, a simulation of the Last Glacial Maximum (~21,000 years B.P.) is analyzed to determine the effect of massive ice sheets on the circulation. Second, a simulation of the mid-Holocene (~6,000 years B.P.) is analyzed to determine the effect of an increased seasonal cycle in the Northern Hemisphere. Third, a simulation with double the amount of atmospheric carbon dioxide is analyzed to determine the effect of high latitude warming on baroclinic wave activity and the general circulation. Fourth, a simulation in which the Earth is completely ice-covered, as has been proposed for the Neoproterozoic (~600-800 million years B.P.; Hoffman *et al.*, 1998; Hyde *et al.*, 2000), is analyzed to determine the impact of rendering surface baroclinicity negligible. While the nature of the Snowball Earth is a matter of some debate (the debate is centred on whether or not the tropical oceans were completely ice-covered or whether open water refugia existed), it is assumed here that the oceans were completely ice-covered; therefore, this simulation is performed with the atmosphere-only model only. Numerical results are included here to provide an example of the dynamics that may have occured during one of the most extreme climates of Earth's past. Results of these four experiments are compared to a control simulation for today's climate.

A brief description of the coupled model is given in the next section, followed by the results and discussion in section 3 and concluding remarks in section 4.

2. Configurations of the model

For brevity, the reader is referred to Bush and Philander (1999) for full details of the numerical model and for a description of the model configuration in the LGM simulation. Key points are that: the atmospheric model (Gordon and Stern, 1982) is spectral with rhomboidal 30 truncation and a 216 second time step; the oceanic model (MOM, version 2; Pacanowski *et al.*, 1991) uses finite-differencing with a resolution of 2.25^o in latitude, 3.75^o in longitude, 15 vertical levels, and a 1-hour time step; and dynamic and thermodynamic coupling between the models is performed once per day of integration time. In the LGM simulation, continental ice sheets are imposed according to reconstructions (Peltier, 1994), sea level is lowered by 120 meters (Fairbanks, 1989), glacial land surface albedo is imposed (CLIMAP, 1981), and atmospheric carbon dioxide is set to 200 ppm.

The model configuration for the mid-Holocene simulation is identical to that of the control simulation with the exception that the orbital parameters of obliquity, eccentricity, and longitude of perihelion are set to those appropriate to 6,000 years B.P. (Berger, 1992). In particular, obliquity at 6,000 B.P. was 24.1^o as opposed to the modern 23.446^o so an increased seasonal cycle is expected. Also, perihelion occured during boreal summertime (today it occurs in austral summertime). The increased CO_2 simulation is identical to the control simulation except that the amount of atmospheric carbon dioxide is doubled. The Snowball simulation assumes flat continental ice over the continents (which are all located in the tropics; see Hyde *et al.*, 1990) and sea ice over all of the oceans; bare surface albedo is therefore set to 0.6 everywhere in the tropics and increases to 0.8 at the poles. A 7% reduction in solar luminosity is assumed, as appropriate for the early Neoproterozoic (Endal and Sofia, 1981).

The coupled simulations are decadal in timescale (specifically, 70 years for the control and CO_2 simulations and 25 years for the others). The simulations are sufficiently long for the atmosphere and the upper ocean to have reached a new radiative equilibrium state and for the oceanic wind-driven circulation to be established. Any possible influence of benthic ocean currents on SST are therefore neglected in these simulations. Model output fields are averaged to produce monthly mean data, from which the following results were obtained. In the discussion to follow we will focus on changes in eddy activity in the Northern hemisphere since this is where most of the changes occur from, for example, topographic forcing in the LGM simulation and radiative forcing in the mid-Holocene simulation. Changes in the Southern hemisphere eddies are small in comparison.

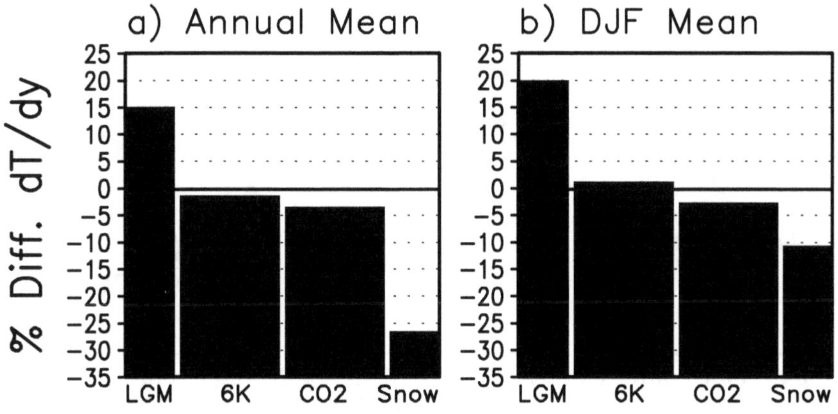

Figure 1. **a)** Climatological, zonally averaged, surface baroclinicities averaged between 30-60N, shown as a percentage difference from the control simulation (the value for which is -0.5^o per degree of latitude) for the Last Glacial Maximum (LGM) the mid-Holocene (6K), the increased CO_2 (CO2) and the Snowball (Snow) simulations. **b)** Same as in **a)** but for the December-January-February means (the value in the control simulation is -0.65^o per degree of latitude).

3. Simulated mean climates and baroclinicity

There is a relatively wide range of baroclinicities in the climatological mean state of these simulations (Figure 1a) with the greatest existing in the LGM simulation (when large ice sheets cool the Northern Hemisphere polar latitudes) and the smallest in the Snowball simulation (when the planet is completely ice-covered).

In the Northern Hemisphere winter (Figure 1b), baroclinicity increases above its climatological value by 30% in the control simulation and by 34% in the mid-Holocene simulation. Winter values are therefore greater than today in the LGM and mid-Holocene simulations and are lesser than today in the CO2 and Snowball simulations.

Since baroclinic eddies develop primarily in wintertime, these differences in mean wintertime baroclinicity should also be reflected in the Northern Hemisphere momentum flux. The total northward momentum transport is $[\overline{VU}]$, where a bar denotes a time average and square brackets denote a zonal average. The following decomposition then follows (e.g. Peixoto and Oort, 1992):

$$[\overline{VU}] = [\overline{V}][\overline{U}] + [\overline{V'U'}] + [\overline{V^*U^*}]. \tag{1}$$

where a prime indicates a departure from the time mean and a star indicates a departure from the zonal mean. Terms on the right hand side therefore represent contributions to the total momentum flux from, respectively, the

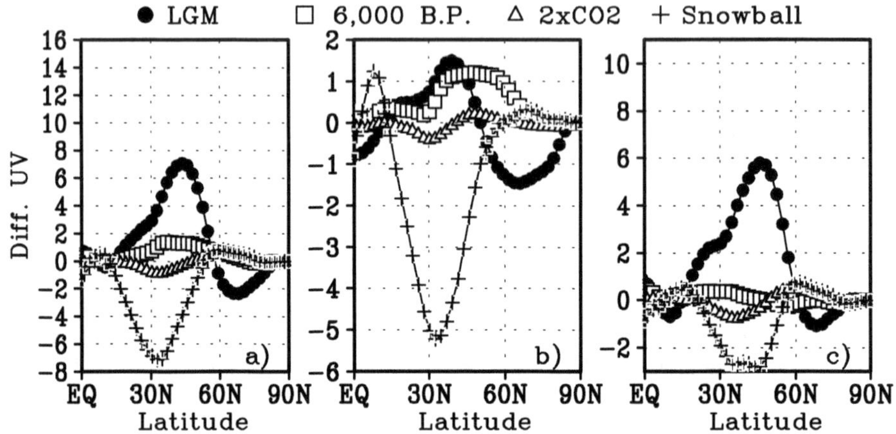

Figure 2. Changes (experiment minus control) in northward momentum transport in the Northern hemisphere by **a)** all motions, **b)** transient eddies, and **c)** stationary eddies. Units are m^2/s^2. Note that the axis scale is different between the three panels. (Peak values in the control simulation are 7.3, 5, and 3.1, respectively.)

mean meridional circulation, the transient eddies, and the stationary eddies. Note that because monthly mean values are used in these diagnostics, all interannual variability is included in the stationary eddy statistics (see Peixoto and Oort (1992)). Although we wish to focus here only on changes in the climatological mean states in the simulations, we note that there are also changes in interannual variability in the three coupled model experiments (the LGM, mid-Holocene, and increased CO_2 simulations). Some differences in the stationary eddy statistics may therefore be attributable to these changes.

The net momentum transport in all simulations (Figure 2) indicates that the LGM and mid-Holocene simulations have greater equatorward flux of easterly momentum than the control, while the CO2 and Snowball simulations have smaller fluxes. A breakdown of the contributions in the LGM and mid-Holocene simulations indicates that it is the stationary eddies that contribute most to the increased flux in the LGM simulation (although transient eddy activity also increases), whereas it is the transient eddies that contribute most to the increase in the mid-Holocene simulation (Figures 2b-c). Additional high latitude momentum fluxes in the LGM simulation are related to splitting of the midlatitude jet by the Cordilleran and Laurentide ice sheets over North America.

Contributions of midlatitude eddies to both heat and momentum transport are readily visualized by the E-P flux vectors, whose positive horizontal and vertical components, F_λ and F_p, represent, respectively, equatorward flux of westerly momentum and poleward heat transport. They may be written as (e.g. Piexoto and Oort, 1992):

$$F_\lambda = -\frac{2\pi R^2 \cos^2(\phi)}{g} \; [U^* V^*],$$

$$F_p = \frac{2\pi R^3 \cos^2(\phi)}{g} \; f[V^* \Theta^*](\partial \Theta_s / \partial p)^{-1}. \tag{2}$$

Here, R is the Earth's radius, g is gravity, f is the Coriolis parameter, ϕ is latitude, Θ is potential temperature and Θ_s is the global mean potential temperature on a pressure surface.

In the control simulation, the baroclinic eddy field generates E-P flux vectors that indicate poleward heat transport in the lower-mid troposphere, with equatorward flux of easterly momentum in the upper troposphere (Figure 3a). The values plotted are annual means, so the entire life cycles of all eddies (i.e. their baroclinic growth, F_p and their barotropic decay, F_λ) is included (e.g. Simmons and Hoskins, 1980; Edmon *et al.*, 1980). Differences between the simulations and the control (Figures 3b-d) indicate greater momentum flux in the LGM and mid-Holocene simulations, and less in the CO2 simulation. E-P fluxes in the Snowball simulation are extremely small and are therefore not shown in this figure. In the LGM simulation, the northern hemisphere jet stream is split north and south by the Laurentian ice sheet over North America and exhibits greater baroclinicity downstream of the ice sheet in the Atlantic storm track (e.g. Hall *et al.*, 1996; Bush and Philander, 1999). Analysis of the separate contributions by transient and stationary eddies (not shown) indicates that stationary eddies dominate in the LGM simulation, whereas transient eddies contribute more in the mid-Holocene simulation. In the increased CO_2 simulation, the contributions from each are comparable.

In the control simulation the greatest convergence of momentum flux is in the upper troposphere of the subtropics on the poleward edge of the tropical Hadley cell (cf. Figure 3a). Increased convergence in this region in the LGM and mid-Holocene simulations increases subtropical subsidence, whereas decreased convergence in the CO2 and Snowball simulations decreases subsidence. Differences in the zonal mean Hadley circulation do show such changes (Figure 4) with magnitudes that are in agreement with the differences in momentum transport (cf. Figure 2), in the sense that the LGM simulation shows the strongest increase in subsidence, the mid-Holocene simulation a lesser increase, the increased CO_2 simulation a decrease, and the Snowball simulation a large decrease. For reference, the Hadley cells in the control simulation have maximum amplitudes of -6.1×10^{10} kg/s and 5×10^{10} kg/s in the Southern and Northern hemispheres, respectively. The Ferrel cells have maximum amplitudes of 2.4×10^{10} kg/s and -2.6×10^{10} kg/s in the Southern and Northern hemispheres, respectively.

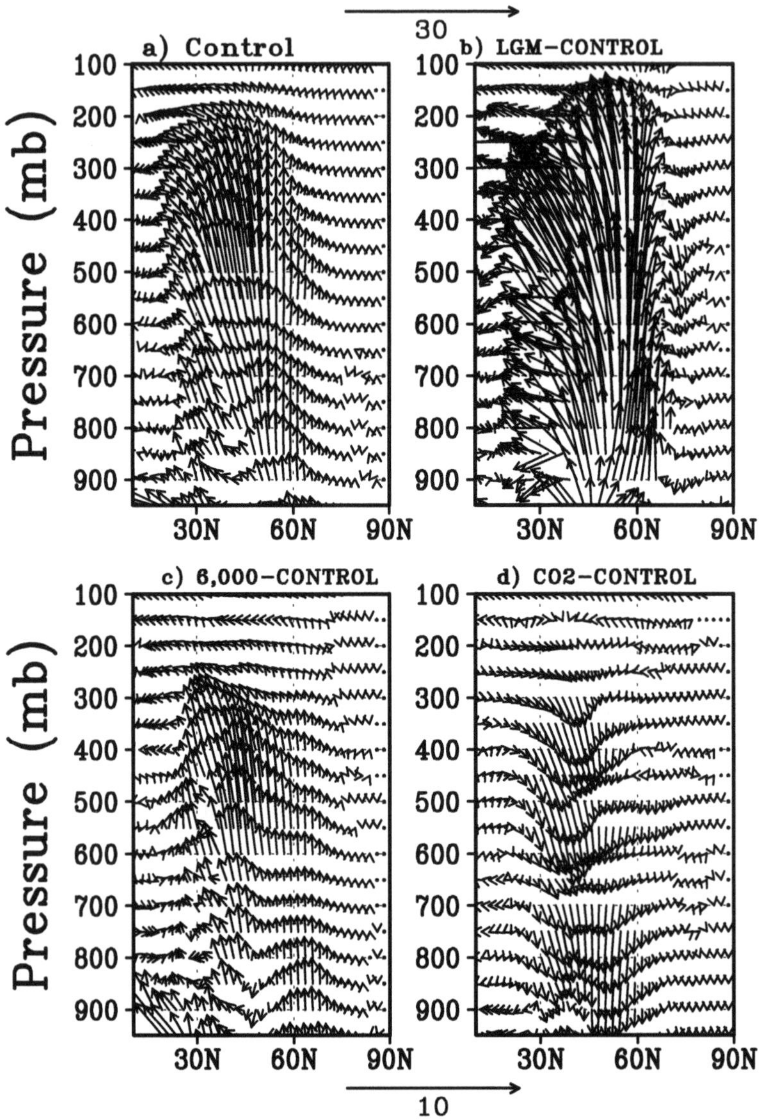

Figure 3. **a)** E-P flux vectors (as defined in the text) in the control simulation, from transient and stationary eddies. Differences in E-P flux are shown for **b)** the LGM, **c)** the mid-Holocene, and **d)** CO2. Vector components have been scaled down by the common factor $\frac{2\pi R^2}{g}$, so that the units are m^2/s^2. The arrow scale for panels **a)** and **b)** is shown above the figure, whereas the scale for panels **c)** and **d)** is shown below the figure.

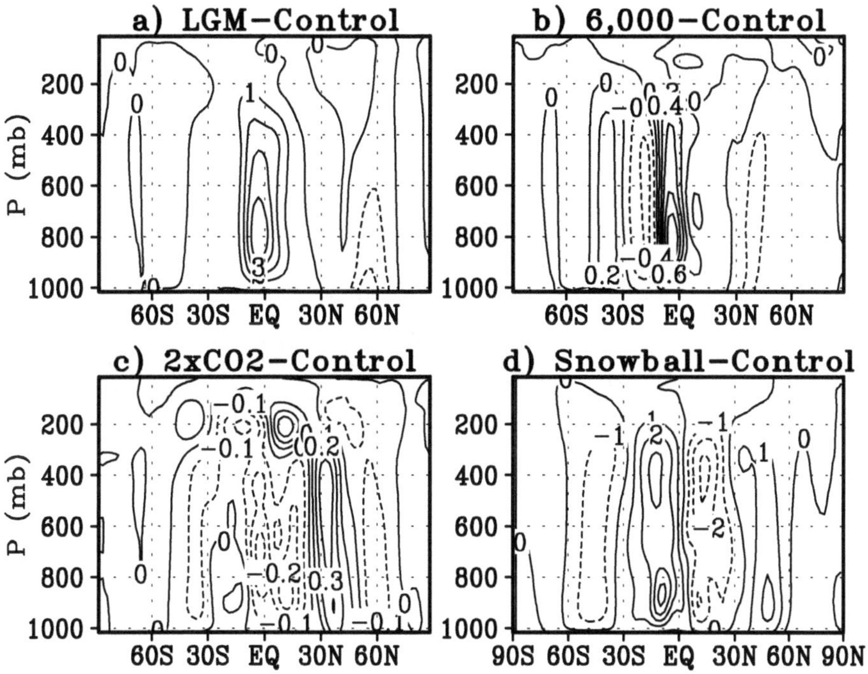

Figure 4. Differences (experiment minus control) in the annual mean Hadley circulation for **a)** the LGM simulation, **b)** the mid-Holocene simulation, **c)** the increased CO_2 simulation, and **d)** the Snowball simulation. The sign convention is such that positive values indicate a clockwise circulation in the plane of the diagram. Units are 10^{10} kg/s.

The meridional pressure gradient between the near tropics and the subtropics is, climatologically, in near geostrophic balance with the zonal wind field. That is, the higher surface pressure of the subtropics, caused by subsidence of the air between the Hadley and Ferrel cells, creates a positive meridional pressure gradient that balances the mean easterly trade winds. (For climatological mean fields, geostrophy holds quite well close to the equator, to within 5 degrees of latitude.) Without any compensating changes in equatorial pressure, these changes in subtropical subsidence should therefore alter the strength of the equatorial easterlies. Since topographic heights have been altered in two experiments (LGM and Snowball), there are concomitant changes in surface pressures. Between the control, mid-Holocene, and increased CO_2 simulations there is a change of less than one millibar in the mean equatorial pressure. In the LGM and Snowball simulations, however, equatorial pressures are 9 millibars higher and 17 millibars lower, respectively, because of the increased and decreased topographic heights. It is assumed that these changes are uniform spatially so that the gradients are not affected by this mechanism.

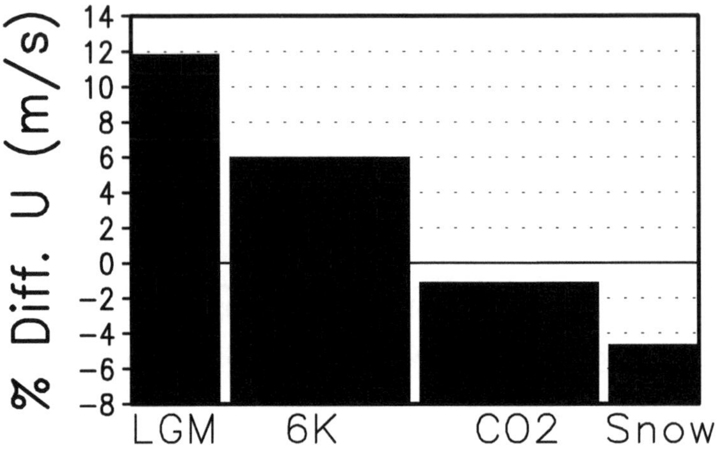

Figure 5. Climatological, zonally averaged, surface zonal wind speed change between 10S and 10N, shown as a percentage difference from the control simulation (the value for which is −4.7 m/s).

Simulated changes in mean zonal wind speed in the tropics are in agreement with what would be inferred from the changes in subtropical subsidence (Figure 5; note that a positive change in zonal wind speed implies stronger trade easterlies). The change in LGM zonal winds is the greatest of all the simulations, despite the fact that the greatest change in momentum transport is in the Snowball simulation. However, in the LGM, mid-Holocene, and increased CO_2 simulations the oceans are allowed to respond to changes in wind speed. There is, therefore, an element of atmosphere-ocean feedback in these simulations that is precluded in the Snowball simulation. Increased zonal wind speeds in the LGM and mid-Holocene simulations increase oceanic upwelling, particularly in the eastern tropical Pacific. This upwelling increases the extent of the Pacific cold tongue which, in turn, increases the zonal pressure gradient along the equator. This amplifies the winds even further, in the type of positive feedback that controls the El Niño Southern Oscillation, leading to a more La Niña-like climatological mean state. In the increased CO_2 simulation, the opposite holds true, and decreased easterlies reduce equatorial upwelling and lead to a more El Niño-like mean state.

The east-west tilt of the mean thermocline in the tropical Pacific therefore increases in the LGM and mid-Holocene simulations, and is smaller in the increased CO_2 simulation (Figure 6). This result is consistent with the changes in mean zonal wind speeds, and has implications for the frequency of interannual variability (Fedorov and Philander, 2000).

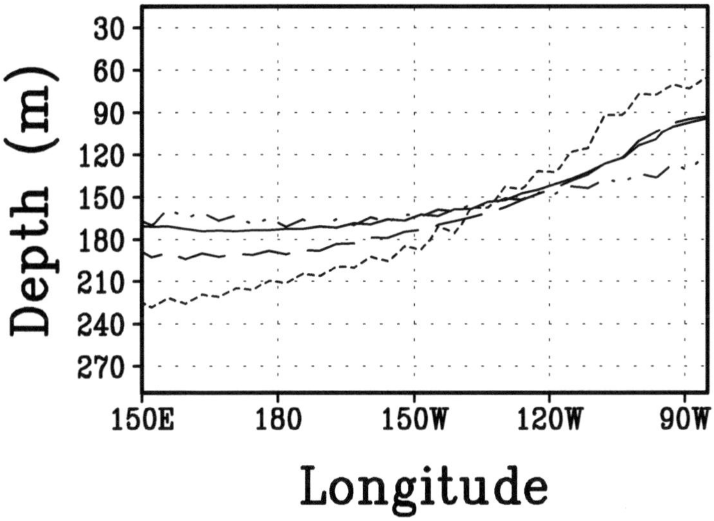

Figure 6. Depth of the 18°C isotherm in the Pacific Ocean in the coupled model simulations. **Solid line:** control simulation; **short dashed line:** LGM simulation; **long dashed line:** mid-Holocene simulation; **dashed-dot-dot line:** increased CO_2 simulation.

4. Conclusions

Simulated changes in the mean state of the tropical Pacific Ocean are consistent with the idea that, through the momentum fluxes associated with their decay phase, midlatitude baroclinic eddies play a role in governing the strength of subtropical convergence in the upper troposphere, subsidence, surface pressure gradients, and hence equatorial trade winds. Moreover, the changes in eddy activity are consistent with the changes in mean state baroclinicity and topographic forcing that are prescribed in each simulation. Atmosphere-ocean interactions in the tropics act to amplify the response beyond what would be expected from the change in momentum flux alone. While evidence exists for stronger trade winds and a steeper thermocline tilt at the LGM, proxy data have yet to be provided for other times such as the mid-Holocene.

References

Andreasen, D., and A. C. Ravelo, Tropical Pacific Ocean thermocline depth reconstructions for the Last Glacial, *Paleoceanography*, 12:395-414, 1997.

Barron, E.J., and W. Washington, The atmospheric circulation during warm, geologic periods: Is the equator-to-pole surface temperature gradient the controlling factor? *Geology*, 10:633-636, 1982.

Becker, E., G. Schmitz, and R. Geprags, The feedback of midlatitude waves onto the Hadley cell in a simple general circulation model. *Tellus*, 49A:182-199, 1997.

Berger, A., Orbital variations and insolation database, IGBP PAGES/World Data Center-A for Paleoclimatology Data Contribution Series # 92-007. NOAA/N GDC Paleoclimatology Program, Boulder CO, USA.

Berner, R.A., A model for atmospheric CO_2 over Phanerozoic time, *Am. J. Sci.*, 291:339-376, 1991.

Bush, A.B.G., Assessing the impact of mid-Holocene insolation on the atmosphere-ocean system. *Geophy. Res. Lett.*, 26:99-102, 1999.

Bush, A.B.G., Simulating climates of the Last Glacial Maximum and of the mid-Holocene: Wind changes, atmosphere-ocean interactions, and the tropical thermocline. *AGU Monograph Series 126 (The Oceans and Rapid Climate Change: Past, Present, and Future)*, 135-144, 2001.

Bush, A.B.G. and S.G.H. Philander, The climate of the Last Glacial Maximum: Results from a coupled atmosphere-ocean general circulation model. *J. Geophys. Res.*, 104:24509-24525, 1999.

Chang, E.K.M., Mean meridional circulation driven by eddy forcings of different time scales. *J. Atmos. Sci.*, 53:113-125, 1996.

Clemens, S.C. and W.L. Prell, Late Pleistocene variability of Arabian Sea summer-monsoon winds and dust source-area aridity: A record from the lithogenic component of deep-sea sediments, *Paleoceanography*, 5:109-145, 1990.

Climate: Long-Range Investigation, Mapping, and Prediction (CLIMAP) Project Members, Seasonal reconstructions of the Earth's surface at the last glacial maximum, *Map and Chart Series MC-36*, Geol. Soc. of Am., Boulder, CO, 1981.

Edmon, H.J., B.J. Hoskins, and M.E. McIntyre, Eliassen-Palm cross-sections for the troposphere, *J. Atmos. Sci.*, 37:2600-2616, 1980.

Endal, A.S. and S. Sofia, Rotation in solar-type stars, I, Evolutionary models for the spindown of the sun. *Astrophys. Jour.*, 243:625-640, 1981.

Fairbanks, R.G., A 17,000-year glacio-eustatic sea level record: Influence of glacial melting rates on Younger Dryas event and deep-ocean circulation, *Nature*, 342:637-642, 1989.

Fedorov, A.V. and S.G.H. Philander, Is El Niño Changing? *Science*, 288:1997-2002, 2000.

Gordon, C.T., and W. Stern, A description of the GFDL global spectral model, *Mon. Weather Rev.*, 110:625-644, 1982.

Hall, N.M.J., P.J. Valdes, and B. Dong, The maintenance of the last great ice sheets: A UGAMP GCM study, *J. Clim.*, 9:1004-1019, 1996.

Haynes, P.H., and T.G. Shepherd, The importance of surface pressure changes in the response of the atmosphere to zonally-symmetric thermal and mechanical forcing, *Q. J. R. Meteorol. Sci.*, 115:1181-1208, 1989.

Held, I.M., and A.Y. Hou, Nonlinear axially symmetric circulations in a nearly inviscid atmosphere, *J. Atmos. Sci.*, 37:515-533, 1980.

Hoffman, P.F., A.J. Kaufman, G.P. Halverson, and D.P. Schrag, A Neoproterozoic snowball earth, *Science*, 281:1342-1346, 1998.

Hyde, W.T., T.J. Crowley, S.K. Baum, and W.R. Peltier, Neoproterozoic 'snowball Earth' simulations with a coupled climate/ice-sheet model, *Nature*, 405:425-429, 2000.

Kim, H.K., and S. Lee, Hadley cell dynamics in a primitive equation model: Part I. Axisymmetric flow. *J. Atmos. Sci.*, 58:2845-2858, 2001.

Kim, H.K., and S. Lee, Hadley cell dynamics in a primitive equation model: Part II. Nonaxisymmetric flow. *J. Atmos. Sci.*, 58:2859-2871, 2001.

Kutzbach, J.E. and B.L. Otto-Bliesner (1982). The sensitivity of the African-Asian monsoonal climate to orbital parameter changes for 9000 years B.P. in a low-resolution general circulation model. *J. Atmos. Sci.*, 39:1177-1188.

Lindzen, R.S., and A.Y. Hou, Hadley circulations for zonally averaged heating off the equator, *J. Atmos. Sci.*, 45:2416-2427, 1988.

Lyle, M.W., F.G. Prahl, M.A. Sparrow, Upwelling and productivity changes inferred from a temperature record in the central equatorial Pacific, *Nature*, 355:812-815, 1992.

Otto-Bleisner, B.L., El Niño/La Niña and Sahel precipitation during the middle Holocene. *Geophys. Res. Lett.*, 26:87-90, 1999.

Pacanowski, R.C., K. Dixon, and A. Rosati, *The GFDL Modular Ocean Model user guide, GFDL Ocean Group Tech. Rep. 2*, Geophys. Fluid Dyn. Lab., Princeton, N.J., 1991.

Pedersen, T.F., Increased productivity in the eastern equatorial Pacific during the last glacial maximum (19,000 to 14,000 yr B.P.), *Geology*, 11:16-19, 1983.

Peixoto, J.P. and A.H. Oort, *Physics of Climate*, American Institute of Physics, New York, 520 pp., 1992.

Peltier, W.R., Ice age paleotopography, *Science*, 265:195-201, 1994.

Pfeffer, R.L., Wave-mean flow interactions in the atmosphere. *J. Atmos. Sci.*, 38:1340-1359, 1981.

Philander, S.G.H., El Niño and La Niña, *J. Atmos. Sci.*, 42:2652-2662, 1985.

Philander, S.G.H., *El Niño, La Niña, and the Southern Oscillation*, Academic Press, New York, 293 pp., 1990.

Prell, W.L., Monsoonal climate of the Arabian Sea during the late Quaternary: A response to changing solar radiation. In *Milankovitch and Climate* (A. Berger *et al.*, Eds.), pp. 349-366. Reidel, Dordrecht, 1984.

Prell, W.L. and J.E. Kutzbach, Sensitivity of the Indian monsoon to forcing parameters and implications for its evolution. *Nature,* 360:647-652, 1992.

Rutter, N.W., Presidential Address, XIII INQUA Congress 1991: Chinese loess and global change, *Quat. Sci. Rev.*, 11:275-281, 1992.

Sarnthein, M., G. Tetzlaff, B. Koopman, K. Wolter, and U. Pflaumann, Glacial and interglacial wind regimes over the eastern subtropical Atlantic and north-west Africa, *Nature*, 293:193-196, 1981.

Simmons, A.J. and B.J. Hoskins, Barotropic influences on the growth and decay of non-linear baroclinic waves, *J. Atmos. Sci.*, 37:1679-1684, 1980.

Sloan, L.C., and E.J. Barron, "Equable" climates during Earth history? *Geology*, 18:489-492, 1990.

Wright, H.E. Jr., J.E. Kutzbach, T. Webb III, W.F. Ruddiman, F.A. Street-Perrott, and P.J. Bartlein (Eds.), *Global climates since the Last Glacial Maximum*, 569 pp, University of Minnesota Press, Minneapolis, 1993.

Xie, S.-P., Ocean-atmosphere interaction in the making of the Walker circulation and the equatorial cold tongue, *J. Clim.*, 11:189-201, 1998.

MEAN AND EDDY DYNAMICS OF THE MAIN THERMOCLINE

GEOFFREY K. VALLIS
GFDL
Princeton University
Princeton, NJ 08544, USA

Abstract. This paper reviews and discusses a selection of developments in the theory of the structure of the main thermocline, and the mesoscale eddies that inhabit it. In classical theories, that is theories that assume a steady, near-laminar flow and that are based on the planetary geostrophic equations, the upper thermocline (below the surface mixed layer) is conservative and advectively dominated — that is, the dominant balance in the thermodynamic equation lies in the advective terms, leading to a ventilated thermocline. Below this the internal thermocline is a diffusive transition region, and in the limit of small diffusivity it becomes an internal boundary layer between the ventilated thermocline and the abyss. The thermocline is, typically, baroclinically unstable and this leads to an upper ocean populated by vigorous mesoscale eddies. The eddies are strongest in regions of western boundary currents, 'mode water' regions, and in the circumpolar current, and in these regions the eddies significantly affect the structure of the main thermocline. Elsewhere, the structure of the upper (ventilated) thermocline is largely determined by mean-flow advection. Lower in the water column, eddies typically tend to thicken the isostads that form the internal thermocline, leading to a complex three-way balance between mean flow, eddy fluxes and diffusion, suggesting that the internal thermocline may have finite thickness even as diffusivity tends to zero. In the circumpolar current eddies are a dominant effect, and qualitatively change the structure of the stratification.

Key words: dynamics, oceanography, thermocline, eddies, turbulence

1. Background

The modern development in our understanding of the thermocline is often considered to have begun with two back-to-back papers in 1959 in the journal *Tellus*. Welander (1959) suggested an adiabatic model, based on the ideal-fluid thermocline equations (i.e. the planetary geostrophic equations, with no diffusion terms in the buoyancy equation), whereas Robinson and Stommel (1959) proposed a model that is intrinsically diffusive. In the latter model [developed further by Stommel and Webster (1962)] the thermocline is a diffusive front that forms at the convergence of two different homogeneous water types, warm near surface fluid and cold abyssal fluid below,

O.U. Velasco Fuentes et al. (eds.), Nonlinear Processes in Geophysical Fluid Dynamics, 141-171.

whose thickness decreases as the diffusivity falls. The diffusive model was developed further by Salmon (1990) who found that numerical solutions of the planetary geostrophic equations can show a tendency to develop into a 'two-fluid' state — a pool of warm subtropical near surface water separated from a cold abyss by a diffusive front, an internal boundary layer, which Salmon associated with the main thermocline. Meanwhile, throughout the 1960s and 1970s the adiabatic model had continued its own development (see Veronis, 1969), culminating in the ventilated thermocline model of Luyten *et al.* (1983) (the 'LPS model') and its continuous extensions (Huang, 1988; Lionello and Pedlosky, 2000). Noting the clear difference between the two classes of theory, Welander (1971) commented that 'interior diffusive regions' may be necessary below an adiabatic upper-ocean thermocline. Samelson and Vallis (1997a) eventually suggested a model in which the upper thermocline is adiabatic, as in the ventilated thermocline model, but has a diffusive base that for small diffusivity constitutes an internal boundary layer. This model differs from the original diffusive models (e.g. the Robinson-Stommel-Webster model) and the model of Salmon in that the upper boundary conditions of the boundary layer are obtained (in principle) from matching the boundary layer to the lowest layer of a ventilated thermocline model. Thus, in this model, the thermocline has two dynamical regimes — an upper 'ventilated' region in which the dynamics is essentially adiabatic (or at least adiabatic below the mixed layer), and a lower diffusive layer. To the extent that the upper layer follows the dynamics of the LPS model of the thermocline, it will display features associated with that model — an eastern shadow zone and a western pool region for example. The relative strength of these two thermocline regimes depends on the geography of the situation: the upper advectively dominated thermocline is a mapping of the horizontal meridional temperature gradient across the subtropical gyre; whereas the lower diffusively dominated thermocline is a mapping of the surface temperature across the subpolar gyre, and taken together they constitute the main thermocline.

The role of mesoscale eddies in all of this has only recently been investigated, and their role is now slowly emerging. That eddies may play a role in ocean circulation has, of course, long been conjectured and even accepted. For example Rhines and Young (1982) pointed out that, because potential vorticity is conserved on parcels save for the effect of mild diffusive processes, it (potential vorticity) will tend to become homogenized by the effect of eddies where the circulation both forms closed gyres and is shielded from the direct effect of surface forcing. Thus, we might expect regions of homogenized potential vorticity in, for example, the subsurface recirculation regions of the subtropical gyre. Because their theory is quasigeostrophic, it takes the vertical structure of the stratification as given and thus is not a theory of the thermocline *per se*. (However, it did anticipate the eastern shadow zone, as well as provide a mechanism for setting subsurface layers into motion that is different from that of the LPS model.) More recently, the possible role of eddies in actually setting ocean stratification has been explored. Vallis (2000b) noted that classical thermocline theories implicitly assume that mesoscale eddies are not important in setting the stratification, and the agreement or otherwise of such theories with observation will be one test of that assumption. Directly testing such theories against observation is of course very difficult, and the use of eddy resolving numerical models as an intermediary is likely to be a *sine qua non* of any such activity. Marshall *et al.* (2002) explicitly asked whether mesoscale eddies might play a role in setting the structure of the stratification in the upper ocean and suggested a simple model of that stratification, and Karsten *et al.* (2002) argued that mesoscale eddies set the stratification in the Antarctic Circumpolar Current (ACC). The stratification of the ACC differs from that of the subtropical gyres because one cannot build on Sverdrup balance to obtain a reasonable theory of the thermocline in the manner, say, of LPS. Because of the absence of meridional boundaries an E-W pressure gradient cannot be maintained over broad regions of the ACC and, by geostrophic balance, a mean meridional flow cannot be sustained. A consequence of this is that, in the absence of mesoscale eddies or their parameterized effects, the isopycnals become almost vertical and the convection is too deep (Vallis, 2000a). This is a highly baroclinically unstable situation, suggesting that if eddies are allowed to form they will have a first-order influence on the stratification. Recent numerical simulations do suggest that eddies may indeed play a role in determining the stratification and heat balance of the ocean, certainly in the ACC and, perhaps to a lesser degree (but still importantly), in the subtropical thermocline.

The rest of this paper expands on the above remarks to form a somewhat subjective review of some of the theoretical and numerical developments in thermocline dynamics. It is not a comprehensive review of all aspects of ocean stratification, or even of the theoretical aspects, and there is no discussion of the observations. Nevertheless, I hope readers will find it useful.

2. The Classical Picture

In this section we review the classical picture, by which we mean the picture described by the planetary geostrophic equations and that takes no account of mesoscale eddies. (Sometimes 'classical' is used in a complimentary way, meaning having stood the test of time, and sometimes in a derogatory way as a euphemism for 'wrong.' Our use here is neutral.)

2.1. EQUATIONS OF MOTION

The planetary geostrophic equations in the Boussinesq approximation are

$$\frac{\partial b}{\partial t} + \boldsymbol{v} \cdot \nabla b = \kappa \frac{\partial^2 b}{\partial z^2} \tag{1a}$$

$$-fv = -\phi_x \tag{1b}$$

$$fu = -\phi_y \tag{1c}$$

$$b = \phi_z \tag{1d}$$

$$\nabla \cdot \boldsymbol{v} = 0 \tag{1e}$$

The notation is standard, saline effects are omitted and a linear equation of state is implicit. Thus, the buoyancy b is given by $b = g\alpha\Delta T$, where α is the coefficient of thermal expansion and ΔT is the temperature perturbation from a reference value, ϕ is pressure divided by a constant density, $\boldsymbol{v}$ is the three-dimensional velocity field and κ is a diffusivity. In oceanography these equations were originally known as the *thermocline equations* (Robinson and Stommel, 1959; Welander, 1959); a presentation from a more atmospheric perspective was given by Burger (1958), and Phillips (1963) and Pedlosky (1987) subsequently gave rather more systematic derivations. The diffusive term on the right-hand-side of the thermodynamic equation is generally regarded as representing a real physical process, namely turbulent diffusion at small scales. Even with this term the equations as written above are of insufficiently high order to be well-posed in a laterally closed domain, and small additional terms are needed in both the thermodynamic and momentum equations if the equations are to be solved in such a domain without numerical boundary layers (Samelson and Vallis, 1997b). The 'ideal' thermocline equations have no dissipative (diffusive or viscous) terms at all in the thermodynamic or momentum (geostrophic) equations, and thus cannot support the presence of lateral boundaries.

The planetary geostrophic equations are valid only for scales larger than the deformation radius and for scales for which the Coriolis parameter varies by an $\mathcal{O}(1)$ amount. The absence of nonlinear terms in the momentum equations means that baroclinic instability is incorrectly described: the equations have an ultra-violet catastrophe in that the growth rate increases monotonically with wavenumber (Colin-de-Verdiere, 1986). However, for scales of order the deformation radius and larger, baroclinic growth rates are *underestimated* (Smith and Vallis, 1998), and in fact are sufficiently small that they may readily be controlled with a modest amount of dissipation — smaller than might be required in a primitive equation model. The upshot of all this is with a grid resolution coarser than the deformation scale (say a grid scale of 1° or greater) steady solutions of the planetary geostrophic equations with small lateral and vertical diffusion are somewhat easier to obtain than corresponding solutions with the primitive equations, and unphysical cross-isopycnal diffusion (the 'Veronis effect') is diminished.

2.2. SCALING

2.2.1. *Advective Scaling*

A scaling for the depth of the thermocline can be obtained from Sverdrup balance in conjunction with the thermal wind equation. These equations are, respectively,

$$\beta v = f\frac{\partial w}{\partial z}, \tag{2}$$

with corresponding scaling

$$\beta V \sim \frac{fW_E}{D_a}, \tag{3}$$

and

$$f\frac{\partial v}{\partial z} = \frac{\partial b}{\partial x} \tag{4}$$

with corresponding scaling

$$\frac{fV}{D_a} \sim \frac{\Delta b}{L}. \tag{5}$$

In these equations Δb is the magnitude of the buoyancy variation and the other scaling variables are denoted with a capital letter. Thus, L is the horizontal scale of the motion, which we take as the gyre or basin scale, and the parameter D_a is the vertical scale where the subscript a denotes an advective scale. The appropriate vertical velocity to use is that due to Ekman pumping, W_E; we will assume *a priori* that this is much larger than the abyssal upwelling velocity, which in any case is zero by assumption at $z = -D_a$. W_E and L are thus given by the geometry and the strength of the wind forcing, whereas U and D_a are part of the solution. Eliminating V from (3) and (5) gives the estimate

$$D_a \sim \left(\frac{W_E f^2 L}{\beta \Delta b}\right)^{1/2} \tag{6}$$

which has its roots in Welander (1959). We can also derive this result by imagining the upper ocean to be a single layer of homogeneous fluid of variable depth h lying over a resting abyss, then (4) can be replaced by $fv = g'\partial h/\partial x$ where g' is the usual reduced gravity and h is, effectively, the depth of the thermocline. This equation scales like

$$fV \sim g'\frac{D_a}{L} \tag{7}$$

and using this and (3) we obtain $D_a \sim (f^2 L W_E/\beta g')^{1/2}$, which is equivalent to (6), because $\Delta b \sim g'$ is just the reduced gravity across the thermocline.

As an aside, we note that if the horizontal scale is large enough so that there is little cancellation in the terms comprising horizontal divergence, then the mass conservation equation, $\partial u/\partial x + \partial v/\partial y = -\partial w/\partial z$, scales like $V/L \sim W/D_a$. Using this with (5) gives

$$D_a \sim \left(\frac{W_E f L^2}{\Delta b}\right)^{1/2}. \tag{8}$$

which is the same as (6) if $\beta \sim f/L$, that is for the planetary scale.

Thus, the depth of the wind-influenced region is proportional to the half-power of the strength of the wind-stress, and inversely proportional to half power of the meridional temperature gradient. The former dependence is reasonably intuitive, the latter perhaps less so. One way to think about this is that as the temperature gradient increases, the associated thermal wind-shear V/D_a correspondingly increases. But if the mechanical forcing is unaltered, then Sverdrup balance can only remain satisfied if the vertical scale of the motion decreases. From a shallow water perspective, that interface displacements tend to fall as the reduced gravity increases is a familiar notion, but the constraining effects of Sverdrup balance are such that the thermocline depth does not fall as rapidly as g' increases [equations (3) and (7)].

Finally, let us estimate the thermocline depth as given by such scalings. Using $W_E \sim 10^{-6}\,\mathrm{m\ s^{-1}}$, $\Delta b = g\Delta\rho/\rho_0 = g\alpha\Delta T \sim 10^{-2}\,\mathrm{m\ s^{-2}}$, $L = 5000$ km and $f = 10^{-4}\ \mathrm{s^{-1}}$ in (6)

$$D \sim 500 \text{ m} \tag{9}$$

However, this is only an estimate, and might easily be in error by some nondimensional number not revealed by the scaling analysis. One hopes any such numbers are $\mathcal{O}(1)$.

2.2.2. *Diffusive Scaling*

In the derivation above the diffusion in the thermodynamic equation plays no role; indeed the thermodynamic equation provides no additional vertical scale. The vertical velocity is imposed by the Ekman layer, and a closed scaling is obtained from the mass conservation and Sverdrup balance equations alone. Deeper in the thermocline, the vertical velocity will diminish and no longer be dominated by the Ekman pumping; it will be affected by upwelling from the abyss and should be part of the solution provided by the scaling rather than an imposed parameter. The steady thermodynamic equation, with diffusion,

$$\boldsymbol{v} \cdot \nabla b = \kappa \frac{\partial^2 b}{\partial z^2} \tag{10}$$

implies the scales

$$\frac{U}{L}, \frac{W}{\delta} \sim \frac{\kappa}{\delta^2}. \tag{11}$$

where δ is a vertical scale. This scaling must be consistent with the Sverdrup relation, (2), and thermal wind, (4), now with scalings

$$\beta V \sim \frac{fW}{\delta}, \tag{12}$$

and

$$\frac{fV}{\delta} \sim \frac{\Delta b}{L}. \tag{13}$$

Assuming that $\beta/f \leq 1/L$ then these relations give the diffusive vertical scale

$$\delta \sim \kappa^{1/3} \left(\frac{fL^2}{\Delta b} \right)^{1/3} \tag{14}$$

and the internal vertical velocity scale

$$W \sim \frac{\kappa}{\delta} \propto \kappa^{2/3} \tag{15}$$

which is a scaling for the strength of the overturning circulation. If the vertical advection term dominates in (11) then we can obtain the scaling

$$\delta \sim \kappa^{1/3} \left(\frac{f^2 L}{\beta \Delta b} \right)^{1/3}. \tag{16}$$

but for most purposes this is not qualitatively different.

Using $f = 10^{-4}\,\mathrm{s}^{-1}$, $L = 5\,10^{-6}\,\mathrm{m}$, $g = 10\ \mathrm{m\ s}^{-2}$, $\kappa = 10^{-5}\,\mathrm{m}^2/\mathrm{s} = 0.1\ \mathrm{cm}^2/\mathrm{s}$, $\Delta b = g\Delta\rho/\rho_0 = g\alpha\Delta T$ and $\Delta T = 10\,K$ gives $\delta \sim 130$ m. These parameter values also give $U \sim 0.3$ cm/s and $W \sim 10^{-5}$ cm/s. Note that W_E (which is typically about 10^{-4} cm/s) is indeed much larger than W.

The scaling above assumes that the length scale over which thermal-wind balance holds is the gyre scale itself. In fact there is another length scale that is more appropriate, namely the horizontal difference across the sloping thermocline, and this leads to a slightly different scaling for the thickness of the thermocline. The depth of the subtropical thermocline is not constant; rather it slopes (see Figure 1). It shoals up to the east simply because of Sverdrup balance, and it may slope up polewards because the curl of the windstress falls (and is zero at the polewards edge of the

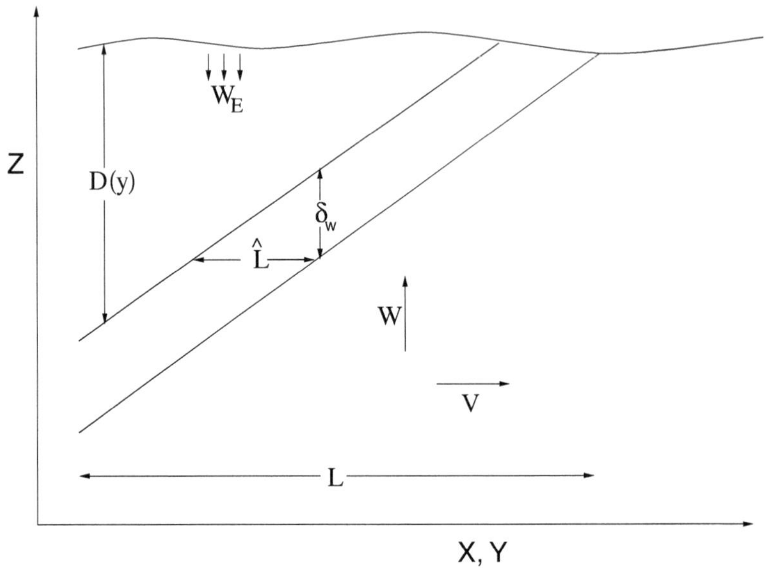

Figure 1. Scaling for the thermocline. If the thermocline slopes, the appropriate horizontal scale is given by $\hat{L}$, which is part of the solution rather than being imposed.

subtropical gyre). Thus, the appropriate horizontal length scale $\hat{L}$ is not the basin scale itself, but is given by

$$\hat{L} = \delta \frac{L}{D}. \tag{17}$$

This is no longer an externally imposed parameter, but must be determined as part of the solution. Using $\hat{L}$ instead of L as the length scale gives, after just a little algebra, the modified diffusive scale

$$\delta_w = \kappa^{1/2} \left(\frac{fL^2}{\Delta b D_a} \right)^{1/2} = \kappa^{1/2} \left(\frac{fL^2}{\Delta b W_E} \right)^{1/4}. \tag{18}$$

Substituting values of the various parameters results in a thickness of about 100–200 m, for both (14) and (18). This is somewhat less than the scaling estimate (6) of the depth of the advective, upper thermocline. Observations, arguably, do not suggest that this is the case. This might be because some other process is thickening the lower thermocline or, perhaps more likely, because all of these scalings are likely to be in error by a nondimensional factor which will differ from case to case.

Note that (18) suggests that the thermocline depth scales as $\kappa^{1/2}$. This scaling is the appropriate one for a wind- and buoyancy-driven ocean, whereas (14) is appropriate if there is no wind forcing, and this expectation is borne out by numerical simulations Vallis (2000a). The vertical velocity, and hence the meridional overturning circulation, now scales as

$$W \sim \frac{\kappa}{\delta_w} \propto \kappa^{1/2} \tag{19}$$

rather than $\kappa^{2/3}$.

2.3. THE LOWER THERMOCLINE AS A BOUNDARY LAYER

If the lower thermocline is indeed a boundary layer, then it is natural to try to apply some of the techniques of boundary layer theory. To do this we first combine the planetary geostrophic equations into a single equation, the 'M-equation'. (This section may be skipped at first reading.)

2.3.1. *The M–equation*

The planetary geostrophic equations can be written as a single partial differential equation in a single variable, although the resulting equation is of quite high order (third, in the absence of friction) and nonlinear. Cross differentiating the geostrophic relations, and using mass continuity, implies the Sverdrup relation

$$\beta v = f\frac{\partial w}{\partial z} \tag{20}$$

or, using geostrophic balance again

$$\frac{\partial \phi}{\partial x} + \frac{\partial}{\partial z}\left(-\frac{f^2}{\beta}w\right) = 0. \tag{21}$$

This equation is automatically satisfied if

$$\phi = M_z \quad \text{and} \quad \frac{f^2 w}{\beta} = M_x. \tag{22}$$

Then straightforwardly

$$u = -\frac{\phi_y}{f} = -\frac{M_{zy}}{f} \qquad \text{and} \qquad v = \frac{\phi_x}{f} = \frac{M_{zx}}{f}, \tag{23}$$

and

$$b = M_{zz}. \tag{24}$$

The thermodynamic equation becomes

$$\frac{\partial M_{zz}}{\partial t} + \left[\frac{-M_{zy}}{f}M_{zzx} + \frac{M_{zx}}{f}M_{zzy}\right] + \frac{\beta}{f^2}M_x M_{zzz} = \kappa M_{zzzz} \tag{25}$$

or

$$\frac{\partial M_{zz}}{\partial t} + \frac{1}{f} J(M_z, M_{zz}) + \frac{\beta}{f^2} M_x M_{zzz} = \kappa M_{zzzz}. \tag{26}$$

This is the 'M–equation,' first derived by Welander (1959). It is analogous to the potential vorticity equation in quasi-geostrophic theory in that it expresses the entire dynamics of the system in a single, nonlinear, advective-diffusive partial differential equation. The Sverdrup relation in the interior is automatically satisfied, and at the surface it is represented by

$$\frac{\beta}{f^2} M_x(x, y, z = 0) = w_E. \tag{27}$$

Equation (26) is rather complicated; analytic solutions are very hard to find, and numerically it is easier to find solutions by integrating the original planetary geostrophic equations. However, it is possible to move forward by appropriately approximating the equation, and the next two sections discuss this.

2.3.2. *A simple one-dimensional model*

A simple special case illustrates clearly the formation of an internal front. If $M = M(y, z)$ then the M-equation becomes

$$\frac{\beta}{f^2} M_x M_{zzz} = \kappa M_{zzzz} \tag{28}$$

which represents the advective-diffusive balance $wb_z = \kappa b_{zz}$. The horizontal advection terms vanish because the zonal velocity (u) and the meridional temperature gradient (b_y) are each zero. We consider, following Salmon (1990), the special case

$$M = xW(z) \tag{29}$$

whence (28) becomes

$$WW_{zzz} = \kappa W_{zzzz}. \tag{30}$$

A closely related equation was put forward by Stommel and Webster (1962) and Young and Ierley (1986), namely

$$[2W - zW_z]\, W_{zzz} = \kappa W_{zzzz}. \tag{31}$$

where the vertical velocity is proportional to the term in square brackets. The similarity of the time-dependent form of these equations (e.g. $W_{zzt} + WW_{zzz} = \kappa W_{zzzz}$) to Burger's equation ($W_t + WW_z = \kappa W_{zz}$) suggests that fronts might form. Appropriate boundary conditions for (30) are a prescribed buoyancy or buoyancy flux and vertical velocity at the upper and lower boundaries, for example

$$\begin{aligned} W &= W_E, \qquad b_z = W_{zzz} = 0 \qquad \text{at top} \\ W &= 0, \qquad b_z = W_{zzz} = 0 \qquad \text{at bottom} \end{aligned} \tag{32}$$

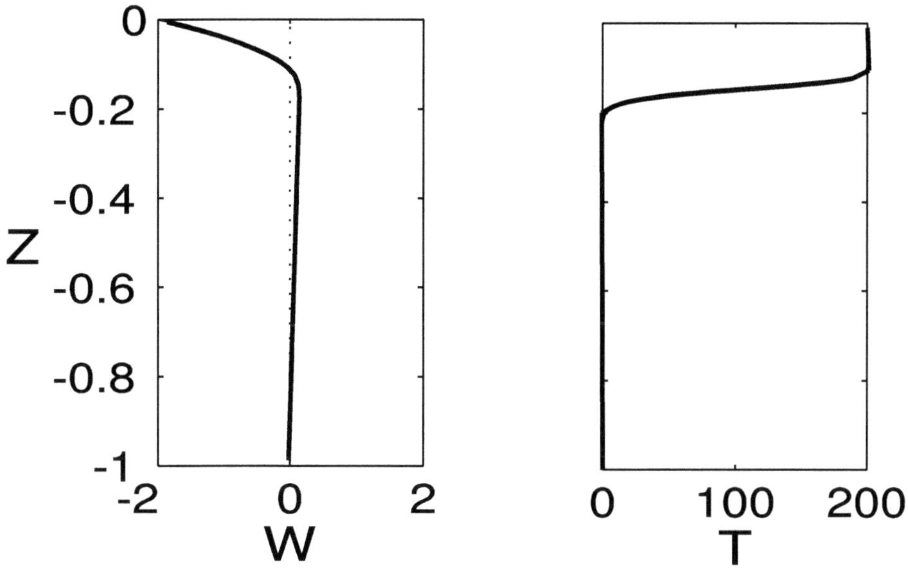

Figure 2. A numerical solution of (31), obtained using Newton's method. Buoyancy flux is zero at top and bottom, and there is an imposed downward vertical velocity at the top. The solution exhibits weak upwelling in the lower part of the domain, and a *front* forms where the vertical velocity passes through zero.

where W_E is the vertical velocity at the base of the top Ekman layer, which is negative for Ekman pumping in the subtropical gyre. We obtain solutions numerically by Newton's method; both (30) and (31) exhibit qualitatively similar solutions, and a typical one is shown in Figure 2.

The important point is that the solutions to (30) and (31) generically produce an interior front, whose thickness goes to zero as $\kappa \to 0$. These equations are rational simplifications of the full three dimensional equations of motion, suggesting that these equations, too, may display frontal behaviour in the limit of small diffusivity. This front occurs where the vertical velocity is zero — where the downwards Ekman pumping and the upwelling from the abyss 'collide', and the water mass properties change rapidly in the vertical.

2.3.3. *Boundary layer analysis*

Following Samelson (1999) let us *assume* that the temperature varies rapidly only in an 'internal boundary layer' of thickness δ; above and below this it is assumed to be at most slowly varying. Thus we write

$$b(x, y, z) = b(x, y, z) + \hat{b}(x, y, \zeta) \tag{33}$$

where $\hat{b}$ is important only in the boundary layer and ζ is a stretched co-ordinate such that

$$z + h(x, y) = \delta\zeta. \tag{34}$$

That is, it is the vertical distance from $z = -h$, scaled by the boundary layer thickness δ, and is presumptively an $\mathcal{O}(1)$ quantity. Note that we allow the depth of the boundary layer to change as a function of horizontal co-ordinate. Then M varies as

$$M = M(x, y, z) + \delta^2 \hat{M}(x, y, \zeta) \tag{35}$$

where $\hat{M} \to 0$ away from the boundary layer. The scaling factor on $\hat{M}$ is needed because $b = M_{zz}$, and so $b \sim (1/\delta^2)M$ in the boundary layer. Since b remains an order one quantity throughout, $\hat{M}$ must be scaled appropriately.

In the boundary layer the derivatives of M become

$$\frac{\partial \hat{M}}{\partial z} = \frac{1}{\delta}\frac{\partial \hat{M}}{\partial \zeta} \tag{36}$$

and

$$\begin{aligned} \frac{\partial \hat{M}}{\partial x} &= \frac{\partial \hat{M}}{\partial \zeta}\frac{\partial \zeta}{\partial x} + \frac{\partial \hat{M}}{\partial x} \\ &= \frac{\partial \hat{M}}{\partial \zeta}\left(\frac{1}{\zeta}\frac{\partial h}{\partial x}\right) + \frac{\partial \hat{M}}{\partial x} \end{aligned} \tag{37}$$

Substituting these into (25) we obtain, for a steady state,

$$\begin{aligned} &\delta\left\{\frac{1}{f}\left(\hat{M}_{\zeta x}\hat{M}_{\zeta\zeta y} - \hat{M}_{\zeta y}\hat{M}_{\zeta\zeta x}\right) + \frac{\beta}{f^2}\hat{M}_x\hat{M}_{\zeta\zeta\zeta}\right\} + \frac{\beta}{f^2}h_x\hat{M}_\zeta\hat{M}_{\zeta\zeta\zeta} \\ &+ \frac{1}{f}\left[h_x\left(\hat{M}_{\zeta\zeta}\hat{M}_{\zeta\zeta y} - \hat{M}_{\zeta y}\hat{M}_{\zeta\zeta\zeta}\right) + h_y\left(\hat{M}_{\zeta x}\hat{M}_{\zeta\zeta\zeta} - \hat{M}_{\zeta\zeta}\hat{M}_{\zeta\zeta x}\right)\right] \\ &\qquad = \frac{1}{\delta^2}\kappa\hat{M}_{\zeta\zeta\zeta\zeta}. \end{aligned} \tag{38}$$

(The horizontal advective terms of order δ^{-1} vanish identically.) Obviously, this equation of itself does not provide much insight even to the most algebraically minded oceanographer. But it is, nevertheless, revealing of its scaling behavior.

If $h_x = h_y = 0$ (i.e. the base of the thermocline is flat), (38) becomes

$$\frac{1}{f}\left[\hat{M}_{\zeta x}\hat{M}_{\zeta\zeta y} - \hat{M}_{\zeta y}\hat{M}_{\zeta\zeta x}\right] + \frac{\beta}{f^2}\hat{M}_x\hat{M}_{\zeta\zeta\zeta} = \frac{1}{\delta^3}\kappa M_{\zeta\zeta\zeta\zeta}. \tag{39}$$

Since all the terms in this equation are, by construction, order one, we immediately see that

$$\delta \propto \kappa^{1/3}, \tag{40}$$

just as in the scaling arguments. On the other hand, if h_x and/or h_y are order one quantities then the dominant balance in (38) is

$$\frac{1}{f}\left[h_x(\hat{M}_{\zeta\zeta}\hat{M}_{\zeta\zeta y} - \hat{M}_{\zeta y}\hat{M}_{\zeta\zeta\zeta}) + h_y(\hat{M}_{\zeta x}\hat{M}_{\zeta\zeta\zeta} - \hat{M}_{\zeta\zeta}\hat{M}_{\zeta\zeta x})\right] = \frac{1}{\delta^2}\kappa M_{\zeta\zeta\zeta\zeta} \tag{41}$$

and

$$\delta \propto \kappa^{1/2}, \tag{42}$$

which again is consistent with the scaling arguments. Thus, if the isotherm slopes are fixed independently of κ (perhaps by the windstress), then as $\kappa \to 0$ an internal boundary layer will form whose thickness is proportional to $\kappa^{1/2}$. We expect this to occur at the base of the main thermocline, with purely advective dynamics being dominant in the upper part of the thermocline, and determining the slope of the isotherms (i.e. the form of h_x and h_y), as in Figure 1.

Interestingly, the balance in the boundary layer equation does not simply correspond to $wb_z \approx \kappa b_{zz}$. Both at $\mathcal{O}(1)$ and $\mathcal{O}(\delta)$ the horizontal advective terms in (38) are of the same asymptotic size as the vertical advection terms. In the middle of the internal boundary layer the thermodynamic balance is thus $\boldsymbol{u} \cdot \nabla_h b + wb_z \approx \kappa b_{zz}$, with all terms in principle important. We might have anticipated this, because the vertical velocity, and the second derivative of buoyancy, passes through zero within the boundary layer. [More discussion is given in Samelson (1999).] Finally, we note that even though the balance at the thermocline base *necessarily* involves diffusion, diffusion (if small) plays a small role in the thermocline heat budget. It is then *not* a significant mechanism for transporting heat or the formation of water masses in the fluid interior; it is not the case for example that a net input of buoyancy at the surface over the subtropical gyre is balanced by an interior diffusive flux across isothermal surfaces. Diffusion is, however, necessary for the maintenance of a meridional overturning circulation, via (15) or (19).

2.4. SUMMARY

The classical line of investigation might be summarized in Figure 3, with a ventilated thermocline lying on top of an internal thermocline. The former describes the structure of isopycnals that outcrop in the subtropical gyre where Ekman pumping is downwards; this merges smoothly into the internal thermocline, which spans those isopycnals that outcrop in the subpolar gyre. Perhaps the most serious shortcoming of this as a conceptual model (rather than a quantitative one, which would need realistic geometry, salinity, etc.) is that no account is taken of mesoscale eddies, and we now turn our attention to this matter.

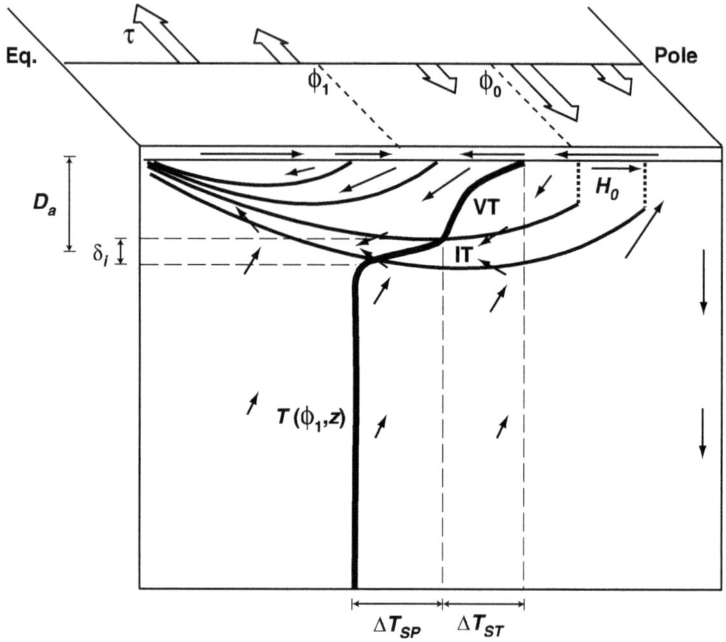

Figure 3. A schematic of the structure of the classical wind- and buoyancy-driven thermocline, according to SV. The medium-weight lines are isopycnals, 'IT' denotes the advective-diffusive internal thermocline, 'VT' denotes the advective ventilated thermocline, and the thick solid line is a temperature profile through the middle of the subtropical gyre. The single arrows represent the meridional circulation, and the double arrows the overlying windstress. Buoyancy forcing may be considered to be low-latitude surface warming and high-latitude surface cooling.

3. Geostrophic Turbulence in the Thermocline

The main thermocline is a region of horizontal temperature gradients and vertical shear; thus, we might expect it to be a region of baroclinic instability — as pointed out by Gill *et al.* (1974), Robinson and McWilliams (1974) and others. Since the time of these papers both observations and simulations with eddying General Circulation Models (e.g. Smith *et al.*, 2000) have indicated the ubiquity of eddies in the midocean and, especially, in the regions of the western boundary currents and their immediate extensions. Furthermore, the scale separation between the first deformation radius ($\sim$ 100 km) and the gyre scale ($\sim$ 1000 km or more) suggests that notions of geostrophic turbulence might be even more applicable in the ocean than in the atmosphere where the scale separation is much weaker. The simplest picture (Rhines, 1977; Salmon, 1980) assumes a uniformly stratified ocean in which a mean shear is maintained at very large scales. Baroclinic instability then leads to the growth of both baroclinic and barotropic modes at scales near the first radius of deformation, followed by an inverse cascade of barotropic energy back to large scales. Dissipative processes are necessary to remove energy in the inverse cascade and enstrophy at small scales.

An interesting variation on this model arises when the stratification is nonuniform, as in the subtropical thermocline. The dominant effect is to inhibit the inverse cascade of energy in the barotropic mode. This is illustrated most clearly in an initial value problem using the quasi-geostrophic equations in a doubly-periodic domain (Smith and Vallis, 2001). Two experiments were carried out differing only in their stratification: in one configuration the stratification N^2 is uniform, whereas in the other the stratification is enhanced near the surface to represent the thermocline. The initial energy is all in the baroclinic modes and confined to very large horizontal scales, representing the energy in the very large scale flow. In the experiment with uniform stratification, the flow of energy largely follows the picture above, with a nonlocal transfer of energy to both baroclinic and barotropic modes at the deformation radius (this is just the usual 'baroclinic instability') followed by an inverse cascade of barotropic energy to larger scales (Figure 4). In the parlance of baroclinic lifecycles, the process is one of 'baroclinic growth and barotropic decay.' It is noteworthy that friction is not a necessary ingredient to this cycle: the cycle still decays barotropically and although energy may pile up at large barotropic scales this does not lead to another round of instability.

If the stratification is thermocline-like, then the transfer proceeds first to the baroclinic mode, and only subsequently to the barotropic mode, where it may then be transferred to larger scales (Figure 5). That is, the transfer to the barotropic mode is inhibited, as if there is a constriction in the pipeline of the baroclinic lifecycle. In statistically steady experiments, this leads to a slightly enhanced level of activity in the baroclinic mode at the scale of the deformation radius. This, coupled with the fact that the ocean is not uniformly subject to baroclinic instability, and therefore that baroclinic eddies may not always find themselves surrounded by other eddies with which to cluster and form an inverse cascade, may be the reason that the baroclinic eddy scales in the ocean are comparable with the first deformation radius. However, we might still expect that the barotropic energy is at a larger scale, perhaps determined by the Rhines scale or frictional processes.

4. Effect of Eddies on the Structure of the Thermocline

We come, finally, to the issue as to whether and how mesoscale eddies substantially affect the structure of the steady thermocline and whether and how classical models, such as those of the previous sections, need to

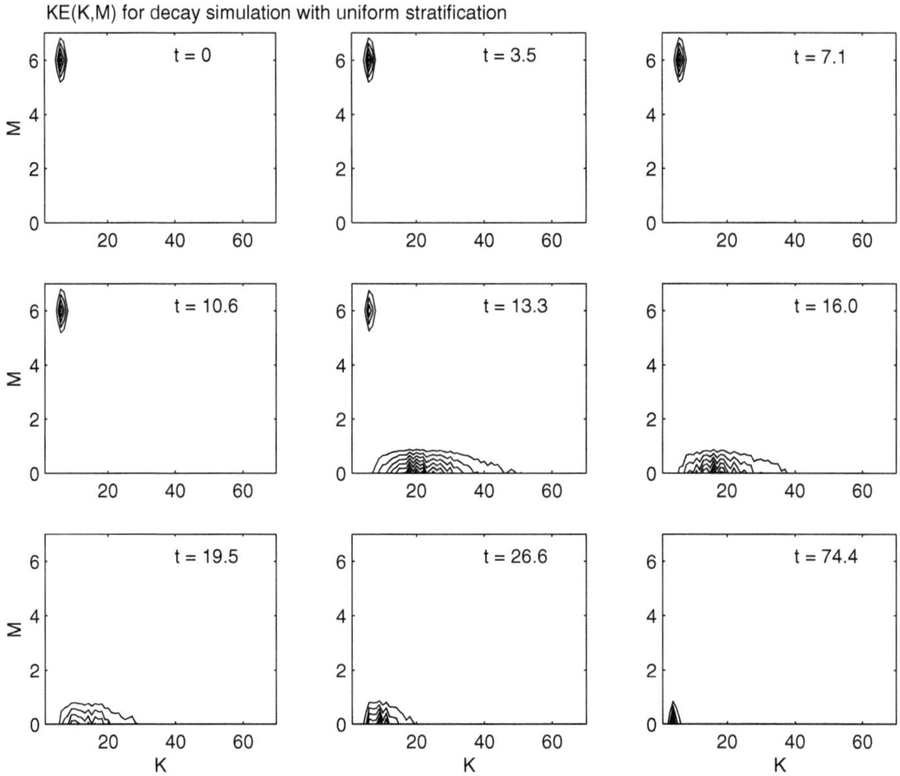

Figure 4. Time sequence of kinetic energy spectra for in a nearly inviscid stratified quasi-geostrophic simulation with uniform stratification. Times are given in terms of eddy turn-around time, τ_{eddy}, and axes are vertical mode number, M, and horizontal isotropic wavenumber, K. Contour values are linear over the range of values at each frame. The first radius of deformation is at wavenumber 15. From Smith and Vallis (1998).

be modified. At the time of writing (2003) this is an open question and, although recent numerical simulations and various theoretical models do suggest that eddies play a role in setting thermocline structure, the picture is not wholly clear.

4.1. CRUDE *A PRIORI* SCALING

Consider first the advectively dominated upper thermocline. An estimate of the importance of mesoscale eddies arises may be roughly estimated by comparing the sizes of the two terms

$$\nabla \cdot \overline{\boldsymbol{u}'b'} \qquad \text{and} \qquad \nabla \cdot (\overline{\boldsymbol{u}}\overline{b}). \tag{43}$$

If we take $b' \sim \delta \boldsymbol{x} \cdot \nabla \overline{b} \sim L'\Delta b/L$, where Δb and L refer to the scales of mean quantities, then the ratio of the magnitude of the mean advection to

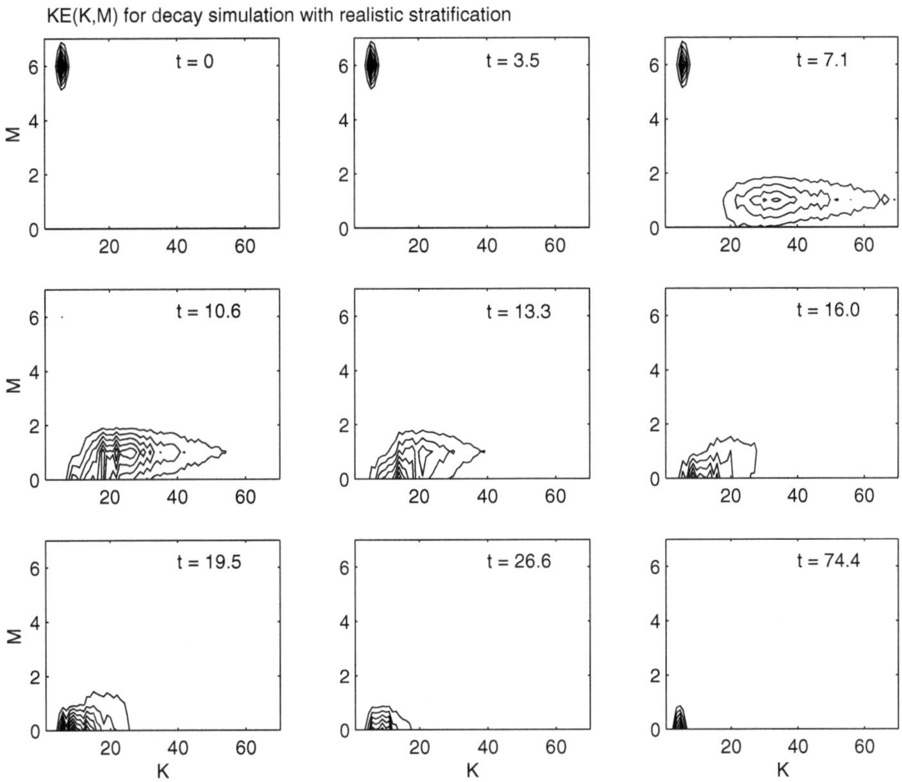

Figure 5. Time sequence of kinetic energy spectra in a nearly inviscid stratified quasi-geostrophic simulation with enhanced stratification in the upper ocean, representing a thermocline. Times are given in terms of eddy turn-around time, τ_{eddy}, and axes are vertical mode number, M, and horizontal isotropic wavenumber, K. Contour values are linear over the range of values at each frame. The first radius of deformation is at wavenumber 24. From Smith and Vallis (1998).

that of the eddy advection is characterized by an eddy Peclet number

$$Pe = \frac{UL}{U'L'} \tag{44}$$

where U' is the eddy velocity scale, U is the velocity scale of the mean flow, and L' and L are the length scales of the eddies and mean flow, respectively. The denominator is just an estimate of the size of an eddy diffusivity.

We can estimate the ratio L'/L if we take L' to be the deformation radius and if we assume, initially, that eddies have only a perturbative effect on the depth of the thermocline. The deformation radius is

$$L_d \sim \frac{ND}{f} \tag{45}$$

where D is the depth of the thermocline and N is the Brunt-Väisälä frequency. An estimate for this is $(\delta b/D)^{1/2}$ where D is the depth of the thermocline and $\Delta b \sim g\alpha\Delta T$ is the change in buoyancy across it. If D is given by (6) then we find

$$L_d \sim \left(\frac{W_E \Delta b L}{\beta f^2}\right)^{1/4} \sim \left(\frac{W_E \Delta b L^2}{f^3}\right)^{1/4} \tag{46}$$

where the second term holds for the planetary scale $\beta \sim f/L$. These are *a priori* estimates for the oceanic deformation scale. With the oceanically reasonable values of $g = 10\,\mathrm{m\,s^{-1}}$, $f = 10^{-4}\,\mathrm{s^{-1}}$, $\alpha = 10^{-4}\,\mathrm{K^{-1}}$, $L = 10^6\,\mathrm{m}$ and $\Delta T = 20$ K we find $L' \approx 50\,\mathrm{km}$ and $L'/L \approx 0.05$, with still smaller values for a (perhaps realistically) larger value of L. Observations suggest a similar value for the value of the deformation radius in the midlatitude ocean. Chelton *et al.* (1998), for example, find that the first deformation radius varies between about 20 km and 100 km in midlatitudes, with values of about 50 km being typical in the subtropical gyres. Eddies themselves tend to be rather larger than this, perhaps reflecting not just a (rather weak) inverse cascade (section 3) but also the tendency of the length scale of maximum instability to be larger than the deformation radius. In the Eady problem, for example, the length-scale of maximum instability is about four times larger than the deformation radius.

Reliable theoretical estimates of the ratio of the eddy to the mean velocity are harder to come by. The simplest assumption of all is to suppose that $U' \sim U$ (Green, 1970; Stone, 1972), although this does not take into proper account the turbulent properties of the flow: for example, the eddy velocity will increase as the extent of any inverse cascade increases. Held and Larichev (1996) do incorporate these properties using the machinery of geostrophic turbulence, but their arguments are not likely to be quantitatively valid in the ocean, because of vertical inhomogeneities in the stratification and horizontal inhomogeneities in the eddy field itself. Except perhaps in the Antarctic Circumpolar Current, it seems unlikely that a fully developed geostrophically turbulent inverse cascade exists in the ocean; in subtropical gyres, for example, eddies will develop in or near the strongly baroclinically unstable western boundary current or its extension (e.g. the Gulf Stream extension, the Kuroshio extension) but then may pass into less eddy rich regions. Eddies then find it more difficult to interact with each other, and the inverse cascade is inhibited.

Observations suggest that the ratio of U'/U ranges from about 1 to 5 in the subtropical gyre, corresponding to ratios of eddy-to-mean kinetic energy from 1 to 30 (e.g. Stammer, 1997). If L/L' is about 5 or smaller, then the eddy Peclet number may drop toward unity or smaller, indicating the importance of eddies. The conclusion of all this is that the eddy Peclet number is probably larger than unity over most of the subtropical gyre (and therefore mean effects locally dominate), but approaches unity in eddy rich regions, such as mode water regions near the western boundary layer. This means we cannot *a priori* eliminate the possibility that eddies play a significant role in setting the structure of the thermocline, but we should also expect (or perhaps at least hope) that the structure of the non-eddying models remains relevant in eddying models, and in the real ocean.

Evidently, the importance of eddies to the heat and momentum budgets of the ocean depends on their size — if the eddies are as large as the mean flow then their effects will be correspondingly large. Green (1970) suggested that the eddy mixing length should be the size of the baroclinic zone, rather than the eddy size, and this was used by Marshall *et al.* (2002) in their arguments. If we additionally take $U' \sim U$, then the eddy Peclet number given by (44) is obviously $\mathcal{O}(1)$ and with this scaling eddies play a first order role in thermocline dynamics, and in ocean dynamics in general. Marshall et al considered the thermodynamic balance of a warm lens created by Ekman pumping and a surface buoyancy flux. In this case, the incoming buoyancy flux must be balanced by the integrated outgoing eddy fluxes over the lens. Then, further assuming that

$$\overline{v'b'} \sim U\Delta b_{\text{lens}} \tag{47}$$

and using also the thermal wind relation (in cylindrical coordinates)

$$f\frac{\partial u}{\partial z} = \frac{\partial \Delta\Delta b}{\partial r}, \tag{48}$$

they obtained a scaling for the depth of the lens

$$h_{\text{lens}} \sim \left(\frac{f}{B}\right)^{1/2} W_E L, \tag{49}$$

where L is the size (e.g. radius) of the lens and B is the surface buoyancy flux. This is actually the same as the classical scaling (8) if we take $B \sim W_E \Delta b$. It is not surprising that we obtain the same result, because we have used the same parameters in the dimensional analysis, but the assumptions underlying the two scalings are quite different. This scaling is supported by laboratory and numerical results, although this is not perhaps a strong argument that the mixing scale in the real ocean is the scale of the baroclinic zone. That would only be the case if the eddy scale itself were the scale of the baroclinic zone, or if eddies were somehow organized on that scale to transfer properties efficiently across the baroclinic zone.

Regarding the lower, diffusive thermocline one may ask whether an internal boundary layer forms even in the presence of vigorous eddies. Scaling arguments themselves are not immediately elucidating on this matter. For example, we might compare the sizes of

$$\nabla \cdot \overline{\boldsymbol{u}'b'} \qquad \text{and} \qquad \kappa \frac{\partial^2 b}{\partial z^2} \tag{50}$$

The ratio of these terms is approximately

$$\frac{U'L'\delta^2}{\kappa L^2}, \tag{51}$$

where δ is a vertical scale. If the eddies have only a perturbative effect on the thickness of the thermocline, then we might use the vertical scale (14) or (18), but the resulting estimate is not very informative [although both scalings suggest a rather weak dependence on the value of the diffusivity itself, and no dependence at all if (18) is used]. We might, nevertheless, expect eddy effects to be *more* important in the lower thermocline than the upper part, at least away from the surface (in the near surface region, eddies have an important diabatic effect, as we see later.) This may seem counter-intuitive since eddies tend to be strongest near the surface; however, over their lifecycle eddies will, even if somewhat inefficiently, tend to barotropize or at least put energy into the first vertical mode, so distributing eddy energy over the entire depth of the thermocline. The intensity of the mean flow, however, falls off fairly rapidly with depth, so that the eddy Peclet number may similarly fall with depth, bringing with it the possibility of a subsurface balance involving eddy, mean and diffusive terms. The precise balance achieved becomes a quantitative issue which probably cannot be addressed by armchair reasoning, and both observations and high resolution numerical simulations will be needed to definitively settle the issue.

In spite of this quantitative uncertainty, if eddies are even moderately active in the internal thermocline then they are likely to lessen the impact of diffusion in setting its structure. To see this, consider a model ocean in which diffusion is small, and in which eddies are not present. As in section 2 this leads to a thin diffusive internal thermocline. If eddies are now allowed to form, and if their leading order effect is to diffuse potential vorticity (or, approximately, thickness) along isopycnals, then the internal boundary layer will generically thicken, as illustrated in Figure 6. The thickness flux is subducted from the ocean mixed layer where thickness is 'generated' by diabatic ocean-atmosphere interactions. Eddies almost certainly play a role in this, but the important point is that there can be subduction of water mass into the internal thermocline. Because the original thickness in the noneddying case was such as to balance diffusion with mean advection, then in a thicker thermocline the diffusive term (proportional to $\partial^2 b/\partial z^2$) becomes correspondingly less important, and an internal boundary layer

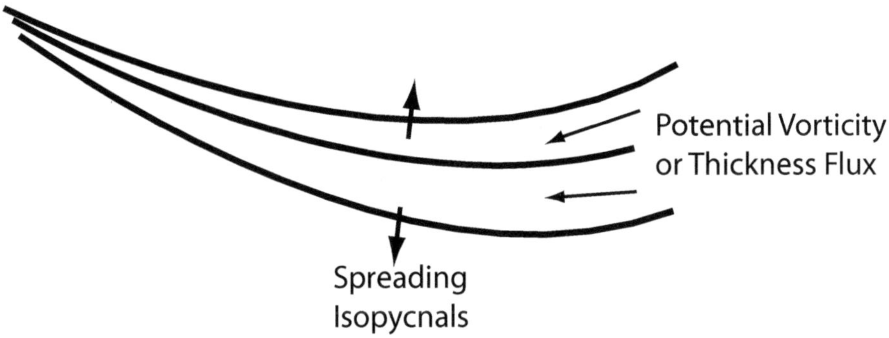

Figure 6. Schematic of the possible thickening of the internal thermocline by eddy effects. If mesoscale eddies tend to diffuse potential vorticity along isopycnals (solid lines) then if β is small they will tend to transport thickness from a region with greater separation between isopycnals to a region with less (in so far as is consistent with a loss of available potential energy), spreading the isopycnals and thickening the internal boundary layer.

need not exist in the sense that the thickness of the internal thermocline remains finite as diffusivity tends to zero. Of course, the eddies might affect the mean flow in such a manner that diffusion remains important and thus, again, the argument is heuristic.

4.2. NUMERICAL RESULTS

We will now briefly describe some numerical calculations of an idealized but eddying ocean. We shall be rather descriptive here — for more detail, see Henning and Vallis (2003). [Radko and Marshall (2003) have also explored the influence of eddies on the thermocline in idealized numerical experiments.] The simulations were carried out in a smaller and more idealized domain and at a rather lower horizontal resolution than is currently 'state-of-the-art', but were integrated for a sufficiently long period (typically $> 10^2$ years at eddying resolution, after a spin-up period of $> 10^3$ years at lower resolution) to allow both the dynamics and the thermodynamics to properly equilibrate.

The numerical model is a standard, primitive-equation Boussinesq z-coordinate model [the Modular Ocean Model (MOM) from GFDL, Pacanowski and Griffies (1999)], although a linear equation of state with no saline effects is used. It was configured in two domains as illustrated in Figure 7. One is an enclosed, rectangular box with wind and buoyancy forcing such as to produce a large subtropical gyre and a smaller subpolar gyre. Specifically, the surface temperature is relaxed back to a temperature that is zonally uniform but which diminishes linearly from low to high latitudes. The other domain was constructed so as to represent in a simple way the

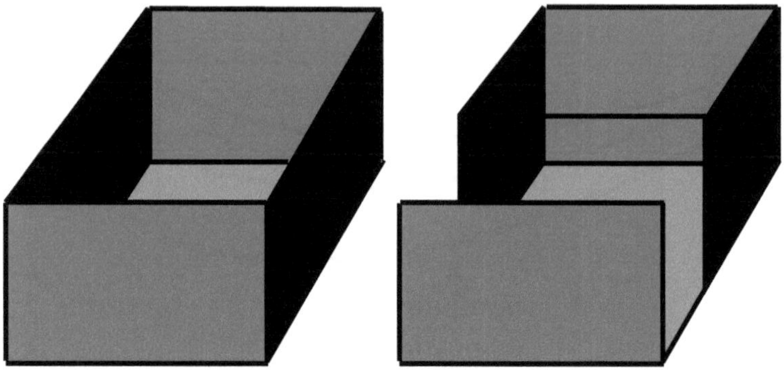

Figure 7. Domains used in primitive equation numerical experiments. Left: A box domain, with wind and buoyancy forcing such as to give a subtropical gyre and a smaller subpolar gyre. Right: a re-entrant channel replaces the subpolar gyre, providing a simple model of the Antarctic Circumpolar Current.

effects of the Antarctic Circumpolar current. Two sets of integrations were performed; one consists of low resolution, non-eddying integrations with a relatively large value of horizontal diffusivity. These integrations evolved into a completely steady state (except for some small oscillations on the decadal to century timescale) typically after an integration period of a few thousand years. The second set consists of eddy resolving integrations, with a horizontal resolution typically of $1/4°$ or $1/6°$, and a significantly lower value of horizontal diffusivity and viscosity. About 25 vertical levels are used in all integrations. These experiments are typically initialized from an equilibrated lower resolution integration, and then run for an additional hundred years and in some cases much longer. They are vigorously eddying, with eddy kinetic energy several times the mean kinetic energy, and a typical snapshot is illustrated in Figure 8.

A section of the subtropical thermocline in a closed basin in eddying and noneddying runs is illustrated in Figure 9. Qualitatively the structure of the stratification is similar in the two simulations, except in and close to the western boundary layer (not shown). The region of 'mode water' in the noneddying simulations is also partially eroded by the eddies, which are particularly vigorous in this region as the available potential energy of the noneddying state is high. The eddies evidently are quite efficient in extracting this available potential energy, resulting in a more uniform stratification and lowered mean available potential energy. A typical term-by-term analysis of the thermodynamic equation through the center of the subtropical gyre (and a little distance from the western boundary layer) is illustrated in Figure 10. This shows that in the upper thermocline the mean advection terms are still dominant, just as in the classical ventilated picture,

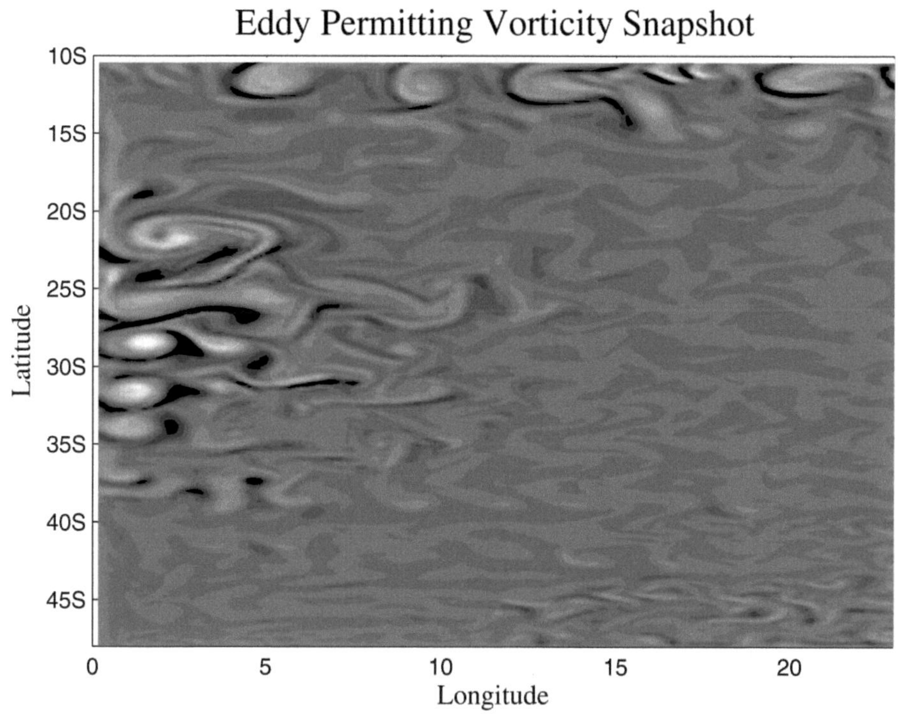

Figure 8. Snapshot of near surface vorticity in an eddying integration.

with eddy terms of secondary importance and diffusive terms negligible. Obviously, if the eddies were still more vigorous it would quantitatively affect this picture, but the upper thermocline would nevertheless remain an advective regime in the sense that explicit diffusion (i.e. the term $\kappa\partial^2 b/\partial z^2$) is small. However, in the near surface region the effect of the eddies is *diabatic* — that is to say, the eddies induce diapycnal fluxes — as can be seen from the transformed Eulerian mean (TEM) diagnostics of Figure 10. This might be expected because this is a region in which interactions with the atmosphere are important. The lower thermocline has a more complicated balance (in the simulations), involving mean flows, eddy terms and explicit diffusion, although the eddy effects are themselves largely adiabatic. Just as in the non-eddying case, a thermocline base that is distinct from the upper advective regime can still be identified. It is slightly thicker in the eddying case but, interestingly, the thickness still scales with the diffusivity according to $\kappa^{1/2}$, much the same way as the noneddying case. It is not known whether this scaling is universal in the eddying case.

Mesoscale eddies tend to have a larger influence in the simulations with a circumpolar channel (Figure 11). The stratification in the channel is quite

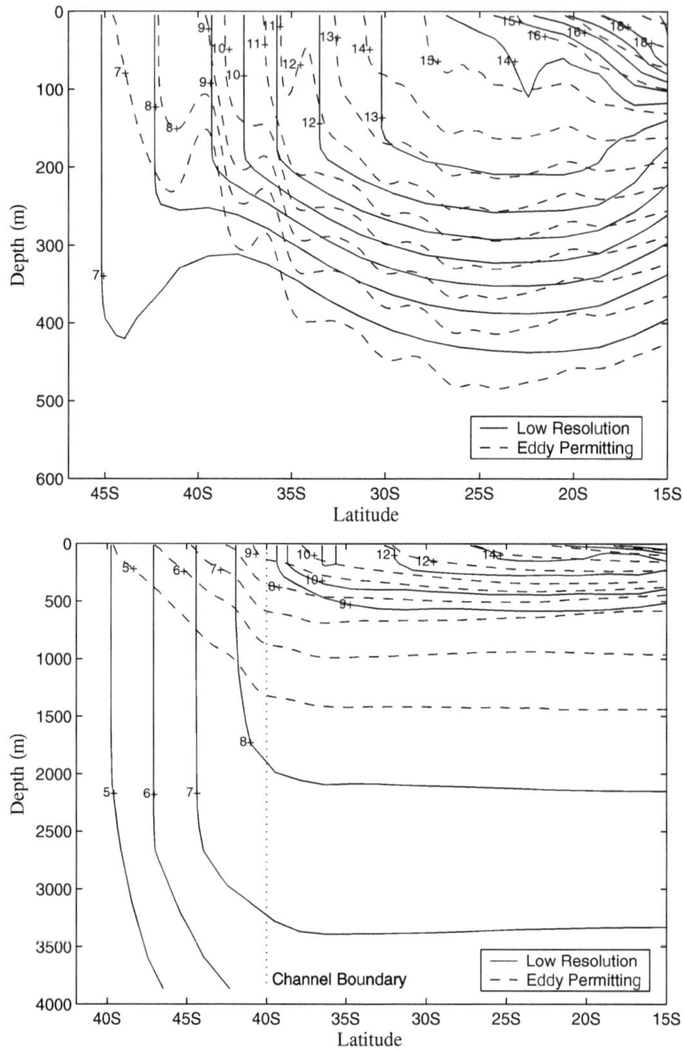

Figure 9. Top: Meridional section of the time averaged temperature in a non-eddying (solid line) and an eddying (dashed lines) simulation in an enclosed box with subtropical and subpolar gyres. Bottom: Similar meridional section, except for an integration with a periodic channel, as in the right-hand panel of Fig. 7.

different with and without eddies and this is at least in part because the non-eddying case has no equivalent of a ventilated thermocline solution, and produces a stratification with near vertical isopycnals that is highly baroclinically unstable. Such instability causes these isopycnals to slump, and the resulting stratification exhibits a balance between the buoyancy input at the surface and its lateral transport by baroclinic eddies [see also

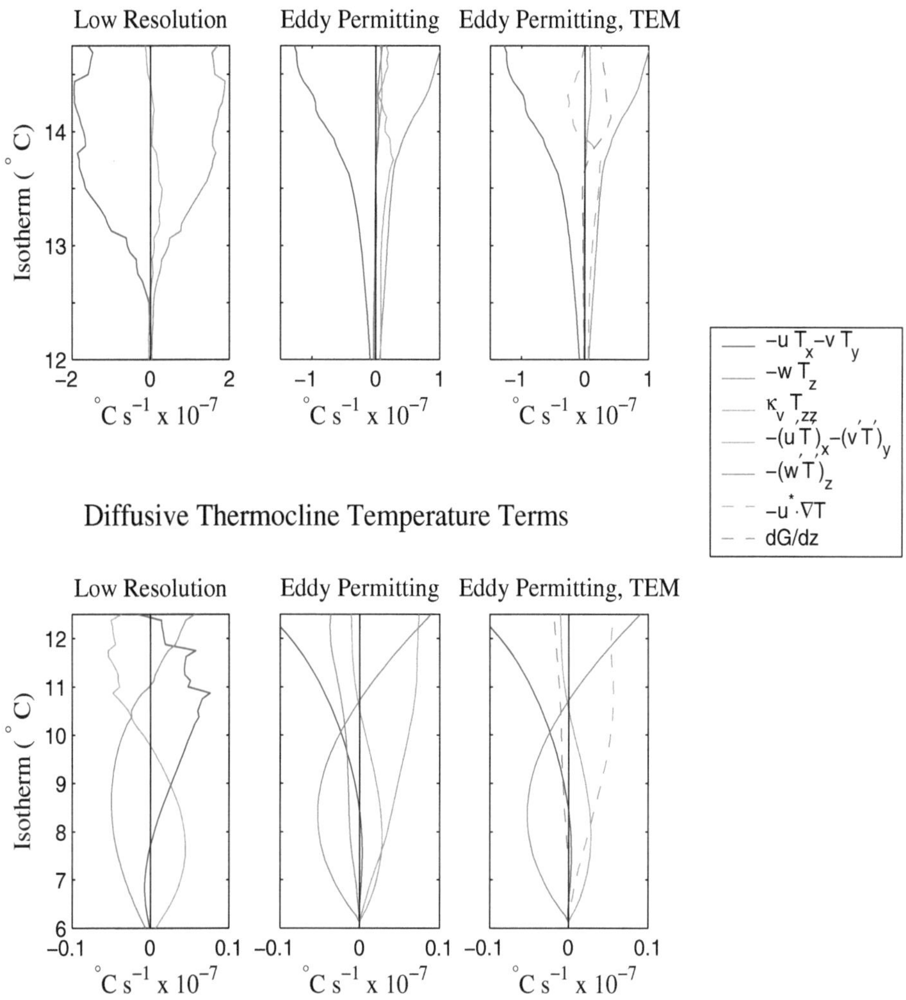

Figure 10. Profile of terms in the thermodynamic equation, plotted as a function of potential temperature, through the middle of the subtropical gyre in eddying and noneddying integrations. The upper portion is for isotherms that outcrop in the Ekman pumping region (ventilated thermocline) and the lower portion is for isotherms that outcrop in the subpolar Ekman suction region (the internal thermocline). The rightmost panels show the transformed Eulerian mean terms: G is the cross-isopcynal eddy buoyancy flux, so that the size of $\partial G/\partial z$ is a measure of the diabaticity of the eddy terms, which evidently is significant near the surface.

Karsten *et al.* (2002)]. Explicit dissipation (the $\kappa\partial^2 T/\partial z^2$ term) plays no direct role in this, at least locally. The presence of the circumpolar channel greatly affects the adjacent subtropical thermocline, thickening its base and leading to a balance in the thermodynamics between mean advection and eddy fluxes, rather than mean advection and diffusion as in the classical subtropical gyre model. [Gnanadesikan (1999) also noted the effect of the ACC on the deep stratification of the ocean elsewhere.] The thickening might be interpreted as a potential vorticity or thickness transport (and if the β effect is small, the two are similar) from the circumpolar channel, similar to that illustrated in Figure 6.

5. Concluding Remarks

Several decades of effort have finally yielded a good, although still not complete, understanding of the subtropical thermocline in a world without baroclinic instability; that is, in a world essentially governed by the planetary geostrophic equations. The picture that has emerged is one of an advectively dominated upper thermocline (the ventilated thermocline) and an advective-diffusive base (the internal thermocline), although important aspects of this quasi-laminar picture are still poorly understood — for example the formation and properties of warm pools of low potential vorticity water ('mode water'). The degree to which this picture should be considered an accurate or quantitative model of the circulation depends to a large degree on the role of mesoscale eddies — do they have only a perturbative effect, or do they completely dominate the circulation, or is it somewhere in between? A definitive answer to this is hard to give, since it will depend at least in part on the results of high resolution numerical simulations as well as observations. The current set of simulations we have described do perhaps allow us a glimpse of the truth, however, and some preliminary conclusions are:

1. In the upper thermocline of the subtropical gyre eddies tend to be most vigorous in or near the western boundary current and in regions of 'mode water.' Eddies may be the dominant factor in setting thermocline structure in these regions, although one should remember that it is the structure of these regions as described by noneddying solutions that at least in part leads to the formation of the eddies.

2. Away from the western boundary current and mode water, the signature of the classical ventilated thermocline can be seen, even in strongly eddying regions, at least in the simulations we have performed so far. The lower thermocline exhibits a complex balance involving eddies,

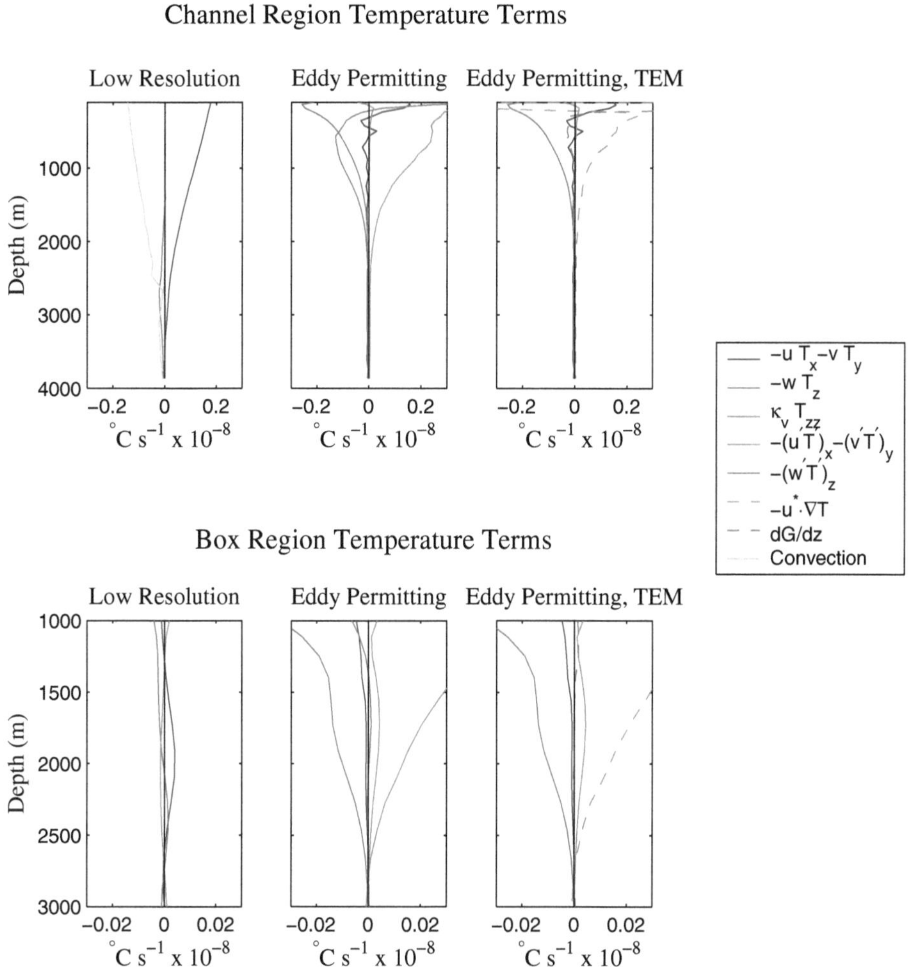

Figure 11. Profile of averaged terms in the thermodynamic equation in the integrations with a circumpolar channel for a low resolution case (left), an eddy permitting case (center) and the eddy permitting case using the TEM form for the eddy terms (right). The upper portion shows the temperature equation terms through the channel region, and the lower portion shows the temperature equation terms in the adjacent subtropical gyre, spanning the depth of those isotherms that outcrop within the channel. Eddy terms are significant nearly everywhere in the channel, and in the lower thermocline of the adjacent subtropical gyre.

mean flow and diffusion. The importance of eddies is a quantitative issue in these regions, and may differ from basin to basin.

3. Eddies are ubiquitous in simulations with a re-entrant channel, and they are the controlling effect in setting the upper ocean stratification there, strongly suggesting that mesoscale eddies are of overwhelming importance in setting the stratification of the Antarctic Circumpolar Current. The effect of such eddies spreads underneath the adjacent subtropical gyres, thickening the thermocline and reducing the importance of diffusivity in setting thermocline thickness.

4. Near the surface eddies induce diapycnal fluxes and have strong diabatic effects, but in the interior their influence is largely adiabatic.

Even as we answer questions, new ones arise and the general effect of mesoscale eddies on the ocean circulation is very much an open problem. Nevertheless, even without fully solving this grand 'problem of turbulence' some more specific problems in ocean circulation may be tractable. Among these are:

1. The thermocline near the western boundary current seems likely to be completely controlled by eddy effects, and we do not have a good understanding of this. Do eddies influence the separation of the western boundary currents? And how does the western boundary current regime transition into the midocean regime, where eddy effects may be secondary?

2. The dynamics of mode water is poorly understood in both noneddying and eddying cases, and it seems likely that an understanding of the latter case will depend on an understanding of the former. In the classical picture the lowest ventilated layer readily forms into a thick thermostad, but this is quite baroclinically unstable and in the eddying simulations we have performed it is apparently partially eroded away. Nevertheless, warm pools of low stratification exist in the real ocean. To understand this we may need to revisit parameterized models, for example Dewar (1986). Perhaps seasonal effects really are important in mode water maintenance?

3. What is the role of explicit diffusion in the interior of an eddying ocean? Eddy effects tend to diminish the role of diffusion in the local balance of terms in the thermodynamic equation, and baroclinic instability can produce its own vertical scale which can be of order 1 km. Yet on more fundamental grounds a finite diffusivity seems necessary to maintain a mean stratification away from a wind-driven layer, and to produce an overturning circulation.

4. How do eddies affect the abyss? The 'barotropization' of energy will lead to large vertical scales and interactions with bottom topography, and indeed eddies may equilibrate via bottom friction. Does deep eddy motion overwhelm the Stommel-Arons circulation? Might the latter still be apparent with very long time averaging?

5. How do mesoscale eddies interact with the mixed layer? Both theoretical reasoning and the numerical results described above suggest that mesoscale eddies have a diabatic effect in regions where they feel the surface. Perhaps the zeroth order effect of this can be modelled with a simple horizontal (and so cross-isopycnal) eddy diffusion, but a host of complicating factors (predicting mixed layer depth, temperature-salinity compensation, convection, predicting the size of any diffusivity) will make this far from simple.

Let me finish with a couple of subjective comments. First, I hope and expect to see much progress in these areas in the next few years, and some of that may of course determine that the above questions are the wrong ones. Second, I am struck by how important it is to have an understanding of the structure of the ocean in the absence of eddies in order to understand the structure of the ocean with eddies. One might take issue with this comment for the ACC, where the stratification seems largely determined by eddies, but the eddy-free basic state still provides a context for understanding eddy influence.

Acknowledgments
This paper reflects my own views and should not be regarded as a comprehensive review of thermocline theory. I am grateful for interactions with Roger Samelson, K. Shafer Smith and Cara Henning, for the comments of John Marshall and an anonymous reviewer, and for the financial support of NSF and NOAA. I would also like to acknowledge the influence of Pedro Ripa on both pure and applied physical oceanography, and this paper is dedicated to his enduring memory.

References

Burger, A. P.: 1958, 'Scale considerations of planetary motions of the atmosphere'. *Tellus* **10**, 195–205.

Chelton, D. B., R. A. deSzoeke, M. G. Schlax, K. E. Naggar, and N. Siwertz: 1998, 'Geographical variability of the first-baroclinic Rossby radius of deformation'. *J. Phys. Oceanogr.* **28**, 433–460.

Colin-de-Verdiere, A.: 1986, 'On mean flow instabilities within the planetary geostrophic equations'. *J. Phys. Oceanogr.* **16**, 1981–1984.

Dewar, W. K.: 1986, 'On the Potential Vorticity Structure of Weakly Ventilated Isopycnals: A Theory of Subtropical Mode Water Maintenance'. *J. Phys. Oceanogr.* **16**, 1204–1216.

Gill, A. E., J. S. A. Green, and A. J. Simmons: 1974, 'Energy partition in the large-scale ocean circulation and the production of mid-ocean eddies.'. *Deep-Sea Res.* **21**, 499–528.

Gnanadesikan, A.: 1999, 'A simple predictive model for the structure of the oceanic pycnocline'. *Science* **283**, 2077–2079.

Green, J. S. A.: 1970, 'Transfer Properties of the Large-Scale Eddies and the General Circulation of the Atmosphere'. *Quart. J. Roy. Meteor. Soc.* **96**, 157–185.

Held, I. M. and V. D. Larichev: 1996, 'A Scaling Theory for Horizontally Homogeneous, Baroclinically Unstable Flow on a Beta-Plane'. *J. Atmos. Sci.* **53**, 946–952.

Henning, C. C. and G. K. Vallis: 2003, 'The Effect of Mesoscale Eddies on Ocean Stratification, parts I and II'. To be submitted to *J. Phys. Oceanogr.*

Huang, R. X.: 1988, 'On boundary value problems of the ideal-fluid thermocline'. *J. Phys. Oceanogr.* **18**, 619–641.

Karsten, R., H. Jones, and J. Marshall: 2002, 'The role of eddy transfer in setting the stratification and transport of a circumpolar current'. *J. Phys. Oceanogr.* **32**, 39–54.

Lionello, P. and J. Pedlosky: 2000, 'The role of a finite density jump at the bottom of the quasi-continuous ventilated thermocline'. *J. Phys. Oceanogr.* **31**, 212–225.

Luyten, J. R., J. Pedlosky, and H. Stommel: 1983, 'The Ventilated Thermocline'. *J. Phys. Oceanogr.* **13**, 292–309.

Marshall, J. C., H. Jones, R. Karsten, and R. Wardle: 2002, 'Can eddies set ocean stratification?'. *J. Phys. Oceanogr.* **32**, 26–38.

Pacanowski, R. C. and S. M. Griffies: 1999, 'The MOM3 Manual'. Technical report, NOAA/Geophysical Fluid Dynamics Laboratory.

Pedlosky, J.: 1987, *Geophysical Fluid Dynamics*. New York: Springer, 2nd edition.

Phillips, N. A.: 1963, 'Geostrophic Motion'. *Rev. Geophys.* **1**, 123–176.

Radko, T. and J. Marshall: 2003, 'Eddy-induced diapycnal fluxes and their role in the maintenance of the thermocline'. Submitted to *J. Phys. Oceanogr.*

Rhines, P. B.: 1977, 'The dynamics of unsteady currents'. In: E. A. Goldberg, I. N. McCane, J. J. O'Brien, and J. H. Steele (eds.): *The Sea*, Vol. 6. J. Wiley and Sons, pp. 189–318.

Rhines, P. B. and W. R. Young: 1982, 'Homogenization of potential vorticity in planetary gyres'. *J. Fluid. Mech.* **122**, 347–367.

Robinson, A. R. and J. C. McWilliams: 1974, 'The baroclinic instability of the open ocean'. *J. Phys. Oceanogr.* **4**, 281–294.

Robinson, A. R. and H. Stommel: 1959, 'The oceanic thermocline and the associated thermohaline circulation'. *Tellus* **11**, 295–308.

Salmon, R.: 1980, 'Baroclinic instability and geostrophic turbulence.'. *Geophys. Astrophys. Fluid Dyn.* **10**, 25–52.

Salmon, R.: 1990, 'The thermocline as an internal boundary layer'. *J. Mar. Res.* **48**, 437–469.

Samelson, R. M.: 1999, 'Internal Boundary Layer scaling in "two-layer" solutions of the thermocline equations'. *J. Phys. Oceanogr.* **29**, 2099–2102.

Samelson, R. M. and G. K. Vallis: 1997a, 'Large-scale circulation with small diapycnal diffusion: The two-thermocline limit'. *J. Mar. Res.* **55**, 223–275.

Samelson, R. M. and G. K. Vallis: 1997b, 'A simple friction and diffusion scheme for planetary geostrohic basin models'. *J. Phys. Oceanogr.* **27**, 186–194.

Smith, K. S. and G. K. Vallis: 1998, 'Linear wave and instability properties of extended range geostrophic models'. *J. Atmos. Sci.* **56**, 1579–1593.

Smith, K. S. and G. K. Vallis: 2001, 'The scales and equilibration of mid-ocean eddies: freely decaying flow'. *J. Phys. Oceanogr.* **31**, 554–571.

Smith, R. D., M. E. Maltrud, F. O. Bryan, and M. W. Hecht: 2000, 'Numerical Simulation of the North Atlantic Ocean at $1/10°$'. *J. Phys. Oceanogr.* **30**, 1532–1561.

Stammer, D.: 1997, 'Global characteristics of ocean variability estimated from regional TOPEX/Poseidon altimeter measurements.'. *J. Phys. Oceanogr.* **27**, 1743–1769.

Stommel, H. and J. Webster: 1962, 'Some properties of the thermocline equations in a subtropical gyre'. *J. Mar. Res.* **44**, 695–711.

Stone, P. H.: 1972, 'A Simplified Radiative-Dynamical Model for the Static Stability of Rotating Atmospheres'. *J. Atmos. Sci.* **29**, 405–418.

Vallis, G. K.: 2000a, 'Large-scale circulation and production of stratification: effects of wind, geometry and diffusion'. *J. Phys. Oceanogr.* **30**, 933–954.

Vallis, G. K.: 2000b, 'Thermocline theories and WOCE: A mutual challenge'. *WOCE Newsletter* **39**, 30–33.

Veronis, G.: 1969, 'On theoretical models of the thermocline circulation'. *Deep-Sea Research* **31** Suppl., 301–323.

Welander, P.: 1959, 'An advective model of the ocean thermocline'. *Tellus* **11**, 309–318.

Welander, P.: 1971, 'Some exact solutions to the equations describing an ideal-fluid thermocline'. *J. Mar. Res.* **29**, 60–68.

Young, W. R. and G. Ierley: 1986, 'Eastern boundary conditions and weak solutions of the ideal thermocline equations'. *J. Phys. Oceanogr.* pp. 1884–1900.

AN OVERVIEW OF THE PHYSICAL OCEANOGRAPHY OF THE GULF OF CALIFORNIA

M.F. LAVÍN AND S.G. MARINONE
Departamento de Oceanografía Física, CICESE
Ensenada, Baja California, México.

Abstract. The Gulf of California (GC) presents several oceanographic features that make it unique among semienclosed seas of similar latitude and dimensions, the most important being strong tidal mixing, some of it close to deep stratification. Three-dimensional numerical model results suggest that tidal mixing may be more important than the thermohaline circulation in causing the long-term residual circulation, which consists of outflow in the upper 200 m and inflow below, plus a seasonally-reversing surface layer. The GC is an evaporative basin, but in the mean it gains heat through the surface. Lacking a sill at the point of connection with the Pacific Ocean (PO), the GC is constantly shaken by a wide spectrum of signals coming from the PO, including tides, subinertial trapped waves of various frequencies and El Niño. The seasonal dynamics and thermodynamics of the GC are dominated by the PO, not by local wind or buoyancy flux. Local processes are important at shorter time scales and in altering the thermohaline characteristics of the upper-layer waters. Tidal currents generate internal tides, packets of solitons, and sea surface temperature fronts from which jets may form. Coastal upwelling also seems to generate jets that separate from capes, especially on the mainland coast. The mesoscale off-shore circulation in the GC consists of a series of basin-wide geostrophic gyres that reach below 1000 m; their effect on the mean and seasonal circulation and thermodynamics of the GC remains to be studied. During summer, the currents in the mainland continental shelf are due to coastal trapped waves, while during winter they are wind-driven. The most important interannual anomalies in the GC are due to El Niño.

Key words: Gulf of California, physical oceanography

1. Introduction

We present in this article an overview of the current understanding of the Physical Oceanography of the Gulf of California (henceforth GC), which has advanced considerably in the last two decades. The normal variety of observational and modelling techniques have been used, including direct and satellite measurements, data bank mining and 3D numerical modelling.

O.U. Velasco Fuentes et al. (eds.), Nonlinear Processes in Geophysical Fluid Dynamics, 173-204.

2. Physiography and water masses

The GC is 1400 km long and its width in the inner region is 150 – 200 km (Figure 1). In this article, the Gulf of California is divided into several provinces, indicated in Figure 1: (a) The entrance zone, in open communication with the Eastern Tropical Pacific Ocean (ETPac) through a line from Cabo San Lucas to Cabo Corrientes ("the outer mouth"). (b) The Southern Gulf of California (SGC) covers from the Cabo San Lucas-El Dorado line ("the inner mouth") to just south of the sills of the large islands. (c) The archipelago, or midriff islands zone has several narrow channels and sills whose maximum depths are between 300 and 600 m (Figures 1 and 2). (d) The Northern Gulf of California (NGC), which has shelf sea the characteristics. (e) The Upper Gulf of California (UGC) is the very shallow (depth <30 m) province north of 31°N.

The laterally-averaged vertical distribution of the water masses in the GC is sketched in Figure 2, together with the water-mass classification of Torres Orozco (1993); the T-S diagram from most of the available data inside the GC is shown in the inset of Figure 1. The water masses of the GC have been reviewed by Bray and Robles (1991) and Badan-Dangon (1998).

The Pacific Deep Water (PDW) has a salinity that increases toward the bottom. The Pacific Intermediate Water (PIW) is characterized by a salinity minimum ($S_{min} = 34.50$, inset in Figure 1) centered at about 900 m. The Subtropical Subsurface Water (StSsW) is found approximately between 150 and 500 m; in the ETPac it has a well marked salinity maximum in the thermocline. The Tropical Surface Water (TSW) is formed from StSsW by upwelling in the ETPac, where its salinity is lowered to 33 by the excess of rainfall over evaporation (Wyrtki, 1966, 1967), but off SW Mexico its salinity exceeds 34. The California Current Water (CCW: $S < 34.5$, $12 \leq T < 18$°C) does not appear in Figure 2 because of its volume inside the gulf; its presence inside the gulf has always been reported close to the mouth.

The Gulf of California Water (GCW) has salinity ≥ 35, but because of its high temperature, it is always found in the upper layers. The ultimate main source water for the GCW is the Subtropical Surface Water of the central South Pacific, either via StSsW or TSW. The most important transformation is probably StSsW + evaporation $\Rightarrow$ GCW (Bray, 1988b; Torres Orozco, 1993; Lavín *et al.*, 1995); it occurs mainly in the northern GC where the process is aided by vertical mixing produced by tidal currents, wind and winter convection.

Some winters, a fraction of the GCW formed in the NGC undergoes bottom water formation, evidenced by bottom salinity maxima in Wagner basin, which extend further south at mid-depth (Alvarez-Borrego and

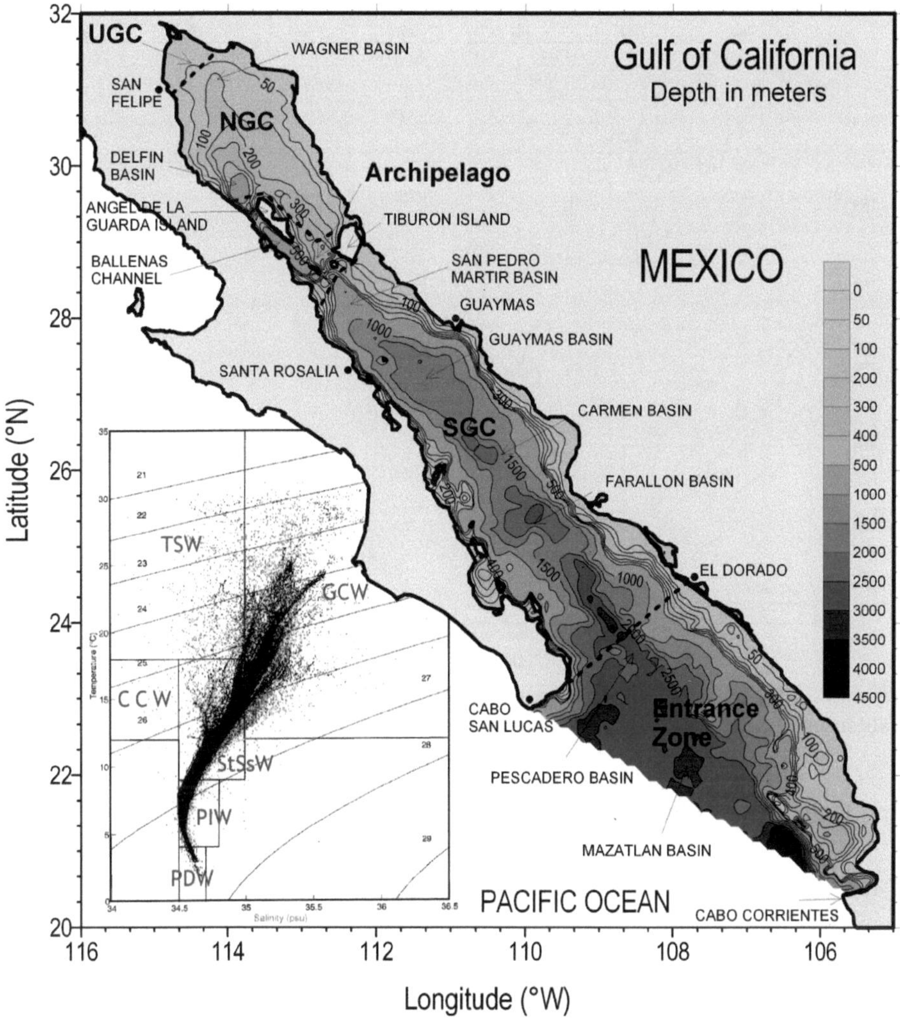

Figure 1. Bathymety of the Gulf of California (depth in meters), with named places and basins. Inset: Temperature-Salinity diagram with all the GC CTD data from 1939 to 1993. SGC, NGC and UGC, see text.

Schwartzlose, 1979; Bray, 1988b; Lavín et al. 1995; López, 1997). The process appears to occur during events of intense northwesterly winds, whose cold dry air enhances evaporation and surface heat loss (Bray, 1988a; Reyes and Lavín, 1997; López, 1997).

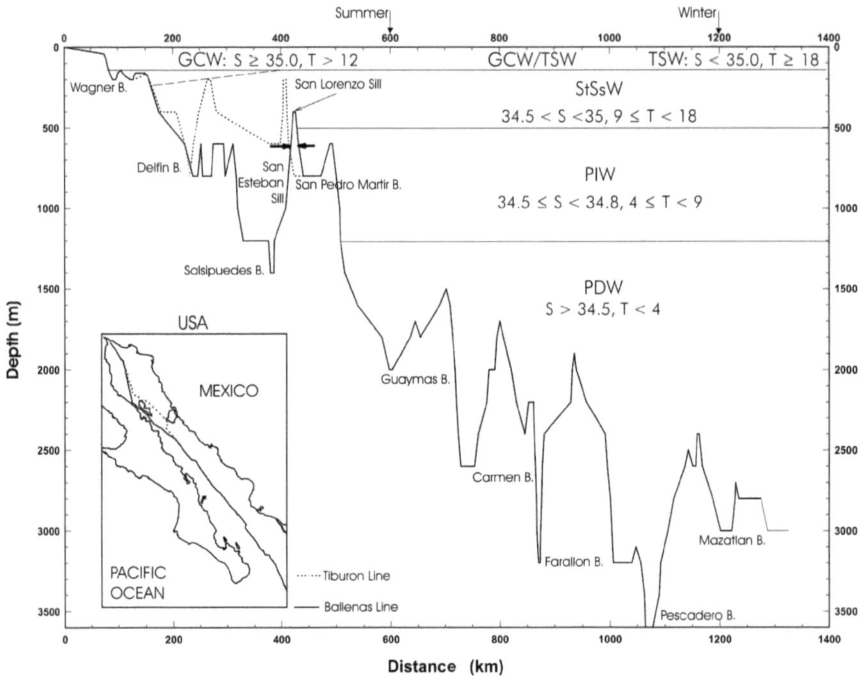

Figure 2. Sketch of the laterally-averaged distribution of water masses in the Gulf of California. Bathymetry along the lines shown in the inset map.

3. Tides and Tidal Mixing

Tides in the GC are produced by co-oscillation with the tides of the Pacific Ocean (Filloux, 1973; Ripa and Velázquez, 1993). The observed cotidal charts of the main tidal constituents inside the GC (Morales and Gutiérrez, 1989) are very well reproduced by barotropic numerical models (Marinone, 2000, and references therein). The semidiurnal components show amplification in tidal height toward the head of the gulf; for M2 from 36 cm in the gulf entrance to 150 cm in the UGC, with a minimum in the central gulf of 5 cm where a virtual amphidromic region is found. The amplification of the semidiurnal frequencies occurs because the GC is almost resonant at those frequencies. By contrast, the diurnal components are basically in phase in the entire gulf and amplitude increases toward the head by continuity.

As a consequence of the different character of the diurnal and semidiurnal components, the tides in the gulf are mixed; mainly semidiurnal in the northern and southern zones and diurnal in the central gulf. Figure 3 shows the form factor, defined as [A(O1)+A(K1)]/[A(M2)+A(S2)], where A(..) is the amplitude of the indicated constituent. There is a large springs-neaps difference in tidal range (Figure 3b and c), especially in the UGC.

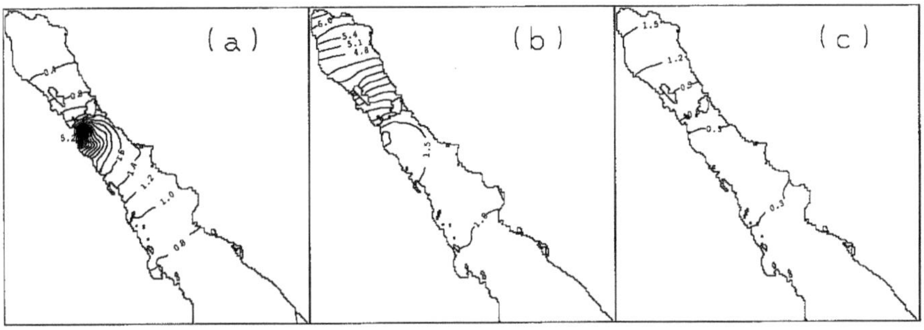

Figure 3. Characteristics of the tides in the Gulf of California, from the three-dimensional numerical model of Marinone (2003). (a) Form factor. (b) Tidal range (m) during spring tides. (c) Tidal range (m) during neap tides.

The tidal ellipses from barotropic models [for comparisons vs. observations, see Argote *et al.* (1995) and Marinone (2000)] are very rectilinear and mainly oriented along the length of the gulf. The orientation of the surface M2 tidal ellipses from the 3D baroclinic numerical model of Marinone (2003) is not very different from that of homogeneous models. In the SGC the M2 tidal currents are only a few cms^{-1}, and they are strongest (60 cms^{-1}) in the archipelago and in the UGC. Tidal currents are fortnightly modulated (Badan-Dangon *et al.*, 1991a; Marinone, 1997), and observations show important variations with depth (Ramírez Manguilar, 2000; Marinone, 2000; López and García, 2003; Jiménez Lagunes, 2003).

The fact that the amphidromic region is not in the center of Guaymas basin means that part of the semidiurnal tidal energy is lost by bottom friction in the northern part. The rate of energy dissipation by bottom friction of the M2 tidal constituent in the GC is around 4.3×10^9 W, from tidal observations (Filloux, 1973). Calculations based on numerical models, made by several authors, range between 3 and 9×10^9 W for M2. Most of the dissipation occurs in the archipelago, especially over the sills, and in the UGC (Argote *et al.*, 1995; Marinone, 1997; Carbajal and Backhaus, 1998; Marinone, 2000).

A fraction of the energy extracted from the tidal currents by bottom friction is used for vertical mixing, therefore tidal mixing is important in the areas where tidal dissipation is high: the UGC, the archipelago and the shelf south of Tiburón Island. Mixing can also occur in the interior of the fluid, by baroclinic processes, like breaking internal waves. Marinone and Lavín (this volume) find that there are conditions for strong internal mixing over the sills and in the surrounding area. Both internal and bottom mixing are tidally modulated, at the diurnal, semidiurnal and fortnightly frequencies, which cause variations in the distribution of sea surface temperature (SST) and stratification around the islands (Paden *et al.*, 1991; Simpson *et al.*,

1994; Argote *et al.*, 1995). The areas of strong tidal mixing are the most biologically productive of the GC (Alvarez-Borrego and Lara-Lara, 1991). *Internal tides and solitons.* As the tide passes over the sills the tidal currents accelerate, and internal hydraulic processes distort the thermocline (Paden *et al.*, 1991) which causes part of the tidal energy to be radiated as internal waves. Filonov and Lavín (2003) reported the presence of semidiurnal, diurnal and quarterdiurnal internal tides, apparently propagating north from San Esteban sill. The semidiurnal internal tide is the most energetic (45 of the energy of the barotropic tide), with wavelengths of 10-40 km and phase speeds of 0.4-0.9 ms^{-1}. The semidiurnal internal wave currents are aligned with the gulf axis and have amplitude of 0.10-0.15 ms^{-1}. Jiménez-Lagunes (2003) found similar results for the semidiurnal and diurnal internal tides in the NGC, and also that inertial oscillations were generated by hurricane Nora in September 1997. The numerical models of Beier (1999) and Marinone (2003) show that internal tides are important enough to alter the smooth flow of the barotropic tide, which explains the apparently random orientation and sense of rotation of the observed M2 current ellipses in the NGC (Ramírez Manguilar, 2000; Marinone, 2000).

Also radiating from the sills are high-frequency soliton-like internal wave packets (Fu and Holt, 1984), with a 1-2 km wavelength, which travel along the thermocline in the same direction as the internal tides, at a speed of roughly 1.2 ms^{-1}. They are generated over the sills mainly in spring tides, and carry an estimated 10 of the barotropic tidal energy. Direct observations of these solitons show vertical displacements of the isotherms of the order of 20-50 m with periods of 40-50 minutes (Badan-Dangon *et al.*, 1991a; Gaxiola *et al.*, 2002).

4. Average and seasonal dynamics and thermodynamics

The upper layers of the GC are strongly seasonal in circulation and in thermohaline structure (sea surface temperature and salinity, water masses, stored heat, surface mixed layer depth, etc.). This behavior is a response to the seasonality of the main forcing agents, namely: the Pacific Ocean, the wind system, and the fluxes of heat and moisture.

4.1. FORCING AGENTS

(i) The Pacific Ocean. The water masses around the entrance of the GC are moved by the ETPac system of oceanic currents, which have a strong seasonal behavior, a reflection of the meteorological seasonal changes at Pacific Ocean scale (Wyrtki, 1965, 1966; Baumgartner and Christensen, 1985; Fiedler, 1992; Strub and James, 2002a). The seasonal movement of the

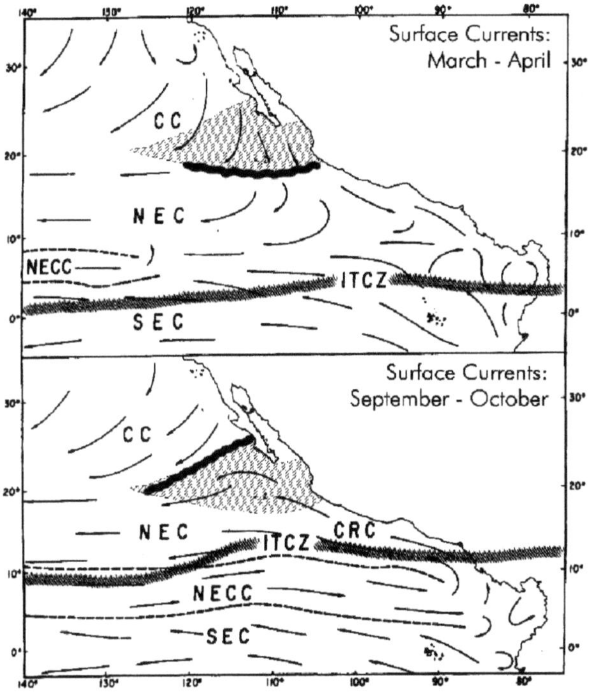

Figure 4. Surface currents in the Eastern Tropical Pacific in (a) March-April and (b) September-October. Arrows indicate the main surface currents: SEC= South Equatorial Current, NECC= North Equatorial Counter Current, NEC= North Equatorial Current, CRC= Costa Rica Current, CC= California Current. ITCZ= Inter-Tropical Convergence Zone. The limits of the California Current in (a) and (b) are marked by a wavy band. Based on Wyrtki (1965) and Baumgartner and Christensen (1985).

Inter-Tropical Convergence Zone (ITCZ) imposes latitudinal displacements of the equatorial current system (Figure 4), which determine in particular how far south the California Current will reach and how far north the Costa Rica Coastal Current will carry TSW (Wyrtki, 1967; Fiedler, 1992).

The sea level of the ETPac shows a seasonal signal, which has been identified in tide gauge data (Ripa, 1997) and in satellite altimeter data (Strub and James, 2002a). This seasonal signal moves toward the pole along the coast, from SW Mexico to Alaska (Strub and James, 2002a).

(ii) Wind. The wind in the GC has a marked seasonal behavior, product of the seasonal changes of atmospheric pressure centers in its vicinity and the channeling effect of the mountain chains in both sides: in autumn, winter and spring the wind blows from the NW with a speed of 8 to 12 ms^{-1}, while in summer it blows from the SE with a mean speed of ≤ 5

ms^{-1}(Roden, 1964; Badan-Dangon *et al.*, 1991b; Merrifield and Winant, 1989; Badan, this volume). Recent wind velocity observations from the NSCAT and QuickScat satellite scatterometers (Parés *et al.*, 2003) suggest that the summer reversals may not occur in the NGC and only for periods of a few days in the SGC, which would explain their absence in quarterly climatologies from ship reports (Fiedler, 1992).
(iii) Surface fluxes. Figure 5 shows the monthly means of heat fluxes in Guaymas basin and in the NGC calculated with bulk formulae (Castro et al., 1994). Marked seasonality is apparent in all cases, which is due to the seasonality of the meteorological and astronomical variables. Solar heating is the largest term, and the largest heat loss is due to evaporation, followed by long wave radiation. The annual fit to the surface heat flux for the entire GC (Castro et al., 1994) is

$$Q_s = \{18 + 20\cos(\omega t - \phi)\} \times 10^{12} W, \tag{1}$$

where ω is the annual frequency and $\frac{\phi}{\omega}$ is June 9. The net heat flux Q_s has, in addition to the seasonal cycle, two important features (Bray, 1988a; Lavín and Organista, 1988; Ripa and Marinone, 1989; Paden *et al.*, 1993): (i) unlike other evaporative marginal seas, the GC gains heat through the surface in the annual mean, and (ii) there are net heat losses in the NGC in November and December.

Evaporation (E) is an important process in the GC, being responsible of the high salinity of the GCW. Precipitation (P) is negligible in the NGC and has a summer maximum in the southern part. The mean monthly values of E-P for the NGC, the archipelago and the SGC, present annual and semiannual variations (Roden, 1958; Ripa and Marinone, 1989; Romero Centeno, 1995; Beron-Vera and Ripa, 2002).

4.2. THERMODYNAMICS

(i) Heat balance. The heat balance for the GC, in the average and in the seasonal time scale, was first made by Castro et al. (1994) and in more detail by Beron-Vera and Ripa (2000). They used CICESE's hydrographic data bank to calculate the heat content of the GC in the upper 400 m, $\mathcal{H} = \int_v \rho c_p T dv$ where V is the volume. The heat balance for the entire gulf is:

$$\frac{\partial \mathcal{H}}{\partial t} = \mathcal{Q}_s + \mathcal{F}_H, \tag{2}$$

where $\mathcal{Q}_s$ is the surface heat flux and $\mathcal{F}_H$ is the heat flux through the mouth. Figure 6a shows the different terms of the heat balance (2). In average $18 \times 10^{12} W$ enter the gulf through the surface and exit trough the mouth. The seasonal variation is very large (larger than the mean), and

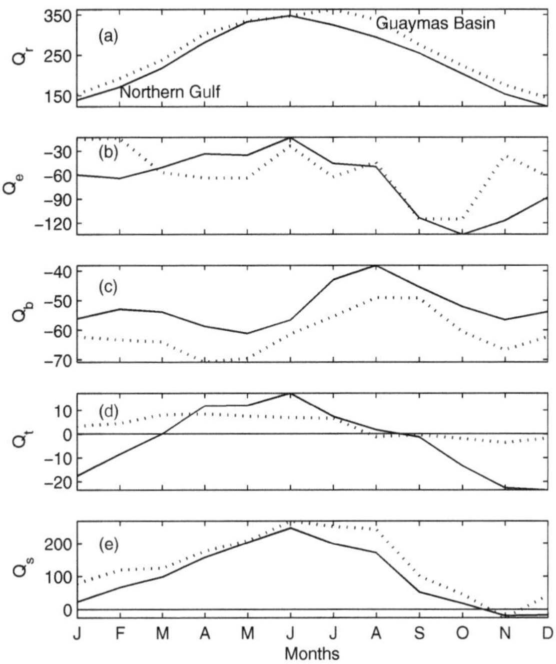

Figure 5. Monthly average surface heat fluxes (in W m^{-2}) in the NGC (continuous line) and in Guaymas basin (dotted). (a) Net short wave radiation, (b) Latent heat flux, (c) Back radiation, (d) Sensible heat flux, (e) Total surface heat flux. Data from Castro *et al.* (1994).

the amplitude of the seasonal horizontal heat flux is twice that through the surface. This means that the seasonal heat exchange between the GC and the PO is more important than the surface heat flux.

(ii) Salt balance. Following Beron-Vera and Ripa (2002), let $<\mathcal{S}> = V^{-1}\int_v Sdv$ be the average salinity of the upper 400 m of the GC, where V is the volume. Then the salt balance is

$$\frac{\partial<\mathcal{S}>}{\partial t} = <\mathcal{S}>(\mathcal{E}-\mathcal{P}) + \mathcal{F}_S, \tag{3}$$

where $\mathcal{E}-\mathcal{P}$ is evaporation minus precipitation and $\mathcal{F}_S$ is the salt flux through the mouth. Figure 6b shows the different terms of equation (3). Although the uncertainties of this balance are larger than those for heat, it is found that the flux of salt through the mouth is more important than evaporation in controlling the mean salinity (Beron-Vera and Ripa, 2002).

(iii) Sea level. The sea level in the GC has a seasonal variation, of amplitude ~15 cm (Figure 7a), with maximum elevation in summer and minimum in winter. Although atmospheric pressure accounts for a little of this seasonal variation, most of it is due to changes in heat and salinity of the water

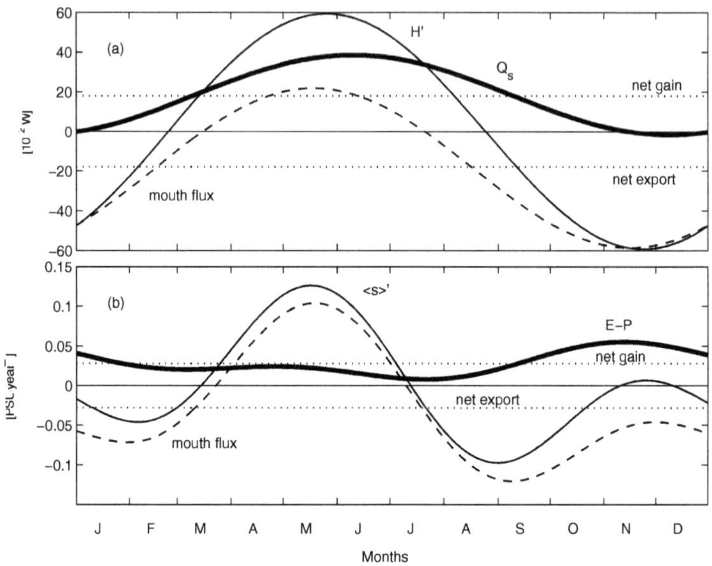

Figure 6. Overall Gulf of California mean and seasonal balances of (a) heat (after Castro *et al.*, 1994), and (b) salt (after Beron-Vera and Ripa, 2002).

column (Roden and Groves, 1959; Ripa and Marinone, 1989; Ripa, 1990 and 1997). Therefore, in accordance with the heat and salinity balances, the seasonal sea level variation in the GC is mainly due to the fluxes of heat through the mouth. This is congruent with the PO forcing the seasonal sea level, and with the heat and salinity balances.

4.3. CIRCULATION

4.3.1. *Seasonal Circulation*

(i) Ripa's model. Figure 7a shows that there are differences in the amplitude and phase of the annual harmonic of the sea level in opposite sides of the GC. The sea level difference is a measure of the geostrophic velocity along the gulf, which is shown in Figure 7b (Ripa, 1997). The high correlation between sea level and geostrophic current apparent in Figure 7, and the lag between the two stations, suggest that there is an intrusion or perturbation forced at the mouth by the PO, which travels counterclockwise around the GC at $1.6ms^{-1}$, which is approximately the phase speed of a first-mode baroclinic wave traveling in a two-layer channel similar to the gulf. The hypothesis that the seasonal forcing of the GC by the PO is in the form of a baroclinic internal Kelvin wave of annual period is due to Ripa (1990, 1997), who also proved that the hypothesis can explain the annual signals of sea level and heat content. This hypothesis suggested that the seasonal

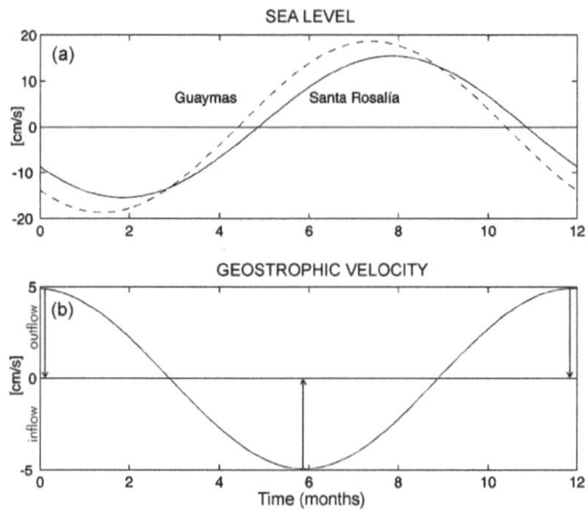

Figure 7. (a) Annual variation of the sea level in opposite sides of the Gulf of California: Guaymas in the mainland (dashed) and Santa Rosalía in the peninsula (solid). (b) Geostrophic velocity calculated from the difference in sea level between the two stations. Adapted from Ripa (1997).

circulation of the GC may be dominated by the PO, in a radical departure from the accepted idea that the wind was the main forcing agent (Roden and Groves, 1959; Bray, 1988a). The geostrophic circulation induced by this wave would be responsible for the seasonal signals of $\mathcal{F}_S$ and $\mathcal{F}_H$.

(ii) Observations. The first descriptions of the surface circulation in the GC, which were based on ship drift reports (U.S. Hydrographic Office, 1947; cited by Roden, 1958 and 1964), suggested influx in summer and outflow in winter, and a cause-effect relationship was suggested between the wind pattern and the surface circulation. The circulation pattern appeared to be supported and extended by the surface salinity distribution, which suggested that high salinity water (GCW) left the GC on the peninsula side and that its influence reached further in winter, while low-salinity water (TSW) was found entering the gulf along the eastern coast (Roden, 1964). Early geostrophic velocity calculations (Roden and Groves, 1959) also showed surface inflow in summer and outflow in winter.

The strength of the seasonal signal of geostrophic velocity, calculated from hydrographic data, was studied by Bray (1998a) for several cross-sections along the length of the gulf and by Marinone and Ripa (1988) and Ripa and Marinone (1989) for the Guaymas basin. A clear seasonal signal was found for the surface transport (with inflow in summer and outflow in winter) and for the wind stress. While Ripa and Marinone (1989) found that the lateral and temporal variability of the geostrophic velocity in Guaymas basin was so large that the lateral structure could not be unraveled, Bray (1988a) found it weakly cyclonic in average, cyclonic in summer with speeds of 0.2 to 0.3 ms^{-1} and anticyclonic ($\sim$ 0.2 ms^{-1}) in spring and fall.

A similar study, but for the inner mouth of the gulf, has been carried out by Castro *et al.* (2000), Castro Valdez (2001) and Mascarenhas *et al.* (2003), but using closely-spaced CTD stations from nine surveys. Clear seasonal patterns are observed in the fields of temperature, salinity and geostrophic velocity. In average there is outgoing flux of salty GCW water close to the peninsula and ingoing flux of fresher water along the mainland coast. The seasonal signal of the geostrophic velocity (37% of the variance) consists of a reversing pattern with inflow (outflow) in the central part and outflow (inflow) mostly on the peninsula side during May (September). The seasonal transports of heat and salinity calculated with these data are in good agreement with those estimated in the balances described above; this is a very important result, since it is an independent estimate of $\mathcal{F}_S$ and $\mathcal{F}_H$, and also gives support to the model of Ripa (1997), in that those fluxes are carried out by the geostrophic currents.

The seasonal circulation in the Northern Gulf is dominated by a seasonally reversing gyre, cyclonic from June to September and anticyclonic from November to April, with speed 0.35 ms^{-1} in both seasons; the transitions appear to last a couple of weeks. Direct evidence for this pattern was obtained from satellite-tracked drifters by Lavín *et al.* (1997a) and from current meter data by Palacios-Hernández *et al.* (2002). The corresponding hydrographic sections show that in summer the strong stratification is characterized by doming isolines, which leads to a low center in dynamic height and cyclonic geostrophic currents with surface speed comparable to that of the drifters, while in winter stratification is much weaker and the isolines are depressed in the center of the basin, which generates a high in dynamic topography. The persistence of the pattern has been proved by Bray (1988a) and Carrillo *et al.* (2002) from geostrophic calculations using hydrographic data banks, and from 14 years of AVHRR data by Soto-Mardones *et al.* (1999).

In winter the geostrophic speed is lower than that of the drifters or the current meters, which means that at least in winter the gyre is partly barotropic and partly baroclinic. The barotropic component appears to be caused by the interaction of the annual internal wave with the bathymetry and the coastline (Beier and Ripa, 1999).

(iii) Numerical models. Ripa (1997) used an analytical, linear, two-layer box model to illustrate the individual effects of the three forcing agents: the Pacific Ocean, the wind and the surface heat flux (Q_s), and to justify the applicability of a laterally-averaged model with laterally-averaged topography. The PO forcing is simulated as an internal Kelvin wave of annual period, trapped within an internal Rossby Radius (30 km), the wind was simulated by a seasonal cycle with NW winds in winter and SE winds in

summer, with maximum speeds of 5 ms^{-1} up-gulf in August, and Q_s was specified from Castro *et al.* (1994). The seasonal cycle of sea level and the seasonal heat balance are well reproduced. The contributions of the PO, the wind and Q_s to the seasonal sea level variability are 6.6 cm, 2.9 cm and 0.9 cm, respectively. The annual amplitude of the laterally-averaged geostrophic velocity is made up of 2.9 cms^{-1} from the PO, 1.0 cms^{-1} from the wind and 0.3 cms^{-1} from Q_s. The seasonal heat flux through the mouth is given by 33 TW from the PO, 10 TW from the wind and 3 TW from Q_s (1TW $= 10^{12}$ Watts). The conclusion is that the seasonal thermodynamics and circulation are not dominated by local processes (wind and Q_s), but driven by the Pacific Ocean.

The lateral structure of the circulation induced by the same three forcings was studied by Beier (1997) with a linear two-dimensional, two-layer baroclinic model with a 70 m thick surface layer and real topography. It is found that the wind also generates an internal Kelvin wave, with amplitude increasing toward the interior of the gulf. The results reproduce the seasonal sea level variation, and give support to the findings of Ripa (1997) regarding the relative importance of the forcing agents. Details of the lateral structure of the circulation are also obtained: the seasonally-reversing gyre in the NGC is reproduced, and in the SGC there is anticyclonic surface circulation in winter and a cyclonic circulation in summer. A non-linear, vertically-entraining version of the model was used by Beier (1999) and Palacios-Hernández *et al.* (2002) to demonstrate that the difference in stratification due to winter mixing is a possible cause for the inequality in the duration of the cyclonic and anticyclonic regimes in the NGC. The simulated seasonal circulation and sea surface temperatures are closer to observations than the model of Beier (1997). This was the first non-linear baroclinic model of the GC that included vertical mixing.

A 3D non-linear numerical model of the seasonal circulation and thermodynamics of the GC with real topography and vertical and lateral mixing has been developed by Marinone (2003) (for more details, see Marinone and Lavín, this volume). While the same sinusoidal seasonal wind is used, the other forcings differ in important ways from those used by Ripa (1997) and Beier (1997, 1999): (a) At the mouth the sea surface elevation is prescribed at tidal (7 harmonics), annual and semiannual frequencies. (b) Observations are used to specify the annual and semiannual variations of the temperature and salinity fields at the mouth cross-section. (c) At the surface, heat and fresh water fluxes were calculated with the model SST and bulk formulae using seasonally-fitted meteorological data.

This model reproduces the annual and semiannual variability of sea level, the balances of heat and salt, and the climatology of the sea surface temperature. In Figure 8 the surface currents predicted by the run with all

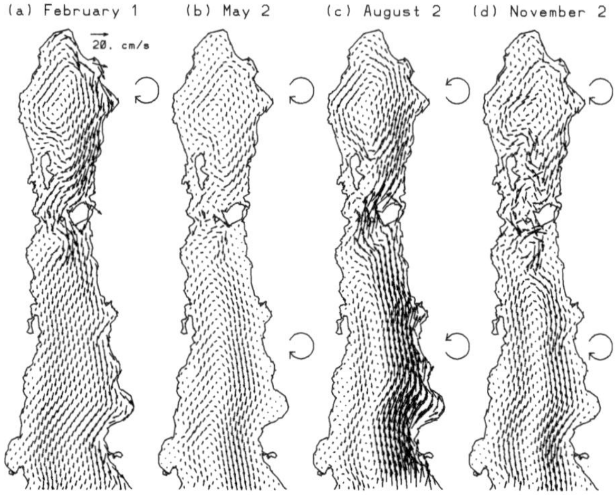

Figure 8. Residual currents for the surface layer (0-10 m) from the 3D model of Marinone (2003), for (a) February 1, (b) May 2, (c) August 2 and (d) November 2. The scale arrow is shown in (a). The gyres on the right of each panel are visual aids indicating the overall sense of rotation in the different sections of the gulf.

the forcings are shown for dates representative of the circulation regimes observed. The seasonally-reversing gyre in the northern GC is obtained. In the Ekman layer in the SGC, one cyclonic period occurs in August, while there are two anticyclonic periods, in May and November. In deeper layers (Marinone, 2003) the circulation is cyclonic during winter and summer, and anticyclonic during autumn and spring. The annual average surface current is shown in Figure 9a, and the first two modes of an Empirical Orthogonal Function (EOF) analysis of the surface circulation is shown in Figures 9b,c,d. The first mode shows a strong annual signal, with anticyclonic surface circulation from June to September, and cyclonic from October to April; this mode is asymmetrical, with strong currents on the mainland side and weak currents on the peninsula side (Figure 9b,d). In the SGC the surface circulation is due to both, the PO and the wind, and it presents a semiannual variability, captured by mode 2 (Figures 9c,d). Therefore the modelled seasonal circulation in the SGC seems to reflect the forcing functions used at the mouth; annual in Beier (1999) and annual plus semiannual in Marinone (2003).

4.3.2. *Annual Average Circulation*

The annual-mean circulation of the GC has been obtained from non-linear models by Beier (1999) and Marinone (2003). The surface circulation predicted by the model of Marinone (2003), shown in Figure 9a, presents an anticyclonic gyre covering the entire NGC, a cyclonic gyre north of Ángel

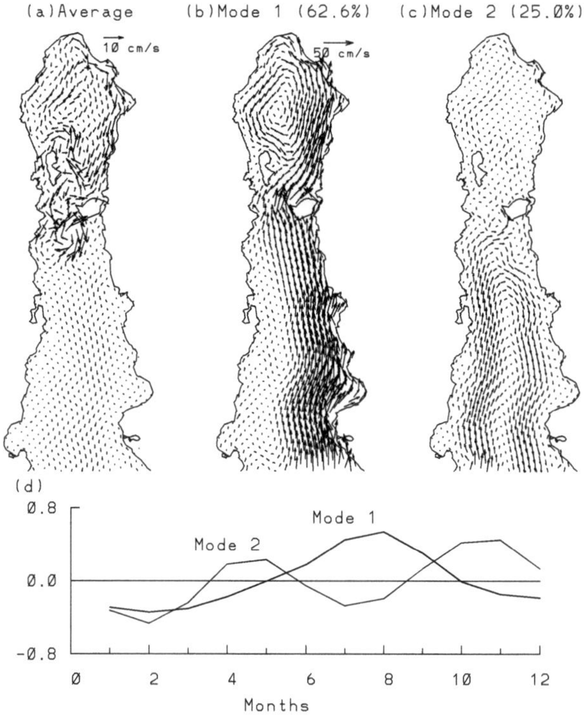

Figure 9. Surface velocity mean and seasonal patterns, from the 3D numerical model of Marinone (2003). (a) Average surface circulation for the surface layer (0-10 m). (b) and (c) Spatial distribution of the first and second EOF modes, respectively. (d) Time variation of the first (thick line) and second modes (thin line). Note that the scale arrows are different for the average and the EOF modes.

de la Guarda island, and anticyclonic circulation south of the San Esteban island. In the SGC a weak outflowing current is present away from the coasts. A net surface outflow from the NGC occurs through the archipelago, which is compensated by a permanent inflow close to the bottom over San Esteban sill (Marinone, 2003; Marinone and Lavín, this volume). A similar mean circulation pattern was obtained with the two-layer non-linear numerical model of Beier (1999), who notes that the two-layer vertical structure is driven by vertical mixing.

The across-gulf average of Figure 9a, shown in Figure 10, produces a two-layer flow. Outflow takes place in the upper layer, which is 200 m thick in the SGC and shallows to 50 m in the NGC. The maximum outflow (0.02–0.03 ms^{-1}) occurs above the sills, centered in a depth of 100 m. Compensating inflow occurs in the rest of the water column, with maximum speeds (0.02–0.03 ms^{-1}) above the sills. This residual flow is responsible of exporting the average surface heat gain and the average excess salt

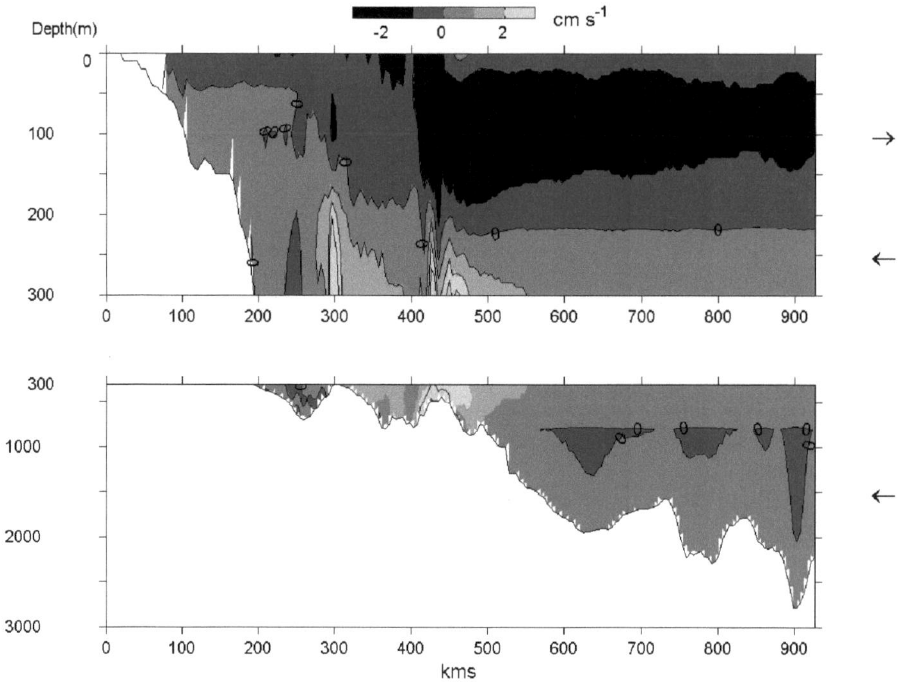

Figure 10. Laterally-averaged residual circulation for the 3D model of Marinone (2003). The top panel shows the upper 300 m, and the bottom panel from 300 m to the bottom. Speed palette on the top of the figure; positive values are into the GC, negative values are out of the GC. The arrows on the right are visual aids indicating the current direction.

apparent in Figure 6. The seasonal behavior of the laterally-averaged flow, described by Marinone (2003), includes a 20-50 m thick seasonally-reversing surface layer which flows in the same direction as the wind; the scheme is practically the same as the three-layer system proposed by Bray (1988a). However, care must be taken when interpreting this for the SGC, because its surface seasonal circulation is in opposite directions on the two sides of the gulf (Figure 8). The two-layer residual flow pattern is supported by direct current measurements in the sill between the Tiburón and Delfín basins, where a strong (0.27 ms^{-1}) inflowing bottom current transporting an average of 0.1 x 10^6 m^3s^{-1} was observed by (López and García, 2003). *Tidal Residuals.* Tidal residual currents in barotropic models are produced by non-linear interaction of the tidal current with the bathymetry, especially in areas where the tidal excursion is comparable to the scale of bathymetric features and around points and capes; for the GC the process has been studied numerically by Marinone (1997), Argote *et al.* (1998), and Carbajal and Backhaus (1998), among others. The tidal residuals for the 10 m thick surface layer produced by the 3D numerical model of Marinone

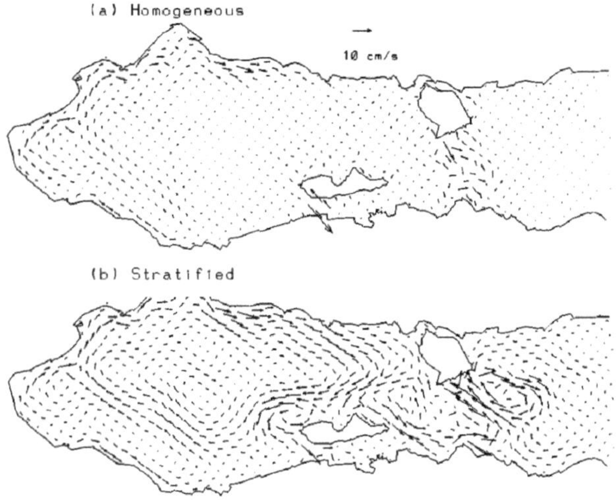

Figure 11. Tidal residual currents for the surface layer (0-10 m) from the 3D numerical model of Marinone (2003), forced with tides only. (a) Homogeneous. (b) Stratified.

(2003), without stratification and forced only with the 7 main tidal harmonics (Figure 11a) show an anticyclonic circulation in the NGC and the UGC, down-gulf coastal currents in the mainland coast of the NGC and in part of the peninsula coast. The speeds are 0.01–0.05 ms^{-1}. In the SGC tidal residual currents are negligible.

In a stratified model with vertical mixing, the generation of residual currents is much more complex. In addition to the interaction of barotropic and internal tides with the bathymetry and with the residual currents, vertical mixing can rearrange the density field, and the resulting pressure gradients induce residual currents, some of which may become geostrophic. The thermal fronts created by tidal mixing in the NGC and around the archipelago (Argote *et al.*, 1995) are a case in point. Figure 11b shows the residual currents for the 10 m thick surface layer from a model run with only the tidal forcing, but with the same initial stratification as the run with all the forcings (Marinone, 2003): in the archipelago and the NGC the residual currents are stronger (0.1 ms^{-1}) than in the homogeneous case, and there is more lateral structure. This structure is similar to that of Figure 9a, the mean residual currents from the model run with all the forcing agents. The lateral average of the 3D annual-mean tidal residual current field produces a distribution very similar to Figure 10, which suggests that the annual mean residual flow field is induced by the tides, most likely through tidal mixing (Marinone, 2003).

Thermohaline circulation. The ocean circulation that results exclusively from the surface fluxes of heat and moisture is called thermohaline circulation. Despite high evaporation, the expected annual-mean thermohaline circulation in the GC is "inverse Mediterranean" (Bray, 1988a), because of the annual-mean gain of heat through the surface. Three-dimensional modelling of this process in isolation by Marinone (2003) produce currents that are one order of magnitude smaller than those shown in Figure 10. The three-layer system proposed by Bray (1988a) came from geostrophic transport profiles in Guaymas basin, and therefore represents the general geostrophic circulation, not only the thermohaline circulation. The three-layer system obtained by Marinone (2003) (Figure 10) seems to be tidally-induced; it is not obvious if they are the same. More realistic surface forcing functions may be necessary to study this subject, but the role of vertical mixing in the mean circulation and thermodynamics seems to be very important.

5. Mesoscale Oceanography

5.1. THE ENTRANCE ZONE

The line from the tip of the Baja California peninsula to Cabo Corrientes is the physical entrance to the GC. However, in most of the recent observational and modelling work the shorter line from the tip of the peninsula to El Dorado in the mainland is used. The main thermohaline characteristic of this province is its transitional character (Griffiths, 1968; Stevenson, 1970; Roden, 1971 and 1972). As suggested by the shaded area in Figure 4, the southern reach of the CC passes the area seasonally, and three surface water masses (TSW, CCW and GCW) intermingle among themselves in the top 150-200 m and with the upper layers of the StSsW immediately below.

In the inner mouth, the GCW is usually found as surface or subsurface layers or cores of high salinity in the upper 100 m. These cores are most frequently found adjacent to the peninsula, but they have also been detected on the mainland side (Roden and Groves, 1959; Roden, 1971; Castro *et al.*, 2000). The distribution of salinity across the inner mouth in December 1993 (Fig. 12a) illustrates the spatial variability of the distribution of the water masses in this area. According to the classification of Torres Orozco (1993), Figure 12 shows the GCW as 20 m thick layers, under the surface on the western side and on the surface on the eastern side, embedded in a surface layer of TSW. This TSW layer occupies most of the upper 50 m across the section up to 60 km west of the peninsula. From 80 to $\sim$400 m depth, there is StSsW, and the salinity around 34.85 at a depth of 150 m is probably the remains of the core of the StSsW, whose maximum has been lost due to the presence of the GCW. At 600 to 800 m there is a salinity minimum under 34.55, which is the core of the PIW.

The annual average hydrography across the inner mouth, obtained by Castro Valdez (2001) from a series of 9 high definition cross-sections (Fig-

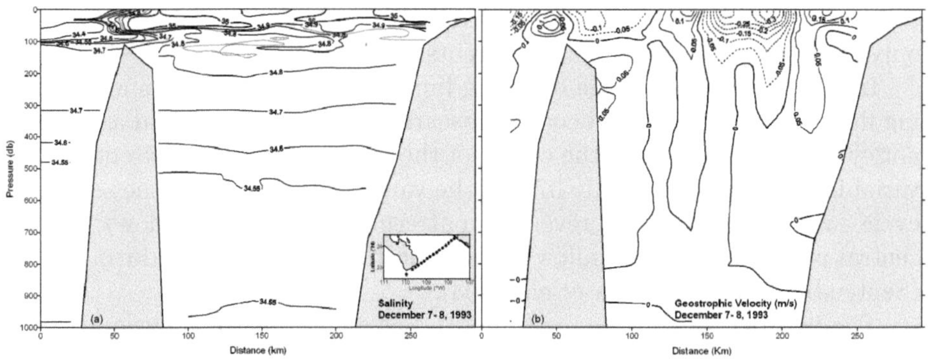

Figure 12. Vertical sections of (a) salinity and (b) geostrophic velocity across the inner mouth in December 1993. The peninsula is on the left side. In (b), common depth between adjacent stations was used as reference; negative values (dashed) indicate flow out of the gulf. No data below 1000 m; maximum bottom depth in this section is 3000 m. A map with the stations is shown in the inset in (a). Data from Godínez *et al.* (2001).

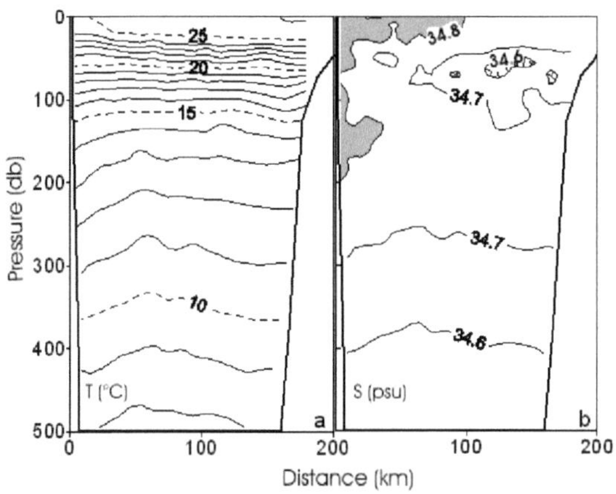

Figure 13. Average distribution of salinity across the inner mouth, from nine high-resolution surveys made from 1992 to 1998. After Castro Valdez (2001); Figure provided by R. Castro.

ure 13), shows that in the mean, the GCW is found as two narrow (~40 km) cores attached to the peninsula, one in the surface and another between 100 and 200 m. On the mainland side the TSW is also concentrated around a core of fresher water.

Very high lateral variability is also found in the geostrophic velocity across the inner mouth, with gyres of length scales from 20 to 100 km and speeds around 0.1 ms^{-1} at depths exceeding 1000 m (Roden, 1971; Collins *et al.*, 1997; Castro Valdez, 2001; Mascarenhas *et al.*, 2003). The lateral variability is evident in the geostrophic velocity across the inner mouth in December 1993, Figure 12b, with alternating inflowing and outflowing regions in the upper 100 m. The maximum depth in these observations is only 1000 m, therefore the deep currents were not detected.

In addition to an average outgoing flux close to the peninsula and ingoing flux along the mainland coast, Mascarenhas *et al.* (2003) find a smaller anticyclonic circulation in the center of the inner mouth. The extent of the variability is reflected in only 37% of the variance being due to the seasonal cycle, which consists of a reversing pattern with inflow (outflow) in the central part and outflow (inflow) mostly on the peninsula side during May (September) (Mascarenhas *et al.*, 2003).

The hydrography in the inner mouth has a clear seasonal cycle (Torres Orozco, 1993), especially in temperature and in mixed layer depth (Castro *et al.*, 2000), which varies from 10 m in summer to around 40 m in winter. A dome is often observed in the deep isotherms across the inner mouth, which is associated to the deep currents and suggests the presence of a gyre (Collins *et al.*, 1997; Mascarenhas *et al.*, 2003). Drastic anomalies occur during El Niño, which consist of deepening of the surface mixed layer and the thermocline up to 50 m from their normal position, an increase in surface temperature of up to 4 °C and a decrease in salinity (0.1-0.2), caused by en extensive invasion of TSW (Lavín *et al.*, 1997b; Castro *et al.*, 2000).

The AVHRR satellite images of this area frequently show the presence of mesoscale gyres, made visible by the temperature contrast between the water masses. More about these gyres below.

5.2. THE SOUTHERN GULF OF CALIFORNIA

In the SGC, the GCW occupies the upper 100 m, the salinity maximum of the StSsW is not discernible, and the salinity minimum of the PIW is slightly lower than in the mouth (Alvarez-Borrego and Schwartzlose, 1979; Robles and Marinone, 1987). Although the distribution of the water masses in the SGC is not as complicated as in the inner mouth, its dynamics is very rich in features.

Upwelling. The wind pattern described above is appropriate for the generation of upwelling, on the mainland coast in winter and on the peninsula in summer. Satellite images, however, only show clearly the winter upwelling on the mainland. It is not clear why the summer upwelling is weaker, but one possible reason is the weakness of the summer winds (Badan, this volume). There are no direct observational studies of upwelling in the GC, although satellite imagery (AVHRR and CZCS) has been used in their study (Badan-Dangon *et al.*, 1985; Santamaría-del-Angel *et al.*, 1994; LLuch-Cota, 2000). Roden (1964) and Carbajal (1993) estimated a mean upwelling rate for the GC of 1 to 3 meters per day.

Geostrophic gyres. Satellite images of the SGC (AVHRR and color) often show a series of gyres about 50-150 km in diameter (Badan-Dangon *et al.*, 1985; Pegau *et al.*, 2002; Navarro-Olache *et al.*, 2003). Their estimated tangential velocity is 25 to 40 cms^{-1} (Emilsson and Alatorre, 1997; Pegau *et al.*, 2002). No study of their life span has been conducted, but there is evidence that the largest ones are geostrophic and reach to over 1500 m in depth; they can be cyclonic or anticyclonic, but no seasonal pattern can be detected with the available hydrographic data (Figueroa *et al.*, this volume). These large gyres are a very important ingredient of the circulation of the GC and probably dominate it at the mesoscale in the offshore region. The only direct measurements of their velocity are those by Emilsson and Alatorre (1997), and maybe the deep currents observed by Collins *et al.* (1997) and Mascarenhas *et al.* (2003) across the inner mouth are due to the presence of gyres of this kind. These gyres do not appear in the baroclinic models of Beier (1997) and Marinone (2003), which are forced by horizontally homogeneous seasonal winds. The first numerical model to produce them is the 3D model of Martínez Alcalá (2002), which uses non-homogeneous, short-term variable winds measured by satellite (NSCAT); the seasonal cycle was not modelled, but the circulation pattern is rich in gyres. These gyres probably have a very important effect on the mean and seasonal circulation and thermodynamics, but the subject has not yet been studied.

Filaments and plumes. Also first observed in satellite images, there are in the GC many smaller-scale features like jets, filaments and plumes, which have a shorter life-span than the gyres (Badan-Dangon *et al.*, 1985; Pegau *et al.*, 2002). They seem to be associated to the basin-wide gyres (appearing on their edges), to wind-induced upwelling and to the thermal fronts induced by tidal mixing over the sills in the archipelago. Like elsewhere, they often occur on the lee of topographic features such as capes, from which they seem to originate. One of these jets, originating north of Guaymas basin from the tidal-mixing zone, was investigated with closely-sampled XBT and CTD stations by Navarro-Olache *et al.* (2003). It was a shallow feature (around 80 m deep), about 30 km wide, with speeds $\sim$0.5 ms^{-1} and a transport around $0.5x10^6$ m^3 s^{-1} . This particular feature lasted some 10 days and stretched for more that 110 km before decaying.

Coastal trapped waves. During summer, the currents in the continental shelf of the SGC are not correlated with the local winds, because internal coastal-trapped waves (CTWs) dominate the velocity variability in scales of several days (Merrifield and Winant, 1989). These CTWs are generated by hurricanes in the ETPac (Christensen at al., 1983; Enfield and Allen, 1983; Zamudio *et al.*, 2002) and travel inside the GC at 150-300 km per day. They are a hybrid of Kelvin and topographic continental shelf waves, and are usually isolated elevations of the sea surface with amplitudes ~10 to 30 cm, and corresponding depressions of the thermocline of 40 to 60 m. The observations of Merrifield and Winant (1989) show currents as fast as 0.5 ms^{-1} in water 100 m deep associated to a 20 cm elevation CTW.

Insight into the behavior and consequences of the CTWs inside the GC is provided by the numerical modeling work of Martínez and Allen (2003a, 2003b). They show that the CTWs travel unhindered along the mainland coast, but upon reaching the archipelago, they split into two waves; one of them enters the NGC and is dissipated there, while the other turns west and returns along the peninsula coast. The returning CTW looses 50% of its amplitude around the sill, and the CTW that enters the NGC carries 10-20% of the incident energy. Most of the dissipation occurs in the sills. The strong nonlinearity of these waves cause partial breaking and a transfer of energy to the bottom of the deep basins.

The mechanism behind the hypothesis of Ripa (1997) that the seasonal lateral heat flux in the GC is due to an internal Kelvin wave (of annual period) is illustrated by Figure 8 of Martínez and Allen (2003a). It shows that on the mainland shelf the currents under an elevation CTW are directed into the gulf, and there is no (or little) return flow after it has passed. This causes a net flux of properties into the GC.

5.3. THE ARCHIPELAGO

The archipelago contains several sills and narrow channels or straits (Figure 1). Its most distinctive oceanographic characteristic is the presence of extensive and strong tidal mixing both by bottom friction and by internal instabilities (Argote *et al.*, 1995; Marinone and Lavín, this volume). As the tide squeezes through the straits, tidal currents are intensified to such an extent that the flow may become supercritical (Badan-Dangon *et al.*, 1991a), leading to turbulence so intense that surface boils can be seen in waters hundreds of meters deep. The upwelled water being colder than the surroundings, the area is highlighted in AVHRR images of the gulf; the lowest SST in the GC is found in this area throughout the year (Badan-Dangon *et al.*, 1985; Paden et al., 1991). The overall distribution and the temporal and spatial variability of the SST in the area north of Guaymas basin is controlled by tidal mixing in the presence of the seasonal heat flux (Paden *et al.*, 1991; Argote et al., 1995). Paden *et al.* (1991 and 1993) suggested that tidal mixing may be responsible for the net surface heat gain, by inhibiting latent heat flux via the low SST; indeed Ballenas channel is the area to the north of Guaymas basin where the least evaporation occurs and therefore the most heat is gained (Paden *et al.*, 1993; Romero Centeno, 1995).

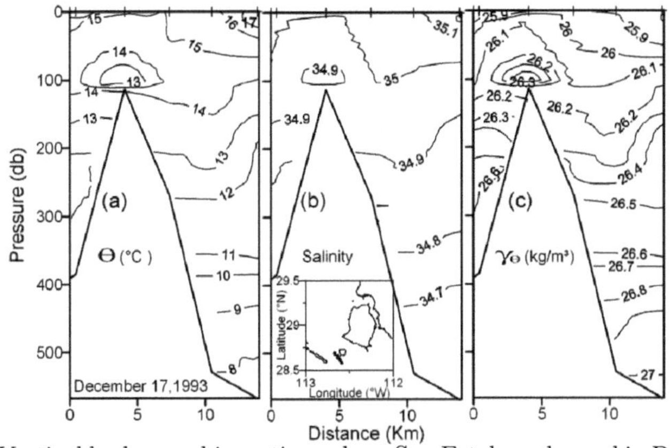

Figure 14. Vertical hydrographic sections along San Esteban channel in December, 1993: (a) potential temperature, (b) salinity, (c) density anomaly. A map with the position of the stations is shown in the inset.

Figure 14 shows vertical sections of temperature, salinity and density anomaly along San Esteban channel made in December, 1993. The isolines present vertical distortions over 100 m, which are part of the generation mechanism of internal tides. There are also thermal and haline surface fronts where the isolines break into the surface. The GCW ($S \geq 35$) occupies the upper 100 m, and underneath only StSsW is present. The effect of tidal mixing in lowering the SST can extend some 400 km to the south of the archipelago (Navarro-Olache *et al.*, 2003), and probably all over the NGC (Paden *et al.*, 1991).

The Ballenas channel is very deep (maximum 1600 m), and has sills both to the north (600 m deep) and south (400 m deep), which give it such unique characteristics that it can be considered an oceanographic province in itself. The water below 600 m in Ballenas channel has a temperature of 11°C and a salinity of 34.8. The deep water in Ballenas channel is well aired (Alvarez-Borrego and Schwartzlose, 1979), which indicates that the residence time is not long.

The interchange of water between the northern and the southern parts of the gulf takes place through the straits of the archipelago. The StSsW that enters to the NGC below 100-150 m is transformed into GCW and then exits in the upper layers. The best communication is through San Esteban Channel, between San Lorenzo and San Esteban islands, because of its depth (600 m) and width. The San Lorenzo channel is 200 m shallower at the threshold than San Esteban channel and very narrow. The channel between San Esteban and Tiburón islands reaches 500 m in depth but is isolated from the SGC below ~350 m by a ridge extending east from San Esteban island. The Infiernillo channel, between Tiburón island and the

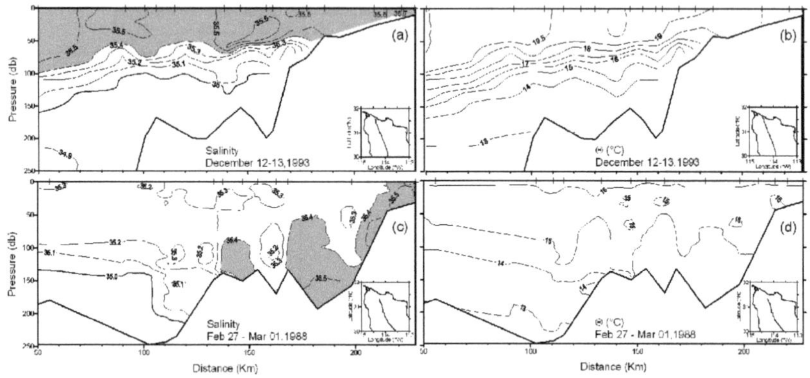

Figure 15. Temperature and salinity sections in the NGC in (a and b) December 12-13, 1993, and (c and d) February 27-March 1, 1988.

mainland is only $5m$ deep. The flow of StSsW into the NGC has been measured with current meters located right on the thresholds: in San Lorenzo sill by Badan-Dangon *et al.* (1991a) for a couple of weeks, in the sill between Tiburón basin and Delfín basin for several months by López and García (2003), and in San Lorenzo sill for over one year by M.L. Argote (Pers. Comm.); they have all found strong (0.3 to 0.5 ms^{-1}) inflowing currents close to the bottom and weaker outflow in the upper layers.

5.4. THE NORTHERN GULF OF CALIFORNIA

The water mass distribution in the NGC consists of GCW in the top 150 m and StSsW underneath (Torres Orozco, 1993; Romero Centeno, 1995). Vertically homogeneous conditions exist in the northernmost fringes of the NGC, down to the 30 m isobath in summer and down to 60 m in winter; the position of the front is imposed in summer by tidal mixing and in winter by tidal mixing and vertical convection (Argote *et al.*, 1995). The salinity (S) and temperature (T) distributions in December of 1993 (Figures 15a,b) show the GCW in the top 100-150 m and the StSsW below. There is a surface mixed layer (S$\sim$35.5, T$\sim$19.5 °C) that is 100 m deep in Delfín basin and shallows to 50 m in the northern edge of Wagner basin, where the thermocline ends as a bottom front. The hydrographic structure in late February-early March 1988 (Figures 15c,d) show a case when the saltiest water is found in the bottom of Wagner basin, evidence that the GCW has gone through vertical convection (Lavín *et al.*, 1995).

The subsequent movement of the convected high salinity water is not well known. Bray (1988b) proposed that they travel south as mid-water gyres, which would then be mixed with the surrounding waters when passing over the sills. Palacios Hernández (2001) and López and García (2003) find

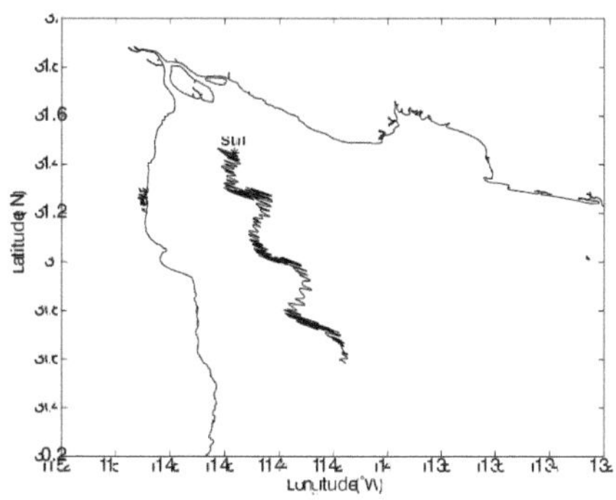

Figure 16. Progressive Vector Diagram from current meter data collected from June 29 to August 22, 1999. Initial position marked by *. Current meter was located 5 m above the bottom in mean bottom depth 15 m. Although the map of the coast and the PVD are in the same scale, the diagram is an artifice and represents no physical trajectory.

evidence that winter water-mass formation events may be able to alter the circulation and thermohaline structure in the entire NGC.

5.5. THE UPPER GULF OF CALIFORNIA

The UGC is the shallow area (depth < 30 m) at the northern end of the GC (Figure 1). It is a tidal, evaporative basin (1 m/year) subjected to strong tidal mixing (Argote *et al.*, 1995; García Silva and Marinone, 2000), which keeps it vertically mixed. During spring (neap) tides, the tidal range is 6.4 m (1.7 m) (Figure 3) and the tidal currents are about 0.5 ms^{-1}(0.1 ms^{-1}). Before the Colorado River was dammed, the UGC was an estuarine environment, at least during the peak river discharge (Carbajal *et al.*, 1997; Lavín and Sánchez, 1999). Today the UGC is an inverse estuary throughout the year (Lavín *et al.*, 1998).

Lavín *et al.* (1998) suggested that gravity currents might be important in the UGC, and be modulated by the spring-neaps cycle, with events occurring during neap tides; recent surveys and moored current meter observations confirm this. A most compelling evidence for the existence, and fortnightly modulation, of gravity currents in the UGC is the progressive vector diagram (PVD) from a current meter deployed in the center of the UGC during the summer of 1999 (Figure 16). During spring tides there is a weak eastward residual flow, but during neap tides there is a strong southward flow.

It has been proposed from the distribution pattern of temperature, salinity and bottom sediments, that the residual circulation pattern of the UGC consists of inflow in the mainland side and outflow in the peninsula side

(Alvarez-Borrego *et al.*, 1975; Carriquiri and Sánchez, 1999). Current meter data obtained in the summer of 1999 suggest that the residual circulation pattern may indeed be as described, driven by the fortnightly gravity-current events on the western side of the UGC. The gravity currents have speeds 0.02 to 0.07 ms^{-1}, and are concentrated in a relatively small portion of the western side of the entrance to the UGC. The heat and salt budgets suggest that the gravity currents play an important role in the flushing of the UGC. Although the largest resuspension of sediments in the UGC occur during spring tides, significant near-bottom suspended sediment fluxes out of the UGC occur only during neap tides, driven by the gravity currents (Alvarez and Jones, 2002).

6. Interannual variability

The strongest interannual anomalies in the GC are due to the ENSO phenomenon (Baumgartner and Christensen, 1985; Bray and Robles, 1991), which is not surprising considering that its entrance is located in the coastal wave guide at only 23 °N. The phase speed of CTWs that carry the oceanic El Niño signal along the ETPac coast is 100-240 km per day (Strub and James, 2002b,c and references therein), so that it can reach the GC entrance in about one month. The anomalies observed just outside and inside the GC during an El Niño (Baumgartner and Christensen, 1985; Robles and Marinone, 1987; Ripa and Marinone, 1989; Marinone, 1989; Torres Orozco, 1993; Soto *et al.*, 1999; Filonov and Tereshchenko, 2000; Castro *et al.*, 2000; Durazo and Baumgartner, 2002; Strub and James, 2002b,c; Lavín *et al.*, 2003) are explained by the properties of the internal CTWs: a 10-30 cm surface elevation, a 50-100 m deepening of the surface mixed layer and the thermocline (and a consequent increase in the heat content of the upper 100-200 m). The surface temperature elevation (up to 3 °C) and surface salinity depression (0.1) are associated to the advective properties of a surface-elevation CTW in the presence of horizontal SST and salinity gradients. As noted above, the currents under surface-elevation CTWs are in the direction of travel of the wave (Martínez and Allen, 2003a,b), so that it looks like a burst or enhancement of the Costa Rica Coastal Current, as proposed by Baumgartner and Christensen (1985). The surface distribution of salinity inside the GC in March and August 1983 (see figures in Lavín *et al.*, 1997b and 2003), in the decay phase of the 1982-83 El Niño, show TSW ($S<35$) almost to the archipelago, while those in January and March 1984 represent the return to normal conditions with TSW retreating from the GC.

The interannual anomalies in the 17-year AVHRR SST record (Soto *et al.*, 1999; Lavín *et al.*, 2003) clearly show that during El Niño (La Niña) the monthly-mean temperature in the GC can be up to 3 °C above (below) normal. The positive anomalies are largest around the archipelago thermal fronts. There are other anomalies that are as large as those produced by a mid-strength El Niño but not associated to ENSO (e.g. summer of 1990 and in the winter of 1994-1995), which can produce noticeable effects in the hydrography of the GC (Palacios-Hernández, 2001).

The trend in the AVHRR SST is large and statistically significant (Lavín *et al.*, 2003), amounting to a surface heating of 1 °C over the 17 years of the AVHRR record. This trend in the GC SST, which has also been observed in the COADS data set (Lluch-Cota, 2000; Bernal *et al.*, 2001), appears to be part of the interdecadal variability, which has been producing a mean SST increase of the entire PO since the mid 1970s.

7. Conclusion

We understand much more about the physical oceanography of the Gulf of California than twenty years ago. However, as knowledge advances and as models and remote sensing become more powerful and usable, it is becoming apparent that more and sustained direct observations are needed in order to test current and future hypothesis and models, and to be able to use this knowledge for practical applications. From the early investigations in the GC it has been recognized that its extreme biological richness is due to physical factors (Gilbert and Allen, 1943; Alvarez-Borrego and Lara-Lara, 1991), but only recently have we began to unravel why and how it occurs. Tidal mixing must be highlighted in this respect.

The seasonal cycle has been submitted to close scrutiny (following the lead of Pedro Ripa), and research in the subject continues, especially in the water exchange across the mouth and between the NGC and the SGC. The study of mesoscale phenomena and processes with shorter time scales (e.g., geostrophic gyres, upwelling, plumes and fronts, gravity currents, etc.), and their influence upon the mean and seasonal circulation and thermodynamics is a standing challenge for observationalists and modelers.

Acknowledgements

This article is a tribute to the intellectual and humanistic inheritance left by Pedro Ripa (1946-2001). This work was supported by CICESE through regular budget, and by CONACYT (México) through contracts 25555-T9712 and 35251T. We thank the colleagues who made available their published or unpublished writings and data, and those that reviewed drafts of this article: A. Martínez, A. Mascarenhas, R. Castro, F.J. Beron-Vera,

P.T. Strub and others. We thank the technical support of C. Cabrera, V. Godínez and A. Ocampo.

References

Alvarez-Borrego S., B.P. Flores-Baez and L.A. Galindo-Bect. Hidrologia del Alto Golfo de California II. Condiciones durante invierno, primavera y verano. *Cienc. Mar.*, **2**, 21-36, 1975.

Alvarez-Borrego, S. and R. S. Schwartzlose. Masas de agua del Golfo de California. *Cienc. Mar.*, **6**, 43-63, 1979.

Alvarez-Borrego, S., and J.R. Lara-Lara. The physical environment and primary productivity of the Gulf of California. In *The Gulf and Peninsular Province of the Californias, Mem. Am. Assoc. Pet. Geol.,* **47**, 555-567, 1991.

Alvarez, L.G. and S.E. Jones. Factors influencing suspended sediment flux in the Upper Gulf of California. *Est. Coast. Shelf Sci.*, **54**, 747-759, doi: 10.1006/ecss.2001.0873, 2002.

Argote M. L., A. Amador, M. F. Lavín and J. R. Hunter. Tidal dissipation and stratification in the Gulf of California. *J. Geophys. Res.* **100**, 16103-16118, 1995.

Argote, M.L., M.F. Lavín and A. Amador. Barotropic residual circulation due to M2 and wind stress in the Gulf of California. *Atmósfera*, **11**, 173-197, 1998.

Badan-Dangon A., C. J. Koblinsky and T. Baumgartner. Spring and sumer in the Gulf of California: observations of surface thermal patterns. *Oceanol. Act.*, **8**, 13-22 pp, 1985.

Badan-Dangon A., M. C. Hendershott and M. F. Lavín. Underway Doppler current profiles in the Gulf of California, *Eos, Trans. Am. Geophys. U.,* **72**, 209, 217-218, 1991a.

Badan-Dangon A., C.E. Dorman, M.A. Merrifield and C.D. Winant. The lower atmosphere over the Gulf of California, *J. Geophys. Res.*, **96**, 16,877-16,896, 1991b.

Badan-Dangon, A. Coastal circulation from the Galápagos to the Gulf of California. In Robinson, A. and K.H. Brink (Eds.) *The Sea*, **11**, 315-343, 1998.

Badan, A. The atmosphere over the Gulf of California. *This volume.*

Baumgartner T. R. and N. Christensen. Coupling of the Gulf of California to large-scale interannual climatic variability. *J. Mar. Res.* , **43**, 825-848, 1985.

Beier, E. A numerical investigation of the annual variability in the Gulf of California, *J. Phys. Oceanogr.,* **27**, 615-632, 1997.

Beier, E. and P. Ripa. Seasonal gyres in the northern Gulf of California. *J. Phys. Oceanogr.,* **29**, 302-311, 1999.

Beier E. Estudio de la marea y la circulación estacional en el Golfo de California mediante un modelo de dos capas heterogéneas. Ph. D. thesis, Department of Physical Oceanography, CICESE, Ensenada, B.C., México, 64 pp, 1999.

Bernal, G., P. Ripa and J.C. Herguera. Oceanographic and climatic variability in the lower Gulf of California: links with the tropics and North Pacific. *Cienc. Mar.*, **27**, 591-617, 2001.

Beron-Vera, F. J. and P. Ripa. Three-dimensional aspects of the seasonal heat balance in the Gulf of California. *J. Geophys. Res.*, **105**, 11441-11457, 2000.

Beron-Vera, F.J. and P. Ripa. Seasonal salinity balance in the Gulf of California. *J. Geophys. Res.* **107**, 10.1029/2000JC000769, 2002.

Bray, N. A. 1988a. Thermohaline circulation in the Gulf of California, *J. Geophys. Res.*, **93**, 4993-5020.

Bray, N. A. Water mass formation in the Gulf of California, *J. Geophys. Res.*, **93**, 9223-9240, 1988b.

Bray, N. A. and J.M. Robles. Physical Oceanography of the Gulf of California. In *The Gulf and Peninsular Province of the Californias, Mem. Am. Assoc. Pet. Geol.*, **47**, 511-553, 1991.

Carbajal, N. Modelling of the circulation in the Gulf of California. Ph.D. Thesis, Institut für Meerskunde, Universität Hamburg, Germany, 186 pp, 1993.

Carbajal, N., A. Souza and R. Durazo. A numerical study of the ex-ROFI of the Colorado River. *J. Mar. Sys.*, **12**, 17-33, 1997.

Carbajal, N. and J.O. Backhaus. Simulation of tides, residual flow and energy budget in the Gulf of California. *Oceanol. Act.*, **21**, 429-446, 1998.

Carriquiry, J.D. and A. Sánchez. Sedimentation in the Colorado River Delta and Upper Gulf of California after a century of discharge loss. *Marine Geology*, **158**, 125-145, 1999.

Carrillo L., M.F. Lavín and E. Palacios-Hernández. Seasonal evolution of the geostrophic circulation in the northern Gulf of California. *Est. Coast. Shelf Sci.*, **54**, 157-173, 2002.

Castro R., M. F. Lavín and P. Ripa. Seasonal heat balance in the Gulf of California, *J. Geophys. Res.*, **99**, 3249-3261, 1994.

Castro R., A. S. Mascarenhas, R. Durazo, C. A. Collins. Seasonal variation of the temperature and salinity at the entrance to the Gul of California, México. *Cienc. Mar.*, **26**, 561-583, 2000.

Castro Valdez, R. Variabilidad termohalina e intercambios de calor, sal y agua en la entrada al Golfo de California. PhD thesis, Facultad de Ciencias Marinas, UABC, Ensenada, México. 121pp, 2001.

Christensen, N., R. de la Paz and G. Gutiérrez. A study of sub-inertial waves off the west coast of Mexico. *Deep Sea Res.*, **30**, 835-850, 1983.

Collins, C.A., N. Garfield, A.S. Mascarenhas Jr., M.G. Spearman and T.A. Rago. Ocean currents across the entrance to the Gulf of California. *J. Geophys. Res.*, **102**, 20,927-20,936, 1997.

Durazo, R. and T. Baumgartner. Evolution of oceanographic conditions off Baja California: 1997-1999. *Progr. Oceanogr.* **54**, 7-31, 2002.

Emilsson, I., and M.A. Alatorre. Evidencias de un remolino ciclónico de mesoescala en la parte sur del Golfo de California. In M.F. Lavín (Editor) Contribuciones a la Oceanografía Física en México. Monografía No. 3, *Unión Geofísica Mexicana*, 173-182, 1997.

Enfield, D.B. and J.S. Allen. The generation and propagation of sea level variability along the Pacific coast of Mexico. *J. Phys. Oceanogr.*, **13**, 1012-1033, 1983.

Fiedler, P.C. Seasonal Climatologies and variability of Eastern tropical Pacific surface waters. NOAA Technical Report **NMFS 109**. U.S Department of Commerce, 65pp, 1992.

Figueroa, M., G. Marinone and M.F. Lavín. Geostrophic gyres in the southern Gulf of California. *This volume.*

Filloux, J.H. Tidal patterns and energy balance in the Gulf of California. *Nature*, **243**, 217-221, 1973.

Filonov, A. and I. Tereshchenko. El Niño 1997-98 monitoring in mixed layer at the Pacific Ocean near Mexico's west coast. *Geophys. Res. Lett.*, **27**. 705-707, 2000.

Filonov, A.E. and M.F. Lavín. Internal tides in the northern Gulf of California. *J. Geophys. Res*, in the press, 2003.

Fu, L.L. and B. Holt. Internal waves in the Gulf of California: Observations from a spaceborne radar, *J. Geophys. Res.*, **89**, 2053-2060, 1984.

García-Silva and S.G. Marinone. Tidal dynamics and energy budget in the Gulf of California. *Cienc. Mar.*, **27**, 323-353, 2000.

Gaxiola-Castro, G., S. Álvarez-Borrego, S. Nájera-Martínez and A.R. Zirino. Internal waves effect on the Gulf of California phytoplankton. *Cienc. Mar.*, **28**, 297-309, 2002.

Gilbert, J.Y. and W.E Allen, The phytoplankton of the Gulf of California obtained by the *E.W. Scripps* in 1939 and 1940, *J. Mar. Res.*, **5**: 89-110, 1943.

Godínez, V.M., M.F. Lavín, J.M. Robles, R. Ramírez and C. E. Cabrera. Datos hidrográficos de la primera campaña del B/O Francisco de Ulloa al Golfo de California (diciembre de 1993). 2001. *Comun. Acad.*, CICESE, Ensenada, B.C., México. *CTOFT20011*, 206 pp., 2001.

Griffiths, R.C. Physical, Chemical, and Biological Oceanography of the entrance to the Gulf of California, spring of 1960. *Spec. Sci. Rep. U.S. Fish Wild Serv.* No. **573**, 43 pp, 1968.

Jiménez Lagunes, A. Análisis de las corrientes de marea y series de temperatura en la parte norte del Golfo de California. Tesis de Maestría, CICESE, Ensenada, B.C., México. 84 pp, 2003.

Lavín M. F. and S. Organista. Surface heat flux in the northern Gulf of California. *J. Geophys. Res.*, **93**, 14033-14038, 1988.

Lavín M. F., G. Gaxiola-Castro and J. M. Robles. Winter water masses and nutrients in the northern Gulf of California. *J. Geophys. Res.*, **100**, 8587-8605, 1995.

Lavín, M.F., R. Durazo, E. Palacios, M.L. Argote, and L. Carrillo. Lagrangian observations of the circulation in the northern Gulf of California. *J. Phys. Oceanogr.*, **27**, 2298-2305, 1997a.

Lavín M. F., E. Beier and A. Badan. Estructura Hidrográfica y Circulación del Golfo de California: Escalas estacional e interanual, *Contribuciones a la Oceanografía Física en México*, Unión Geofísica Mexicana, Monografía No. 3: 141-171, 1997b.

Lavín, M.F., V. Godínez and L.G. Alvarez. Inverse-estuarine features of the Upper Gulf of California. *Est. Coast. Shelf Sci.*, **46**, 769-795, 1998.

Lavín, M.F., and S. Sánchez. On how the Colorado River affected the hydrography of the Upper Gulf of California. *Cont. Shelf Res.*, **19**, 1545-1560, 1999.

Lavín, M.F., E. Palacios-Hernández and C. Cabrera. Sea surface temperature anomalies in the Gulf of California. *Geofís. Int.*, in the press, 2003.

Lluch-Cota, S.E. Coastal upwelling in the Eastern Gulf of California. *Oceanol. Act.*, **23**, 731-740, 2000.

López, M. A numerical simulation of water mass formation in the northern Gulf of California during winter. *Cont. Shelf Res.*, **17**, 1581-1607, 1997.

López, M. and J. García. Moored Observations in the Northern Gulf of California: a strong bottom current. *J. Geophys. Res.* **108**, 3048, doi:10.1029/2002JC001492, 2003.

Marinone, S. G. Una nota sobre la variabilidad no estacional de la región central del Golfo de California. *Cienc. Mar.*, **14**, 117-134, 1988.

Marinone, S.G. and P. Ripa. Geostrophic flow in the Guaymas Basin, central Gulf of California. *Cont. Shelf Res.*, **8**, 159-166, 1988.

Marinone, S.G. Tidal residual currents in the Gulf of California: is the M2 tidal constituent sufficient to induce them?. *J. Geophys. Res.* **102**, 8611-8626, 1997.

Marinone, S.G. Tidal Currents in the Gulf of California: Intercomparison among two- and three-dimensional models with observations. *Cienc. Mar.*, **26**, 275-301, 2000.

Marinone, S.G. A three dimensional model of the mean and seasonal circulation of the Gulf of California. *J. Geophys. Res.*, submitted, 2003.

Marinone, S.G. and M.F. Lavín. Residual circulation and mixing in the large islands region of the central Gulf of California. *This volume.*

Martínez Alcalá, J.A. Modeling studies of mesoscale circulation in the Gulf of California. PhD thesis, Oregon State University, 173 pp, 2002.

Martínez, J.A. and J.S. Allen. A modeling study of coastal-trapped wave propagation in the Gulf of California. Part 1: response to remote forcing. *J. Phys. Oceanogr.*, in the press, 2003a.

Martínez, J.A. and J.S. Allen. A modeling study of coastal-trapped wave propagation in the Gulf of California. Part 2: response to idealized forcing. *J. Phys. Oceanogr.*, in the press, 2003b.

Mascarenhas Jr., A.S., R. Castro, C.A. Collins and R. Durazo. Seasonal variation of geostrophic velocity and heat flux at the entrance to Gulf of California, Mexico. *Unpublished manuscript*, 2003.

Merrifield, M.A. and C.D. Winant. Shelf circulation in the Gulf of California: a description of the variability. *J. Geophys. Res.*, **94**, 18,133-18,160, 1989.

Morales, R.A. and G. Gutiérrez. Mareas en el Golfo de California. *Geophís. Int.*, **28**, 25-46, 1989.

Navarro-Olache, L.F., M.F. Lavín, L.G. Alvarez-Sánchez and A.R. Zirino. Internal structure of SST features in the central Gulf of California. *Deep Sea Res.*, submitted, 2003.

Paden, C.A., M.R. Abbott, and C.D. Winant. Tidal and atmospheric forcing of the upper ocean in the Gulf of California, 1, Sea surface temperature variability. *J. Geophys. Res.*, **96**, 18,337-18, 359, 1991.

Paden, C.A., C.D. Winant, and M.R. Abbott. Tidal and atmospheric forcing of the upper ocean in the Gulf of California, 2, Surface heat flux, *J. Geophys. Res.*, **98**, 20,091-20, 103, 1993.

Palacios Hernández, E. Circulación de la región norte del Golfo de California: estacional y anomalías. PhD thesis, CICESE, Ensenada, México, 117pp, 2001.

Palacios-Hernández, E., E. Beier, M.F. Lavín, and P. Ripa. The effect of winter mixing on the circulation of the Northern Gulf of California. *J. Phys. Oceanogr.*, **32**, 705-728, 2002.

Parés-Sierra, A., A. Mascarenhas, S.G. Marinone and R. Castro. Temporal and spatial variation of the surface winds in the Gulf of California. *Geophys. Res. Lett.*, **30**(6), 1312, doi:10.1029/2002GL016716, 2003.

Pegau W.S., E. Boss and A. Martinez. Ocean color observations of eddies during the summer in the Gulf of California. *Geophys. Res. Lett.* **29**, doi:10.1029/2001GL014076, 2002.

Ramírez-Manguilar, A.M. Análisis armónico de datos de corrientes en la región norte del Golfo de California de noviembre de 1994 a febrero de 1996. BSc thesis, Facultad de Ciencias Marinas, UABC, Ensenada, B.C., México, 56 pp, 2000.

Reyes A. C. and M. F. Lavín. Effects of the autumn-winter meteorology upon the surface heat loss in the Northern Gulf of California. *Atmósfera*, **10**, 101-123, 1997.

Ripa, P., and S.G. Marinone. Seasonal variability of temperature, salinity, velocity, vorticity and sea level in the central Gulf of California, as inferred from historical data. *Quart. J. Roy. Met. Soc.*, **115**, 887-913, 1989.

Ripa, P. Seasonal circulation in the Gulf of California, *Ann. Geophys.*, **8**, 559-564, 1990.

Ripa, P. and G. Velázquez. Modelo unidimensional de la marea en el Golfo de California. *Geofís. Intl.*, **32**, 41-56, 1993.

Ripa, P. Towards a physical explanation of the seasonal dynamics and thermodynamics of the Gulf of California, *J. Phys. Ocean*, **27**, 597-614, 1997.

Robles J. M. and S. G. Marinone. Seasonal and interannual thermoaline variability in the Guaymas Basin of the Gulf of California. *Cont. Shelf Res.*, **7**, 715-733, 1987.

Roden, G.I. Oceanographic and meteorological aspects of the Gulf of California. *Pacific Science*, **XII**, 21-45, 1958.

Roden, G.I., and G.W. Groves. Recent oceanographic investigations in the Gulf of California, *J. Mar. Res.*, **18**, 10-35, 1959.

Roden, G. I. Oceanographic aspects of Gulf of California. In: T. H. Van Andel and G. G. Shor Jr. (Eds.) Marine Geology of the Gulf Of California: A symposium. *Am. Assoc. Pet. Geol. Mem.*, **3**, 30-54 pp, 1964.

Roden, G.I. Aspects of the transition zone in the Northeastern Pacific. *J. Geophys. Res.*, **76**, 3462-3475, 1971.

Roden, G.I. Termohaline structure and baroclinic flow across the Gulf of California entrance and in the Revilla Gigedo Islands region. *J. Phys. Oceanogr.*, **2**, 1777-1803, 1972.

Romero Centeno R. D. L. Comportamiento de los campos hidrográficos y flujos de calor y masa en el Canal de Ballenas. M. Sc. thesis, Department of Physical Oceanography, CICESE, Ensenada, B.C., México, 126 pp, 1995.

Santamaría-del-Angel, E., S. Alvarez-Borrego and F.E. Müller-Karger. Gulf of California biogeographic regions based on coastal zone color scanner imagery. *J. Geophys. Res.*, **99**, 7411-7421, 1994.

Simpson, J.H., A.J. Souza, and M.F. Lavín. Tidal mixing in the Gulf of California, in *Mixing and Transport in the Environment,* edited by K.J. Beven, P.C. Chatwin, and J.H. Millbank, pp. 169-182, John Wiley, New York, 1994.

Soto-Mardones L. A., S. G. Marinone and A. Parés-Sierra. Time and spatial variability of sea surface temperature in the Gulf of California, *Cienc. Mar.*, **25**, 1-30, 1999.

Stevenson, M.R. On the physical and biological oceanography near the entrance of the Gulf of California, October 1966-August 1967. *Bull. Inter-Am. Trop. Tuna Comm.*, **14**, 389-504, 1970.

Strub, P. T. and C. James. Altimeter-derived surface circulation in the large-scale NE Pacific Gyres. Part 1. seasonal variability. *Progr. Oceanogr.*, **53**, 163-183, 2002a.

Strub, P. T. and C. James. Altimeter-derived surface circulation in the large-scale NE Pacific Gyres. Part 2: 1997-1998 El Niño anomalies. *Progr. Oceanogr.*, **53**, 185-214, 2002b.

Strub, P. T. and C. James. The 1997-1998 oceanic El Niño signal along the southeast and northeast Pacific boundaries-an altimetric view. *Progr. Oceanogr.*, **54**, 439-458, 2002c.

Torres Orozco, E. Análisis Volumétrico de las masas de agua del Golfo de California. M. Sc. Thesis. CICESE, Ensenada, B.C, México, 80 pp, 1993.

Wyrtki, K. Surface Currents in the Eastern Tropical Pacific. *Bull. Inter-Am. Trop. Tuna Comm.*, , Vol. **IX**, No. 5., 269-303, 1965.

Wyrtki, K. Oceanography of the Eastern Ecuatorial Pacific Ocean. *Oceanogr. Mar. Biol. Ann. Rev.*, **4**, 33-68, 1966.

Wyrtki, K. Circulation and water masses in the Eastern Ecuatorial Pacific Ocean. *Intl. J. Oceanol. & Limnol.,* **1**, 117-147, 1967.

Zamudio, L., H.E. Hurlburt, E.J. Metzger and O.M. Smedstad. On the evolution of coastally trapped waves generated by Hurricane Juliette along the Mexican west coast. *Geophys. Res. Lett.*, **29**. doi: 10.1029/2002GL014769, 2002.

THE ATMOSPHERE OVER THE GULF OF CALIFORNIA

A. BADAN
Departamento de Oceanografía Física, CICESE
Ensenada, Baja California, México

Abstract. The Gulf of California is also a gulf in the lower atmosphere, delimited by the mountains of Baja California and by the Sierra Madre, open to the eastern tropical Pacific to the south and spreading into the Great American Desert of the southwestern United States. The Gulf of California lies asymmetrically under its southwestern portion, and a well-defined Marine Boundary Layer (MBL), the lower atmosphere over the gulf proper, develops over water but dissipates rapidly over land. The flow within the MBL is forced principally by the along-gulf pressure gradient from the Great Basin High over the southwestern United States and by blocking of the isobaric flow by the Baja California mountains; the flow for most of the year is a northwesterly low-level jet with speeds of 8-12 m sec^{-1}, balanced at the surface by friction in the along-gulf direction, but geostrophically across the gulf, and capped by inversions that slope down from west to east. Moisture in the MBL is kept below 6-8 g kg^{-1} by the cold winds drawn from the desert. Modulation of the high pressure over the desert by upper level synoptic activity causes the typical 3 to 6-day wind events. In late spring or early summer, a monsoon sets up as a thermal low develops over the southwestern United States and reverses the along-gulf pressure gradient. Most of the southerly flow now takes place over the lowlands off the Sierra Madre, so the winds in the MBL over the gulf appear weaker and more variable, but remain as a low-level jet in cross-gulf geostrophic balance, under a weaker inversion that slopes down from east to west. The Gulf of California warms considerably; moisture increases dramatically within the MBL to about 21-24 g $kg^{-1}$and spills over the lowlands to the east, feeding the sometimes intense rainfall against the Sierra Madre. Moisture is also driven onto the southwestern North American desert as a source for the summer rains. This sometimes occurs as wind pulses that result possibly from hydraulic self-adjustment of the MBL to changes in the forcing of the flow, and are often a signature of the onset of the monsoon. The monsoon acts like a short tropical summer, which contrasts with the much longer subtropical 'winter', and gives the region's atmosphere its asymmetric, pseudoseasonal character.

Key words: Gulf of California, Marine Boundary Layer, Mexican monsoon

1. Introduction

A significant amount of research on the physical oceanography and meteorology of the Gulf of California has accumulated over the years and a better

O.U. Velasco Fuentes et al. (eds.), Nonlinear Processes in Geophysical Fluid Dynamics, 205-212.

picture of the hemispheric importance of a system once dismissed simply as a fascinating oceanic laboratory of regional importance is beginning to emerge. It is now clear that some of the unique oceanographic features of this marginal sea are parallelled in the lower atmosphere; intense interactions between these two fluid bodies and with their surroundings explain some of the physical characteristics of this unique province and its role in determining some of the climatic conditions of continental North America. This note is a descriptive synthesis of the most relevant features of the lower atmospheric layers over the Gulf of California, explained in terms of possible mechanistic connections to the surrounding environment and its interactions with the oceanic element. Pedro Ripa possessed great ability to extract the simplest physics from existing sets of data and in this way contributed enormously to the understanding of the Gulf of California; some of the knowledge here is his, especially that pertaining to the heat balance of the gulf and its seasonal fluctuations (see Ripa, 1997, and references therein; also the review on the oceanography of the gulf by Lavín and Marinone, 2003, *this vol.*).

Fundamentally, the atmosphere over the Gulf of California can be understood by considering that it is orographically constrained, subjected to the varying influence of four climatic régimes, and forced by the large-scale extratropical pressure fields that surround the region. The lower layers of the atmosphere acquire their signature from the need to maintain thermodynamic equilibrium with the ocean, with fluxes of heat (usually from the atmosphere to the ocean, in spite of the considerable evaporation) and moisture (largely from the ocean to the atmosphere) sustaining a distinct Marine Boundary Layer (MBL). The oceanic gulf exports the excess heat to the Pacific through a reverse-mediterranean circulation, but in summer, heat gain through the gulf's mouth overwhelms the exchange. The atmosphere disperses the moisture it has gained from the gulf by the dissipation of the MBL over land, by convection, and by providing the precipitable moisture for the summer rains in northwestern Mexico and the southwestern United States, all reflected in the prevalent divergence of moisture flux over the gulf. Winds in the MBL blow mostly along the gulf, as a low-level jet balanced down-gulf by surface friction at the surface, and geostrophically by a pressure setup across the gulf.

2. The setting and general conditions

The Gulf of California is also a gulf in the lower atmosphere, delimited by the mountains of Baja California and by the Sierra Madre over continental Mexico, a path about 300 km wide that effectively channels atmospheric motions and moisture below 800 m (Hales, 1974). The Baja California mountains are less elevated than the Sierra Madre but closer to the oceanic gulf, and thus have a more appreciable influence on the lower atmosphere. The atmospheric gulf is open to the eastern tropical Pacific to the south and reaches onto the Great American Desert plateau of the southwestern United States. The Gulf of California lies asymmetrically under its southwestern two-thirds, and a well-defined Marine Boundary Layer (MBL), the lower atmosphere over the Gulf proper, is sustained over water but dissipates within a short distance over land (Badan-Dangon, *et al.*, 1991). The Gulf of California lies at the transition of at least four different climatic régimes, none of which dominates the region entirely, but all of which contribute to the conditions that prevail at any given time (Hastings and Turner, 1965). Most of the Peninsula is a low-latitude west-coast desert. The gulf marks the western end of the mid-latitude easterlies, and is shielded from the cool oceanic conditions of the Pacific by the Peninsula of Baja California, so is warmer in summer and cooler in winter than the open coast to the west. To the north the westerlies flow with embedded cyclones that modulate the anticyclone normally residing over the western United States. To the south, the eastern Tropical Pacific is the region of confluence of the California Current, a portion of the Subtropical gyre of the North Pacific, and the Mexican Current, the poleward branch of the equatorial return flow that connects the Equatorial Countercurrent to the north Equatorial Current. Conditions there are tropical, with possibilities of instability and deep convection, and the source region for the Mexican monsoon; it is the region where hurricanes and tropical storms that cross the eastern North Pacific and sometimes enter the Gulf of California, may trigger some of the moisture pulses that travel poleward along the gulf at the onset of the monsoon and during the summer (Hales, 1972; Badan-Dangon *et al.*, 1991; Douglas and Leal, 2003). The Gulf of California is the 'only evaporative basin of the Pacific Ocean' (Roden, 1958), but is a net exporter of heat (Castro *et al.*, 1994).

Atmospheric motions over the Gulf of California are related to the locations of the hemispheric pressure centers, such as the Aleutian low, the North Pacific High, the Bermuda High, and the evolution of the Intertropical Convergence Zone (ITCZ). The climate over the Gulf of California during most of the year is that of a mid-latitude 'winter' condition that dominates from late September to late April. The flow within the MBL is forced principally by the along-gulf pressure gradient resulting from the Great Basin High that resides most of the year over the desert southwestern United States and by the blocking of the isobaric flow by the mountains of Baja California; the flow that results is a northwesterly low-level jet, capped by a main and several secondary inversions (Badan-Dangon *et al.*, 1991), balanced at the surface by friction in the along-gulf direction, and geostrophically by a pressure set-up in the cross-gulf direction (Overland, 1984), with the inversion sloping down from west to east, away from the Baja California peninsula (Candela, *et al.*, 1984).

Winds in the MBL during the long mid-latitude winter are typically 8-12 m $sec^{-1}$and most intense near the peninsula of Baja California and over the gulf. The moisture beneath the inversion is kept low, about 6-8 g kg^{-1}, by the cold, dry winds drawn from the Great Sonoran desert; the inversion is stable and any possibility of convection is further suppressed by upper level subsidence (Badan-Dangon *et al.*, 1991). Although the Gulf of California gains heat on average from the atmosphere, these northwesterly winds can also extract considerable heat locally from the ocean and induce wintertime convection over its shallower northern end (Reyes and Lavín, 1997). Periodically during the winter, the low-pressure troughs embedded in the westerlies obliterate the high pressure over the desert and its resulting pressure gradient, and the winds over the gulf diminish or cease for a few days; as the troughs pass over to the east, the pressure gradient rebuilds and the winds resume. This modulation by upper level synoptic activity is the cause of the subdiurnal variability of gulf winds during most of the year. Each northwesterly wind event, being strongest over the gulf as opposed to over the lowlands near the Sierra Madre, is usually accompanied by strong coastal upwelling off the mainland coast of the central and southern gulf, with plumes of cool water that upwell on the lee side of each major cape, flow along the coast to the next cape, and veer offshore to cross the gulf as a narrow jet that bifurcates into a mushroom pair of counterrotating eddies; each major northerly wind event adds to the forcing of a sequence of gulf-wide gyres of alternating rotation sense that occupy the central and southern gulf, beginning with an anticyclone in the upper Guaymas Basin, immediately south of the large islands (Badan-Dangon *et al.*, 1985).

3. The monsoon

In late spring or early summer, the Mexican Monsoon (Douglas *et al.*, 1993) sets up as a thermal low develops over the southwestern United States, especially over Arizona, Nevada, and New Mexico, which blocks the penetration of subtropical fronts and reverses the along-gulf pressure gradient for the much shorter summer season. The Sierra Madre is farther from the gulf than a Rossby radius of deformation (Overland, 1984), so most of the southerly summer flows tend to be trapped away from the oceanic gulf, over the lowlands that border the Sierra Madre (i.e. Mitchell, *et al.*, 2003); the winds in the MBL over the Gulf of California are now weaker, near 2-5 m sec^{-1}, and more variable; but the flow retains the structure of a low-level jet in cross-gulf geostrophic balance, confined by a weaker inversion that slopes down from east to west. Summer upwelling is consequently much weaker, but thinner plumes and gyres sometimes develop off Baja California. The summer inversion is weaker, but the MBL is thicker (200-300 m) and moisture over the ocean increases dramatically to 21-24 g kg^{-1}, extracted from the Gulf of California or advected from the Pacific. The Gulf of California warms considerably through exchanges with the atmosphere, but especially by advection from the open Pacific, a process conjectured by Ripa (1997) as possessing a Kelvin wave-like structure. The increased sea-surface temperature probably sustains the elevated moisture within the MBL, which spills over the lowlands to the east. In addition, the inversion is now often unstable and the high humidity and intense solar heating at the top of the MBL favours deep convection, causing the sometimes intense precipitation, both over the foothills of the Sierra Madre and along the eastern shore of the Gulf of California. Moisture is also driven by the low-level jet onto the desert of southeastern Arizona and southwestern New Mexico, as an important source for the monsoonal summer rains (Douglas *et al.*. , 1993). This advection of moisture sometimes occurs in the form of wind and moisture pulses, that result most probably from an hydraulic self-adjustment of the MBL to changes in the forcing of the flow, and are often a first expression of the onset of the monsoon. The monsoon provides the region with a a short tropical summer which is more a departure from the longer subtropical winter than a true seasonal fluctuation. This asymmetrical pseudoseasonal signature has often been reported as a seasonal signal superimposed on mean conditions more typical of the subtropical winter.

The North American or Mexican monsoon is quite predictable and appears related to the evolution of the gulf's sea-surface temperatures (Mitchell, *et al.*, 2002). The early summer warming is indicated by the progression of the 26°C isotherm along the Gulf of California in about one month, that temperature being the threshold for the onset of deep convection and the beginning of the rains. Eventually, a tongue of water warmer than 29°C extends into the gulf, as additional heat storage in the upper layers of the gulf result from both solar heating and advection (Castro *et al.*, 1994). An increase of the gulf's sea-surface temperature leads to an increase of precipitable vapour which, in turn, drives the increased convection, in part by releasing latent heat at the base of the clouds. Strong deep convection, expressed by low values of outgoing longwave radiation, results from a combination of modes of instability; it follows the increase in sea-surface temperatures, suggesting that the convection is driven by the thermodynamics of the gulf. Hence, the atmosphere should also reflect the interannual variations in oceanic conditions (i.e. Higgins, 2003).

4. Diurnal fluctuations and sporadic events

Diurnal winds are important around the Gulf of California, with typical standard deviations close to 2 m sec^{-1}(Badan-Dangon, 1991). Configured like a rather typical sea breeze, they are predominant nearshore, weaker in the middle of the gulf, and poorly correlated from one location to another, reflecting their localized forcing effects. Diurnal winds are strongest onshore in the afternoon reaching close to 4 m sec^{-1}, but near-stagnant or slightly offshore at nighttime, as expected from the asymmetry of the thermal forcing. Diurnal winds are stronger on the eastern side of the gulf, reflecting the wider adjacent lowlands, and their hodographs usually rotate clockwise because of rotation, although at a few locations on the western side of the gulf, the winds rotate in a counterclockwise sense, possibly from topographic steering by the more abrupt topography of the nearby Baja California mountains. A clear effect of the sea breeze off the gulf is that it helps carry moisture from the MBL onshore, a considerable contribution to the moisture flux divergence over the Gulf of California, especially during the day (Berbery, 2001). Moisture convergence can occur over the slopes of the mountains that border the gulf's MBL. During nighttime a moisture flux divergence is observed on the slopes of the hills and some convergence occurs on the coast. The northern Gulf of California shows mostly divergence, hence is a source of moisture, usually carried into southern Arizona by a low-level jet. In general, there is a meridional moisture flux along the Gulf of California that feeds the precipitation in Arizona, New Mexico and the western Sierra Madre.

Strong sporadic wind events can sometimes modify the otherwise regular behaviour of the wind over the Gulf of California. During the winter, sudden bursts of violent wind, accompanied by a sharp drop of temperature and increased relative humidity are known as 'nortes' (Ives, 1962). Other 'toritos' can occur when cold, dry westerly winter winds are forced through the mountain passes and funneled down some of the many canyons on Baja California. These winds are probably of small overall dynamical significance, but have a considerable influence on the transport of sand and can disrupt navigation and fishing. Yet, these intense winds destroy the MBL temporarily, allowing drier air aloft to come into direct contact with the surface of the ocean; a quick calculation shows that large amounts of heat, several hundred W m^{-2}, can be thus extracted from the gulf.

5. Future research

A proper investigation of the atmosphere over the Gulf of California should consider the entire atmospheric gulf, extending to the foothills of the Sierra

Madre, and into the desert valleys of the southwestern United States; this should explain many of the asymmetries in the atmospheric conditions over the gulf itself and the cross-gulf differences in the moisture fluxes drawn from it. As is often the case, large data sets remain unexplored, such as those of Candela *et al.* (1984, 1985), which probably contain answers to numerous problems being posed about the gulf's region by programs such as the North American Monsoon Experiment (NAME). A thorough review by Higgins *et al.* (2003) provides an excellent description of the evolution of the North American monsoon, of the scope of experiments such as NAME, and an overview of the abundant literature on the subject. Especially, the teleconnections of the Gulf of California to some of the sources of interannual and intraseasonal variation, such as the ENSO cycle, have been addressed, (i.e. Baumgartner and Christensen, 1990; Bernal *et al.*, 2002), but other important fluctuations whose effect are less well known, such as the Pacific Decadal Oscillation or the Madden-Julian Oscillations, or even the multidecadal variations in North Atlantic sea-surface temperatures, may be relevant in conducting the gulf's behaviour. In addition to their direct effects, some of these fluctuations modulate each other, resulting in a wide spectrum of long-term variability. Recently, Cavazos *et al.* (2002) have extracted at least two three-cell intraseasonal modes that are related to the mature phase of the North American monsoon, which emerge as anomalous midtropospheric patterns with large amounts of moisture over the southwestern United States and northwestern Mexico, possibly related to sea-surface temperature anomalies in the North Pacific and to convection off southern Mexico. Finally, the Marine Boundary Layer over the Gulf of California poses many elegant problems in geophysical hydraulics, such as the self-adjustment to criticality of the MBL first reported by Botella (1996).

Acknowledgements

I thank Miguel Lavín and Tereza Cavazos for many generous comments.

References

Badan-Dangon, A., C. J. Koblinsky, and T. R. Baumgartner. Spring and summer in the Gulf of California: observations of surface thermal patterns. *Oceanolog. Acta*, 8:13–22, 1985.

Badan-Dangon, A., C. E. Dorman, M. A. Merrifield, and C. D. Winant. The lower atmosphere over the Gulf of California. *J. Geophys. Res.*, 96:16,877–16,896, 1991.

Baumgartner, T. R. and N. Christensen, Jr. Coupling of the Gulf of California to large-scale interannual climatic variability. *J. Mar. Res.*, 43:825–848, 1985.

Bernal, G., P. Ripa, and J. C. Herguera. Oceanographic and climatic variability in the lower Gulf of California. *Cienc. Mar.*, 27:591–617, 2001.
Berbery, E. H. Mesoscale moisture analysis of the North American monsoon. *J. Climate*, 15:121–137, 2001.
Botella, J. *Tesis de Maestra en Ciencias* CICESE, 1996.
Candela, J., A. Badan-Dangon, and C. D. Winant. Spatial distribution of lower atmospheric physical variables over the Gulf of California. A data report, vol. 1. Summer 1983. *SIO Ref.* 84–33, *Scripps Inst. of Oceanogr.* La Jolla, Calif.: 211 pp, 1984.
Candela, J., A. Badan-Dangon, and C. D. Winant. Spatial distribution of lower atmospheric physical variables over the Gulf of California. A data report, vol. 2. Winter 1984. *SIO Ref.* 85–11, *Scripps Inst. of Oceanogr.* La Jolla, Calif.: 303 pp, 1985.
Castro, R., M. F. Lavín, and P. Ripa. Seasonal heat balance inn the Gulf of California. *J. Geophys. Res.*, 99:3249–3261, 1994.
Cavazos, T., A. C. Comrie, and D. M. Liverman. Intraseasonal variability associated with wet monsoons in southeast Arizona. *J. Climate*, 15:2477–2490, 2002.
Douglas, M. W., R. A. Maddox, and K. Howard. The Mexican monsoon. *J. Climate*, 6:1665-1677, 1993.
Douglas, M. W. and J. C. Leal. Summertime surges over the Gulf fo California: Aspects of their climatology, mean structure, and evolution from radiosonde, NCEP reanalysis, and rainfall data. *Wea. Forecasting*, 18:55–74, 2003.
Hales, Jr., J. E. Surges of maritime tropical air northward over the Gulf of California. *MON. Wea. Rev.*, 100:298–306, 1972.
Hales, Jr., J. E. Southwestern United States summer monsoon source - Gulf of Mexico or Pacific Ocean?. *J. App. Meteor.*, 13:331–342, 1974.
Hastings, J. R. and R. M. Turner. Seasonal precipitation regimes in Baja California, Mexico. *Geogr. Ann.*, 47A:204–223, 1965.
Higgins, R. W., A. Douglas, A. Hahman, E. H. Berbery, D. Gutzler, J. Shuttleworth, D. Stensrud, J. Amador, R. Carbone, M. Cortez, M. Douglas, R. Lobato, J. Meitin, C. Ropelewski, J. Schem, S. Schubert, and C. Zhang. Progress in Pan American CLIVAR research: The North American monsoon system. *Atmósfera*, 16:29–65, 2003.
Ives, R. L. The "pestiferous winds" of the upper Gulf of California. *Weatherwise*, 15:197–201, 1962.
Lavín, M. F. and S. G. Marinone. An overview of the physical oceanography of the Gulf of California. *This volume*, 2003.
Mitchell, D. L., D. Ivanova, R. Rabin, T. J. Brown, and K. Redmond. Gulf of California sea-surface temperatures and the North American monsoon: Mechanistic implications from observations. *J. Climate*, 15:2261–2281, 2002.
Mitchell, D. L., D. Ivanova, and K. Redmond. Onset of the 2002 North American monsoon: Relation to Gulf of California sea-surface temperatures. *83rd. AMS Ann. Meeting*, Long Beach, CA, 9-13 February 2003. (available on CD).
Overland, J. E. Scale analysis of marine winds in straits and along mountainous coasts. *Mon. Wea. Rev.*, 112:2530–2534, 1984.
Reyes, A. C. and M. F. Lavín. Effects of the autumn-winter meteorology upon the surface heat loss in the northern Gulf of California. *Atmósfera*, 10:101–123, 1997.
Ripa, P. Towards a physical explanation of the seasonal dynamics and thermodynamics of the Gulf of California. *J. Phys. Oceanogr.*, 27:597–614, 1997.
Roden, G. I. Oceanographic and meteorological aspects of the Gulf of California. *Pac. Sci.*, 12:21–45, 1958.

RESIDUAL FLOW AND MIXING IN THE LARGE ISLANDS REGION OF THE CENTRAL GULF OF CALIFORNIA

S. G. MARINONE AND M. F. LAVÍN
Departamento de Oceanografía Física, CICESE
Ensenada, Baja California, México

Abstract. A three-dimensional numerical model is used to describe mixing and tidal and residual flow in the archipelago of the central Gulf of California. The model is forced at the gulf's opening by prescribing the sea surface elevation (tidal, annual and semiannual) and the climatological (annual and semiannual) fields of temperature and salinity on the cross-section at the mouth. At the sea surface, forcing is with the climatological annual variations of the wind and of the heat and freshwater fluxes. The tidal currents dominate the instantaneous circulation, which floods and ebbs twice a day with speeds as large as 60 cm s^{-1}; the currents present a strong fortnightly modulation, with springs tidal currents being twice than those during neaps. The tidal flow over bottom features (like sills) generate internal tidal currents, which although less energetic than the barotropic currents, imprint their spatial variability on the total current field. Strong mixing is caused by the tidal flow, both by friction against the bottom and in the interior of the fluid. Mixing is tidally modulated, at the diurnal, semidiurnal and fortnightly frequencies. The strongest mixing occurs over and close to the sills, even during neap tides. The residual circulation in the upper layers reverses sign along the year and is rich in gyres over basins and sills. Over San Pedro Basin, a permanent anticyclonic gyre is produced. In the San Esteban and San Lorenzo sills the near-bottom current is always up-gulf, which causes a vertical structure that consists of two layers in winter and three layers in summer. The latter consists of a surface layer that flows into the northern gulf, a middle out-flowing layer, and the permanent near-bottom inflow. The residual currents are fortnightly modulated, with faster currents in springs than in neaps. The dynamics of the currents are such that in the transversal direction the dominant balance is geostrophic with some contribution of the advective terms, while in the along gulf direction all the terms are equally important, reflecting the non-linear and diffusive character of the area.

Key words: tidal currents, seasonal circulation, tidal mixing, Gulf of California

1. Introduction

The Gulf of California (GC) is a semi-enclosed sea between mainland Mexico and the Baja California peninsula (Figure 1); it is approximately 150 km wide and 1100 km long, with mean depths ranging from about 200 m in the

O.U. Velasco Fuentes et al. (eds.), Nonlinear Processes in Geophysical Fluid Dynamics, 213-236.

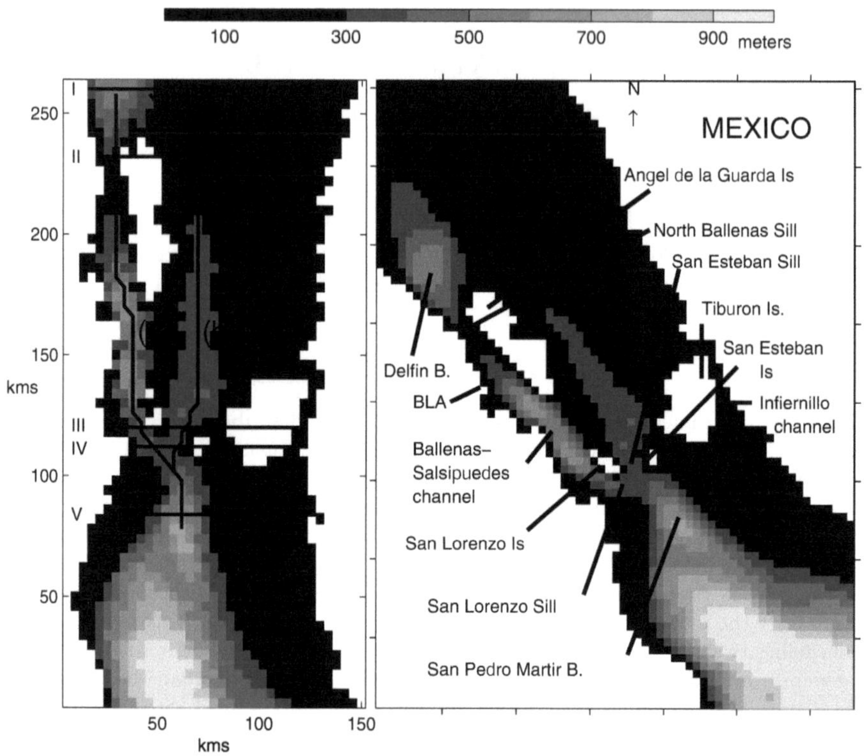

Figure 1. Bathymetry of the Gulf of California. The horizontal and vertical lines indicate the transversal and longitudinal sections along the gulf where some results are presented. BLA stands for Bahia de los Ángeles.

northern gulf to a maximum of 3600 m in the mouth. Between the shelf-like northern province and the deep southern province there is an archipelago containing sills, channels, basins, and islands of different sizes which reduces the communication between the northern and the southern gulf. Two large islands, Ángel de la Guarda and Tiburón, and several smaller islands, from which San Esteban and San Lorenzo protrude, constitute the archipelago. Two important time scales of variability of the GC are tidal (diurnal, semidiurnal, fortnightly) and seasonal (annual, semiannual).

The tides are in co-oscillation with the Pacific Ocean and the semidiurnal constituents are near resonance, with amplitudes at the head 4 times those at the mouth (Marinone, 2003); the tidal currents increase accordingly, becoming dominant in the northern region and more so in the midriff islands zone, where extensive mixing is produced (Argote *et al.*, 1995; Paden *et al.*, 1991).

The seasonal variability of the GC is due to the seasonality of the main forcing agents: the Pacific Ocean at the gulf's entrance (Ripa, 1997), the monsoonal wind regime (Badan *et al.*, 1991a), and the air-sea heat exchange (Beier, 1999; Marinone, 2003). In addition, Paden *et al.* (1991) suggest that tidal mixing also presents a considerable seasonal signal. Ripa (1990, 1997) proposed that the Pacific Ocean forces the gulf in the annual frequency by means of an internal baroclinic Kelvin wave of annual period which enters the GC on the eastern coast and then traverses cyclonically around the entire coastline. The hypothesis has been proved to explain the seasonal circulation and the balances of temperature and salinity (Beier, 1997; Palacios-Hernández *et al.*, 2002; Berón-Vera and Ripa, 2000 and 2002).

The non-linear interaction of the tidal and the residual currents with the bathymetry cause richly varied residual flow patterns, which includes gyres, coastal currents and other features (e.g. Zimmerman, 1980). The best-documented feature of the circulation in the GC is the large-scale seasonally reversing gyres in the northern gulf. They have been documented by Lavín *et al.* (1997) with satellite tracked drifters, by Carrillo *et al.* (2002) from geostrophic calculations, and by Palacios-Hernández *et al.* (2002) with current meters. The cyclonic gyre lasts about 4 months (June to September) and the anticyclonic one about 6 months (November to April). In the southern gulf, there are no equivalent observations, but estimates from ships drift and from the distributions of temperature and salinity indicate surface outflow during winter and inflow during summer, mass conservation requiring a compensating flow at depth (Bray, 1988). At the mouth of the gulf, different studies, mainly in winter, show currents of 10 cm s^{-1} that are cyclonic with a large vertical extension, reaching depths greater than 1000 m (Collins *et al.*, 1997; Castro, 2001).

There are several numerical-modelling studies of the tidal and residual circulation in the GC. Barotropic models were used by Argote *et al.* (1995, 1998) and by Marinone (1997). The first baroclinic model of the circulation in the GC was that of Carbajal (1993), which was forced locally by the wind. Beier (1997) used a two-layer linear model to explore Ripa's (1990, 1997) proposed forcing at the mouth by the Pacific Ocean. The model reproduced the observed annual surface height amplitude, the seasonal heat balance and the seasonally reversing gyres in the northern province of the GC. In the southern province, he found that the circulation is cyclonic (anticyclonic) in summer (winter) with inflow in the continental (Baja California) side. Beier's model was improved by Palacios-Hernández *et al.* (2002) by adding vertical mixing and the advective terms, which allowed the model to reproduce the annual-average heat balance and the annual-mean residual flow. Marinone (2003) has recently presented a full three-dimensional baroclinic model of the GC; it includes the tides, the seasonal circulation and thermodynamics, and vertical and lateral mixing. In addition to the seasonal wind stress and surface heat fluxes, the model is forced at the mouth, with annual and semiannual variations of the thermohaline structure. This model reproduces the seasonally-reversing gyre in the northern GC, while for the southern gulf, the residual circulation, below the Ekman layer, presents a pattern that is cyclonic during winter and summer, and anticyclonic during autumn and spring.

There are surprisingly few studies focusing on the oceanography of the very interesting region between the midriff islands. It is through this zone that the northern and southern provinces intercommunicate, and it is there that tidal currents and tidal mixing are strongest. It is one of the few places in the world where deep stratification is affected directly by tidal mixing (Simpson *et al.*, 1994), producing intense thermal fronts (Badan *et al.*, 1985; Argote *et al.*, 1995). Internal tides and internal solitons are produced by the tide flowing over the sills (e.g. Badan *et al.*, 1991b; Filonov and Lavín, 2003). In this area the tidal currents interact strongly with the background currents, and they also feed energy to these currents by means of tidal rectification and by currents produced by the distortion of the density field at tidal mixing fronts.

Considering that some recent observational studies in the area are producing very interesting results, and more observations are being carried out at present, the objective of this paper is to use the outputs of the numerical model of Marinone (2003) to describe the characteristics of tidal mixing and of the residual circulation in the zone of the archipelago. The seasonal and fortnightly variability will be described by analyzing two fortnightly cycles at the phases of the seasonal cycle (February and July) when the residual circulation is strongest but in opposing direction.

2. Model

The numerical model used is the layerwise vertically integrated Hamburg Shelf Ocean Model (HAMSOM). The model equations are solved semi-implicitly with fully prognostic temperature and salinity fields, thus allowing time-dependent baroclinic motion. The model domain has a mesh size of 2.5' x 2.5' ($\sim$3.9 km $\times$ $\sim$4.6 km) in the horizontal, and twelve layers in the vertical with nominal lower levels at 10, 20, 30, 60, 100, 150, 200, 250, 350, 600, 1000, and 4000 m. The model is described in detail in several papers (e.g. Backhaus, 1985; Marinone *et al.*, 1996), to which the reader is referred. The implementation of the model for the GC is described by Marinone (2003), who focused on the mean and seasonal global residual circulation, after proving that the model reproduces the main seasonal signals of the surface temperature, the heat balance and the tidal elevation.

Forcing. Several runs were performed forcing the model with the most important tidal constituents (M2, S2, N2, K2, K1, O1, P1, Ssa, and Sa) and varying the coefficient of bottom friction (C_d) until the best agreement with observed tidal harmonics at several tidal stations around the gulf was obtained. This was achieved with a constant value for C_d = 4.4 x 10^{-3} (Marinone, 1997). The temperature and salinity fields are prescribed at the gulf's entrance cross-section using the available historical database. Boundary gridpoints are interpolated for each month of the year by means of objective analysis and then least square fitted to obtain the annual and semiannual harmonics. At the sea surface a simple up- and down-gulf sinusoidal seasonal wind is imposed with amplitude of 5 m s^{-1}, with maximum up-gulf in August. Heat and fresh water fluxes were calculated with bulk formulae as in Castro *et al.* (1994) using the monthly meteorological data, from 6 stations around the gulf (also fitted to seasonal functions), and the calculated model SST. Finally, at the open boundary radiation conditions were implemented (Orlanski, 1976) on the dynamical variables and a sponge was also applied in the first 10 grid rows of the domain.

The model is started from rest, with a time step of 300 s. It becomes periodically stable in three years; the results presented here were obtained from the fourth year. While Marinone (2003) analyzes the model results by alternate choices of forcing agents, the results shown in this article come from the model that is forced with all of them, namely: (a) tides, (b) climatological hydrography at the mouth, (c) winds, and (d) heat and fresh water fluxes at the sea-air interface.

Definitions. The coordinate axes are defined with the gulf's axis along the y coordinate (positive toward the NW) and the x axis pointing across the gulf toward the mainland. The total instantaneous horizontal velocity field $\mathbf{u}(x, y, z, t) = (u, v)$, sampled every hour, can be separated into a barotopic and a baroclinic fields. The barotropic velocity field $\mathbf{U}(x, y, t) = (U, V)$ is defined by

$$\mathbf{U} = \frac{1}{H + \eta} \int_{-H}^{\eta} \mathbf{u} \, dz, \tag{1}$$

where η is the surface elevation, and H is the bottom depth. The internal or baroclinic current field $\mathbf{u}_i = (u_i, v_i)$, is defined as the difference between the instantaneous and barotropic currents: $\mathbf{u}_i = \mathbf{u} - \mathbf{U}$. The low-frequency, or residual, current $\mathbf{u}_r = (u_r, v_r)$ is obtained by passing the velocity fields three times through a 25 hour running average. Therefore residual currents also depend on time; the fortnightly, semiannual and annual variations are included.

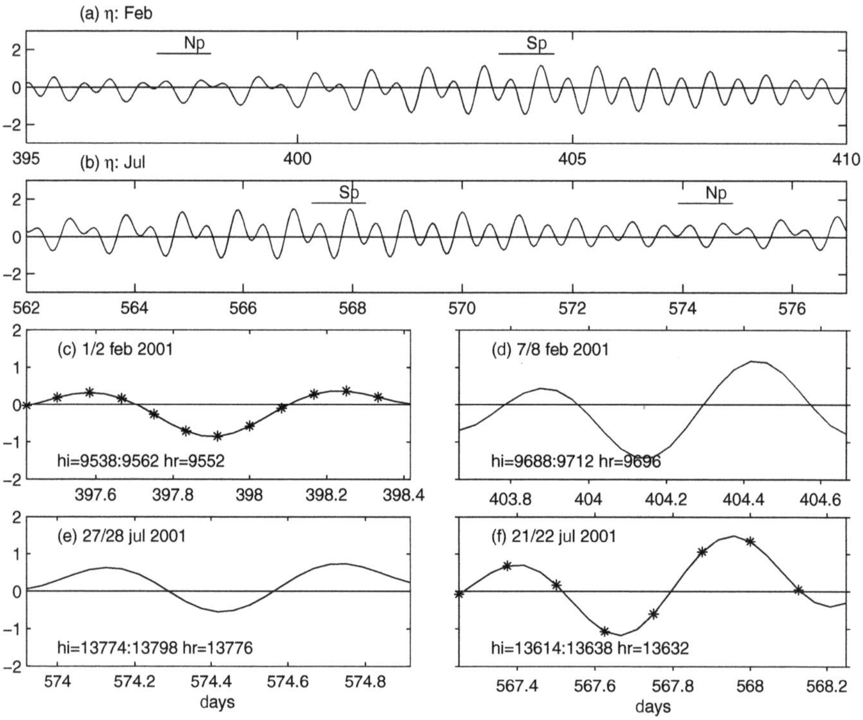

Figure 2. Time series of the surface elevation for (a) February and (b) July. Time is in Julian days starting in January 2001. The horizontal lines labeled Sp and Np stands for spring and neap tides and corresponds to a 25 hour period where results are shown and are expanded in (c) for neap and (d) for spring in February, and (e) for neap and (f) for spring in July. The * show the exact times for which some results will be shown. In c, d, e, and f, hi (instantaneous hour) and hr (residual hour) stands for the hour in which the results were sampled.

3. Results and discussions

The modelled residual circulation of the gulf presents different periods of cyclonic and anticyclonic circulation along the year, as described above. Based on the results of Marinone (2003), two fortnightly periods representative of these regimes were selected for analysis, in February and July. The time series shown in Figures 2a and b are the tidal elevations for the selected fortnightly periods at a grid point close to Bahía de los Ángeles, and is representative of the time series in this area. For each of these Springs-Neaps periods (henceforth abbreviated Sp-Np), 25-hour time series were selected during neap and spring tides, marked by horizontal lines in Figures 2a and b, and labelled Sp and Np, respectively. The time series are 25 hours long,

TABLE I. Temporal average and standard deviation, in a 25 hour cycle, of the root mean square of the magnitude over the horizontal area of the different components of the velocity field. Ins for the instantaneous currents, BT stands for the vertically integrated velocity, BC for the baroclinic currents, and Res for the residual or low frequency currents. The rms of the BT currents is independent of depth, however the reported value correspond to the rms calculated only with the wet gridpoints of the indicated layer. The residual current rms is from one day and therefore it has no standard deviation

	Depth (m): layer	Ins	BT	BC	Res
Feb neaps	0-10: 1	13.9 ± 5.7	11.2 ± 4.2	8.4 ± 0.5	9.2
	30-60: 4	12.1 ± 4.4	10.9 ± 4.2	4.6 ± 0.4	5.3
	150-200: 7	10.6 ± 4.6	10.3 ± 4.4	3.4 ± 0.6	2.6
	350-600: 10	8.7 ± 3.2	8.5 ± 4.0	4.5 ± 0.7	2.9
Feb springs	0-10: 1	32.1 ± 12.9	27.8 ± 12.4	12.0 ± 1.3	10.7
	30-60: 4	30.7 ± 12.4	27.2 ± 12.7	9.3 ± 1.2	7.1
	150-200: 7	28.8 ± 12.6	27.6 ± 12.8	8.5 ± 2.1	5.7
	350-600: 10	23.1 ± 8.7	23.0 ± 10.9	10.9 ± 2.2	4.5
Jul neaps	0-10: 1	16.4 ± 6.1	12.0 ± 4.9	10.0 ± 0.7	9.8
	30-60: 4	13.7 ± 4.7	12.0 ± 5.0	5.0 ± 0.5	5.3
	150-200: 7	12.2 ± 5.3	12.0 ± 5.1	4.0 ± 0.5	4.1
	350-600: 10	10.0 ± 3.5	10.0 ± 4.4	5.0 ± 0.8	2.5
Jul springs	0-10: 1	31.4 ± 11.8	25.8 ± 11.7	13.0 ± 1.0	10.7
	30-60: 4	29.6 ± 11.7	26.0 ± 11.9	9.0 ± 0.8	7.3
	150-200: 7	27.9 ± 12.1	26.0 ± 12.1	8.0 ± 1.6	5.8
	350-600: 10	22.1 ± 8.6	21.8 ± 10.4	11.0 ± 2.0	4.0

in order to cover from LLW (low low water), to LHW (low high water), to HLW (high low water), to HHW (high high water), and finally to LLW again(as shown in Figures 2c, d, e, and f) The 25-hour period will be called "tidal cycle" when integrating or averaging certain quantities.

3.1. INSTANTANEOUS CURRENTS AND MIXING

The instantaneous current fields are interesting because little is known in this respect from direct measurements, and because they are the source of mixing. In the area under study the instantaneous currents are larger than the residual currents, by an order of magnitude in some places. The pattern is illustrated in Figure 3 for the July springs 25-hour barotropic

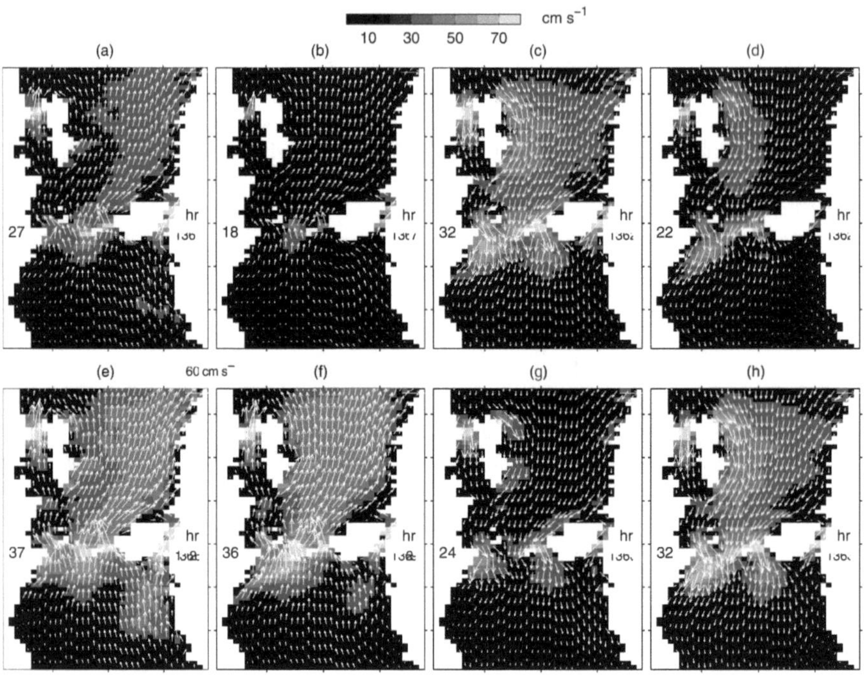

Figure 3. Time series of the vertically integrated velocity during July at spring tides. The numbers at the central and left of each frame is the rms magnitude in cm s^{-1}. The corresponding phase of the tide is shown in Figure 2f. Only one every four arrow vectors are shown. Speed scale is given by the background color and the length of the arrow.

current time series, sampled every 3 hours (see Figure 2f); in neap tides the currents are about half in magnitude and the spatial structure is the same. The numbers on the left center of each frame is the root mean square of the magnitude in cm s^{-1}. The flow is dominated by the tidal signal, so that only the twice-daily ebb and flow can be detected. The rms of the speed for the area, averaged over the tidal cycle, is $\sim$28 cm s^{-1} (in some sites it reaches 80 cm s^{-1}); by contrast, that for the residual flows is only $\sim$10 cm s^{-1} (see Table I). The barotropic currents suggest a funneling effect in all the area, but most clearly over constrictions such as channels and sills. For example, over San Esteban Sill, speeds reach $\sim$120 cm s^{-1} at springs. The lower speeds are in the southern area, where the depths are larger and speed are $\sim$40 cm s^{-1} at most. The associated transport oscillates during neaps from $\sim \pm 12$ Sv (1 sverdrup = 10^6 m^3 s^{-1}) to $\sim$ 30 Sv during springs. (The residual transport is $\sim$50 times smaller.)

The three dimensional total instantaneous current field corresponding to Figure 3g (i.e. hour 13632 see also Figure 2f) is shown for layers 1(0-

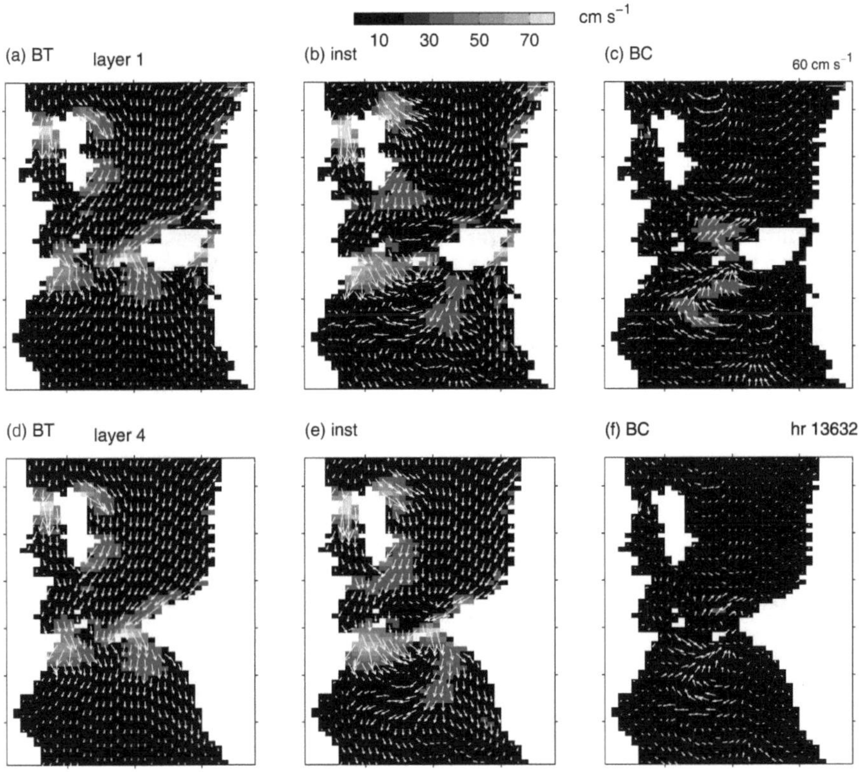

Figure 4. Snapshots of the barotropic (BT), instantaneous (inst), and baroclinic (BC) currents for the layers 1, 4, 7, and 10, respectively. The time corresponds to hour 13632, which is during spring tides in July. Only one every four arrow vectors are shown.

10m), 4(30-60m), 7(150-200m), and 10(350-600m) in Figures 4b, e, h, and k, respectively, together with its separation into barotropic (Figures 4a, d, g, and j) and baroclinic components (Figures 4c, f, i, and l). Table I shows the 25-hour average and standard deviation of the rms of the magnitude of the different velocity fields. At the time Figure 4 was sampled, the flow is ebbing (see Figure 3g) and the barotropic velocity field shows a southward flow that follows the coastlines in all layers. The **u** fields have more horizontal variability than those of **U**, with very large horizontal gradients, especially in the upper layers (Figures 4b and e); the spatial variability diminishes with depth. The variability is due to the baroclinic field $\mathbf{u}_i$ (Figures 4c, f, i, and l), which shows a series of patches of high baroclinic speeds, especially in the upper layers, which according to Beier (1999) and Filonov and Lavín (2003), may be due to internal tides.

To the west of Tiburón Island (henceforth, TI), the surface total current

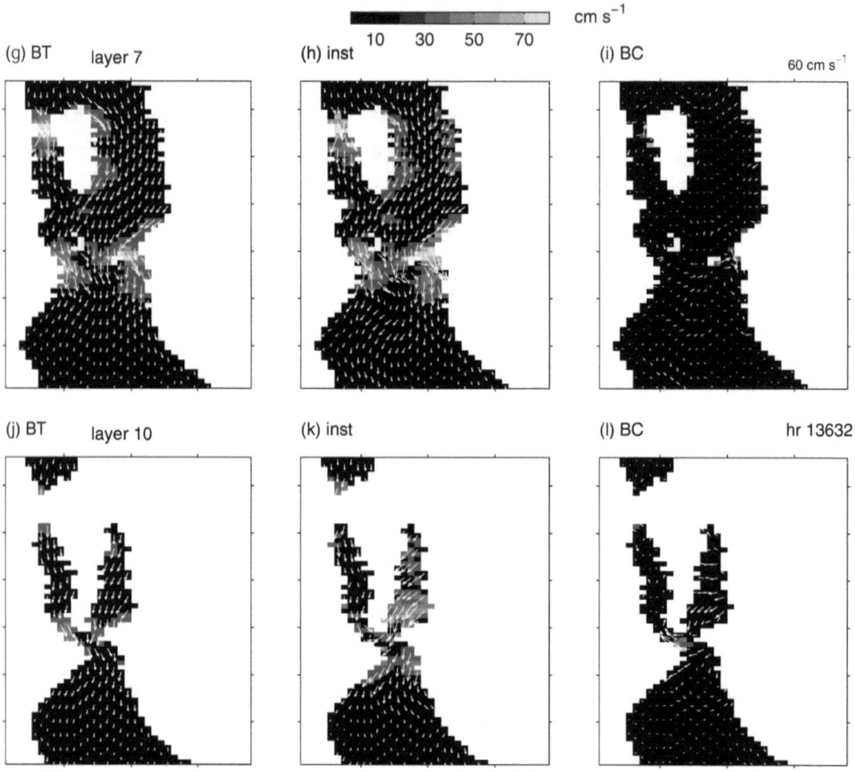

Figure 4. *Continued*

field (Figure 4b) shows a patch of low current flowing to the east, against the ebbing barotropic flow (Figure 4a). The reason for this minimum speed patch is that the surface baroclinic flow is strongly flooding at this time on this patch (Figure 4c); meanwhile, in layer 10 the baroclinic flow is ebbing (Figure 4l). This suggests an internal tide of mode 1, generated by the tidal barotropic flow over San Esteban Sill, as proposed by Filonov and Lavín (2003). Conspicuously strong baroclinic flows are found over San Lorenzo and San Esteban sills, and over San Pedro Basin, where the flow shows very large divergence/convergence; a permanent anticyclonic gyre will be shown to form in the latter. Table I shows that in general, the barotropic and instantaneous currents are twice as energetic as the internal currents and about three times larger than the low frequency currents present at the time of sampling of Figure 4. The time evolution of the instantaneous, $\mathbf{u}$ and $\mathbf{u}_i$ currents varies little from what is shown in Figure 4, with the exception that the direction reverses with the flood of the tide.

Transversal sections. Figure 5 illustrates the distribution of the along-

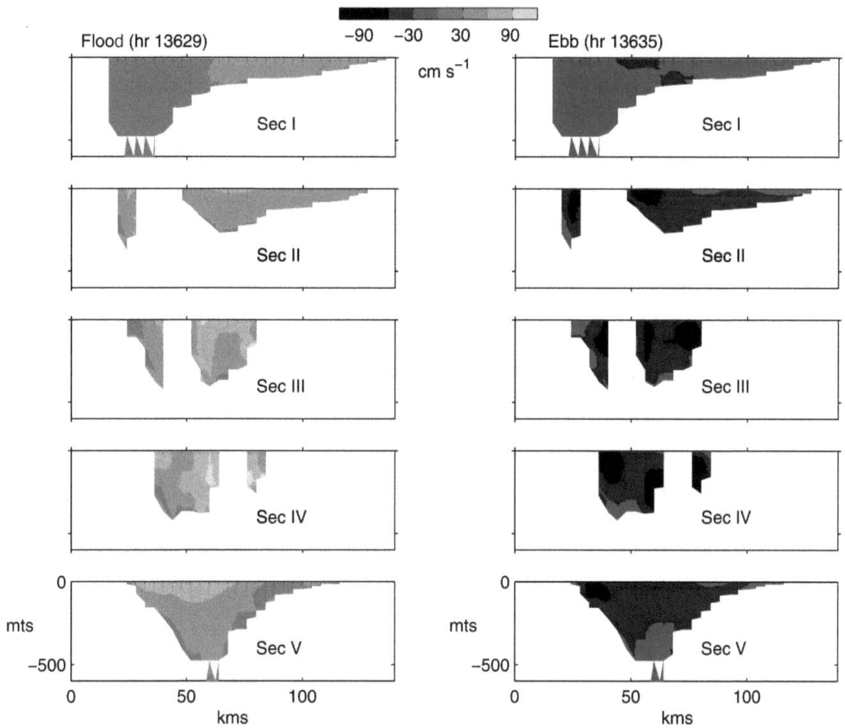

Figure 5. Snapshots of the along gulf instantaneous currents for the transversal sections I to V (see Figure 1 for their location) for flood (left) and ebb (right) currents during spring tides in July.

gulf current in Section I through V at spring tides in July during maximum flood and ebb phase (corresponding to Figures 3f and 3h); at these phases of the tide, the currents flow northward and southward all across every section, respectively. However, there is much spatial shear which is due to the low frequency currents (see below in Figure 10) and the baroclinic flow, which at the particular time of this figure reveals up- and down- gulf currents depending on position. The flow is most intense in the narrower sections III and IV close to San Esteban and San Lorenzo islands and over the sills. At section V the current is stronger at the surface and from the center to the peninsula side.

Dynamics of the instantaneous currents. A budget analysis of the different terms of the momentum equations for the area in general shows, with some exceptions, that the largest terms are those of the across-gulf geostrophic balance. In the longitudinal direction, the Coriolis and pressure gradient terms are equal to or larger than the other terms, i.e. all terms are equally important almost everywhere. In general all the forces are larger during springs than during neaps and there is almost no difference between winter and summer. The vertical and horizontal diffusive terms are smaller but not insignificant, reflecting vigorous mixing specially over the sills.

Mixing. From the first hydrographic measurements in the GC (Sverdrup, 1941), the area of the archipelago was identified as of strong tidal mixing, due to the low surface temperatures found there; the advent of satellite infrared measurements in the early 80s showed this in a very graphic way (Badan *et al.*, 1985). The coldest water is usually located over the sills, but the entire area has cool surface water. Thermal fronts are usually sharp close to the sills, weaker further away and distorted by intermediate scale flows like jets and gyres.

Energy for mixing in this area comes mainly from the tidal flow. There are two principal mechanisms for the transfer of energy from the mean flow to turbulent mixing: friction against the bottom and internal mixing by the breaking of internal waves (Kelvin-Helmholtz, internal jumps and bores).

Bottom-friction mixing in the GC was investigated by Argote *et al.* (1995) with a barotropic model of the M2 tide and by García and Marinone (2000) with several tidal constituents. They used $D_f = C_d(u^2 + v^2)^{\frac{3}{2}}$ and found that the largest amount of energy is dissipated in the archipelago and in the shallow part of the upper north of the gulf. Calculations with this model show the same results, the largest amount of energy by bottom friction occur in the archipelago area with and average of ~ 0.1 W m^{-2} with a $C_d = 3 \times 10^{-3}$; this is twice the amount dissipated in the northern gulf and 20 times larger than in the central and southern gulf.

There are no published measurements or numerical investigations of internal mixing in the area under study. A bulk measure of the degree of internal mixing is the Froude number, defined as

$$Fr = \sqrt{(\partial u/\partial z)^2/N^2}, \tag{2}$$

where $N^2 = -\frac{g}{\rho}\frac{\partial\rho}{\partial z}$ is the Brunt-Väisälä frequency. Fr is the square root of the inverse gradient Richardson number Ri. For continuously stratified parallel flows the inequality $Ri > \frac{1}{4}$ guarantees stability (LeBlond and Mysak, 1978); thus, $Fr < 2$ is a sufficient condition for stability. A flow with $Fr > 2$ is not necessarily unstable, but very likely to become unstable leading to turbulence and mixing.

As an example of the evolution of the distribution of Fr over a tidal cycle, Figure 6 shows the time series of Fr along the Ballenas-Salsipuedes vertical section (see location in Figure 1), for 24-hours every two hours, during neap tides in February (for spring tides Fr exceeds 2 everywhere, and is not shown). The figure clearly shows values larger than 2 appearing first over the sills (San Lorenzo and North Ballenas) and then, as the

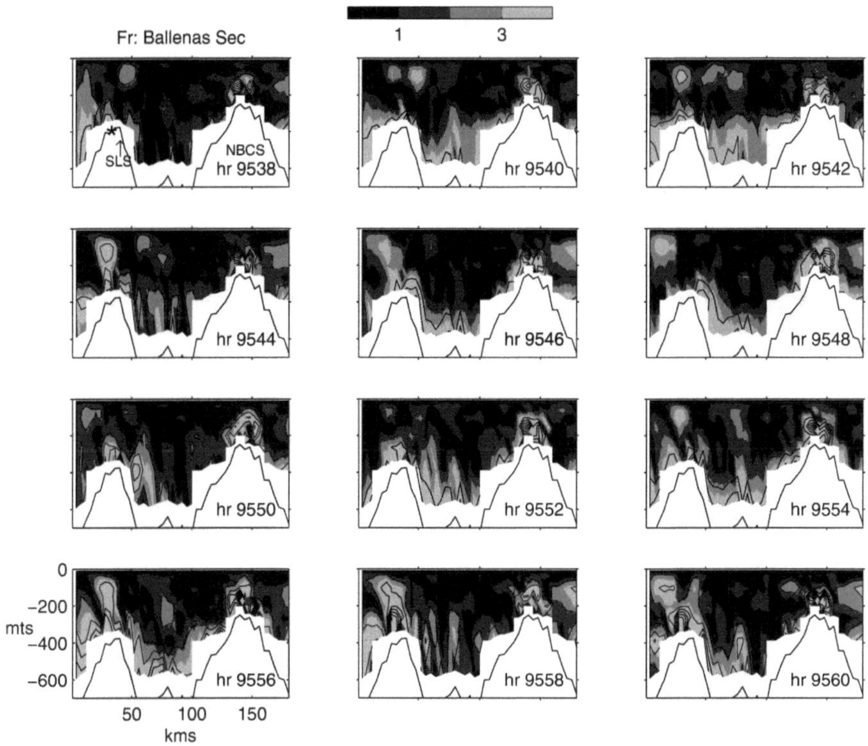

Figure 6. Time series of Fr number along the Ballenas section during a tidal cycle at neap tides in February. See Figure 2c for the phase of the tide.

mechanical energy increases, the mixing occurs in more places including the deeper reaches of the channel.

Figure 7 shows Fr vertically and time averaged over the selected 25-hour periods; the distributions for February (top panels) and July (bottom panels) and for neap (left panels) and spring tides (right panels) are shown. In general and for the whole area, Fr varies from values less than two to much larger than 2, and for a particular area, it is larger during spring tides than during neap tides. Over and close to the sills it is always large. In agreement with the previous figure, over San Esteban, San Lorenzo, and North Ballenas channel sills, Fr is larger than 2 even during neap tides. During spring tides the areas of large Fr extend all around Ángel de la Guarda Island (henceforth, AGI) and to the south of San Esteban and San Lorenzo sills. These changes are reflected in satellite infrared images (Soto *et al.*, 1999) where the coldest SST appear in these areas.

Filonov and Lavín (2003) reported in the area to the east of AGI internal tides with velocity amplitudes of 10-15 cm s^{-1}, and they estimate that

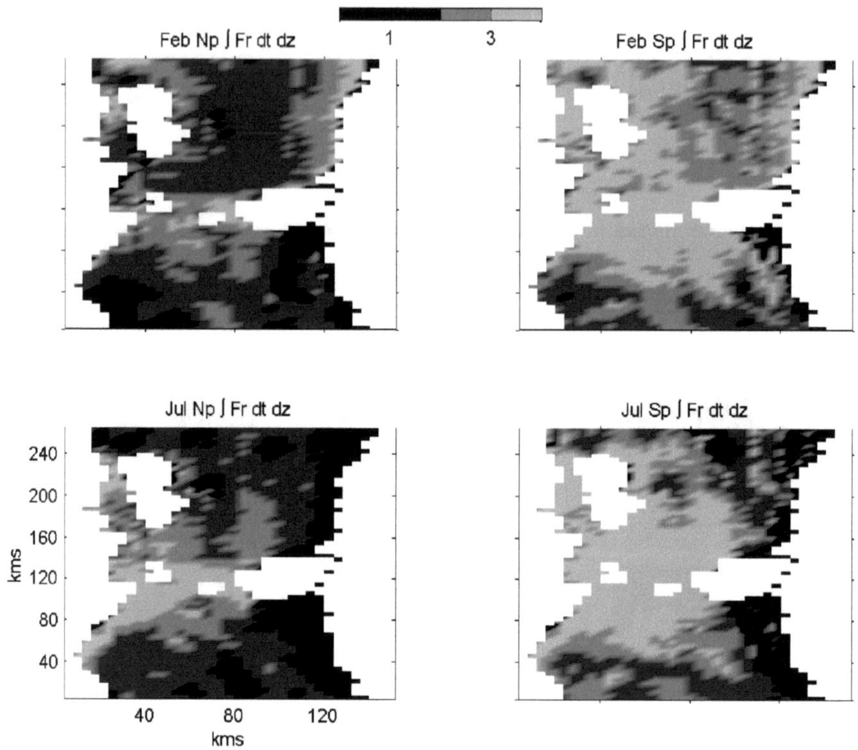

Figure 7. Froude number vertical and time averaged over a tidal cycle during neap and spring tides for February and July, respectively.

about 45% of the barotropic M2 tidal energy is transferred to the internal tide. As a measure of the internal energy, Figure 8 shows the average of $\sqrt{u_i^2 + v_i^2}$ over the selected 25-hours time series (neap and spring tides in February and July) and over the water column. Even at neap tides there are high energy levels around the sills (San Lorenzo, San Esteban, and North Ballenas channel). During spring tides the area with high energy covers a much wider area, including San Pedro, Tiburón, and Delfín basins. The July distributions (Figure 8, bottom panels) show wider coverage of the area with high internal energy than in February (Figure 8, top panels), especially during neaps. This internal energy is mainly produced by the barotropic tide as it moves over sills and basins, and then it is radiated away as internal tides and solitons (Filonov and Lavín, 2003). Although the spatial distributions of the high Fr shown in Figure 7 also show the Sp-Np and the seasonal variation, they are constrained to a smaller area around the sills than the distribution of the rms of the speed of the internal

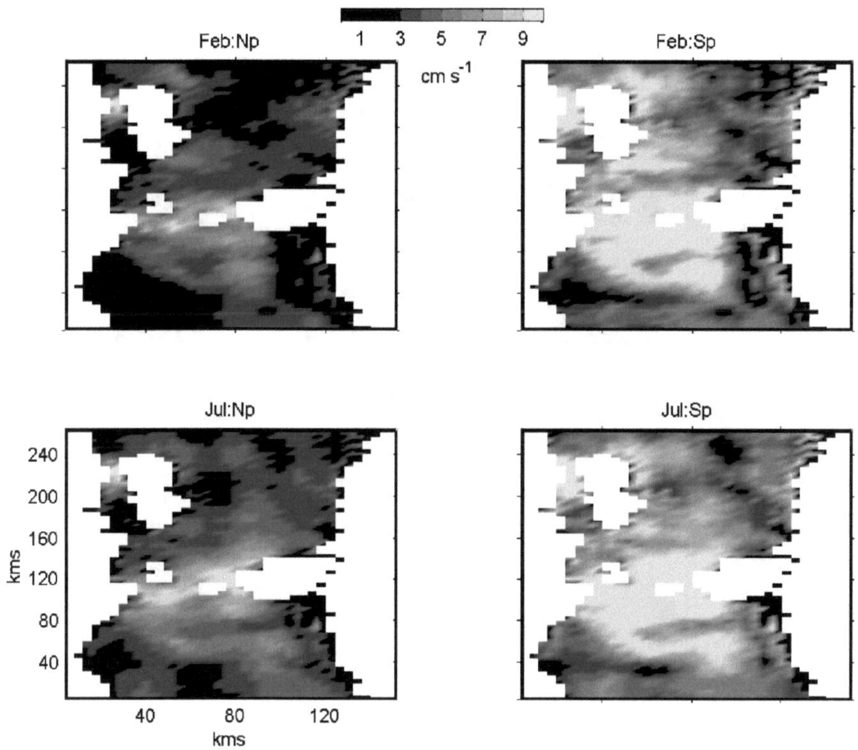

Figure 8. Time and vertical average of the magnitude of the internal currents during a tidal cycle at neap and spring tides in February and July.

currents (Figure 8), which suggests that baroclinic flow causes mixing only close to the sills.

3.2. RESIDUAL CURRENTS

After low-pass filtering the time series of $\mathbf{u}$, the low-frequency signal $\mathbf{u}_r$ was sampled in the middle of the Sp and Np periods shown in Figure 2, for both February and July. Figure 9 shows the horizontal residual currents for model layers 1(0-10m), 4(30-60m), 7(150-200m) and 10(350-600m) during spring tides for February (top panels) and July (lower panels). The corresponding residual currents during neap tides have very similar spatial patterns, but they are weaker (by a factor of 2) and smoother than those during spring tides; they are not shown, but this result means that the residual flow has a fortnightly modulation, with more energy being transferred from the tidal currents to the residual flow during spring tides than during neap tides (this modulation will be demonstrated below in Figure 12).

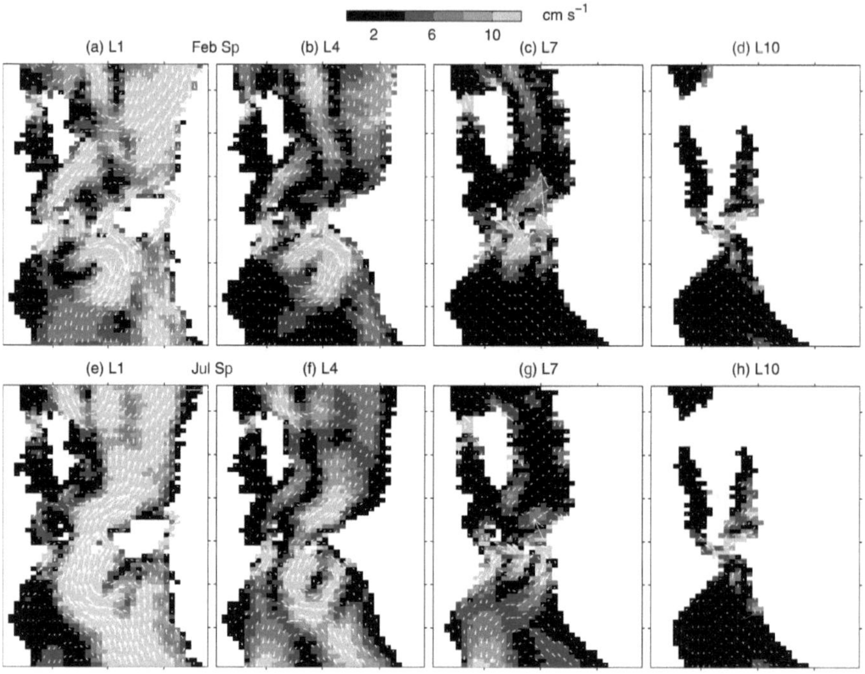

Figure 9. Snapshots of the residual currents for layers layers 1(0-10m), 4(30-60m), 7(150-200m), and 10(350-600m) for February (top) and July (bottom) during spring tides (phase correspond to Figures 2d and 2f. Only one every four vectors are shown.

The surface residual flow in February (Figures 9a), when the wind is blowing from the NW, enters the area over the mainland shelf as a wide well organized southward flow. In the area opposite AGI, it separates into three branches: (i) one branch turns back to the north and goes on to form an anticyclonic gyre in the NW of AGI; (ii) another branch crosses Tiburón basin and flows south parallel to the SE coast of AGI, then goes on to Salsipuedes Channel, and out over San Lorenzo and San Esteban Sills; (iii) the third branch goes on flowing parallel to the mainland coast until TI, where it separates into a fast flow in Infiernillo channel (between TI and the mainland) and another that follows the western coast of TI and then exits through the channel between TI and San Esteban Island. After flowing over the sills, the three branches join back, turn to the east and form an anticyclonic gyre over San Pedro Basin. In the 150 to 200 m layer (Figure 9c) the flow over San Esteban Sill is the reverse of that at the surface, while that over San Lorenzo Sill is still flowing out of the archipelago. In Tiburón basin, the flow is to the NW in the deepest part, while off the mainland coast it is to the SE, like in the surface layers. In the 350-600 layer (Figure 9d) the flow over the sills is toward the interior of the archipelago. It is to be noted that there is almost no residual flow in Ballenas Channel; however, over NBS the residual flow organizes in a small cyclonic gyre (this is better appreciated when zooming and plotting all the arrows).

The surface residual flow in July (Figure 9e), with the wind blowing from the SE, also enters the area over the mainland side. Over San Pedro basin an anticyclonic gyre is traced before flowing across the channels between the islands Tiburón, San Esteban and San Lorenzo. Some water also flows into Salsipuedes Channel over San Lorenzo Sill, which later turns east to join the main inflowing current. The main inflowing residual current flows over Tiburón Basin and in the mainland shelf, leaving an area of very weak residuals adjoining the eastern coast of AGI. In the 30-60 m layer (Figure 9f), the flow in Salsipuedes Channel and over San Lorenzo sill is out of the area; this outgoing flow can be traced up to the east of AGI. The main ingoing flow occurs like in the surface layer, in the channels between the islands of San Lorenzo, San Esteban and Tiburón. The residual flow in the 150-200 m layer (Figure 9g) is very similar to that just described, except that the main flow over the mainland shelf is not present. In the 350-600 m layer (Figure 9h) the flow over the sills is toward the interior, but in Tiburón basin the flow is in the opposite direction. Therefore, in the bottom layer over San Esteban and San Lorenzo sills the currents are always up-gulf (Figures 9d and h).

A cyclonic gyre is always present NE of AGI, while an anticyclonic gyre is found over San Pedro basin. The latter is clearly connecting, during July, the northern and southern parts of the gulf, while during February it appears to be isolated. Marinone (2003) has shown that this gyre is due to tide-induced mixing processes by means of model simulations with and without tides and stratification. Without stratification, the residual tidal currents over San Pedro basin reach only 1-2 cm s^{-1}, while in Figure 9 they reach around 35 cm s^{-1}. When the tides are switched off, no gyre develops at all over San Pedro basin.

López and García (2003) reported at the NE of AGI a strong bottom current from November 1997 to March 1998, with a mean speed of $\sim$25 cm s^{-1}. Here we found strong currents as well, but with speed about half the observed magnitude, maybe due to the low vertical resolution of the model.

More details of these circulation patterns can be appreciated in Fig, 10, which shows the vertical distribution of the along-gulf residual currents for the across-gulf sections I to V (see Figure 1 for their location). On the left panels are the February springs data corresponding to Figure 10a-e, and on the right panels are those corresponding to July springs (Figure 10f-j); as before, during neap tides the residual currents are weaker and present a

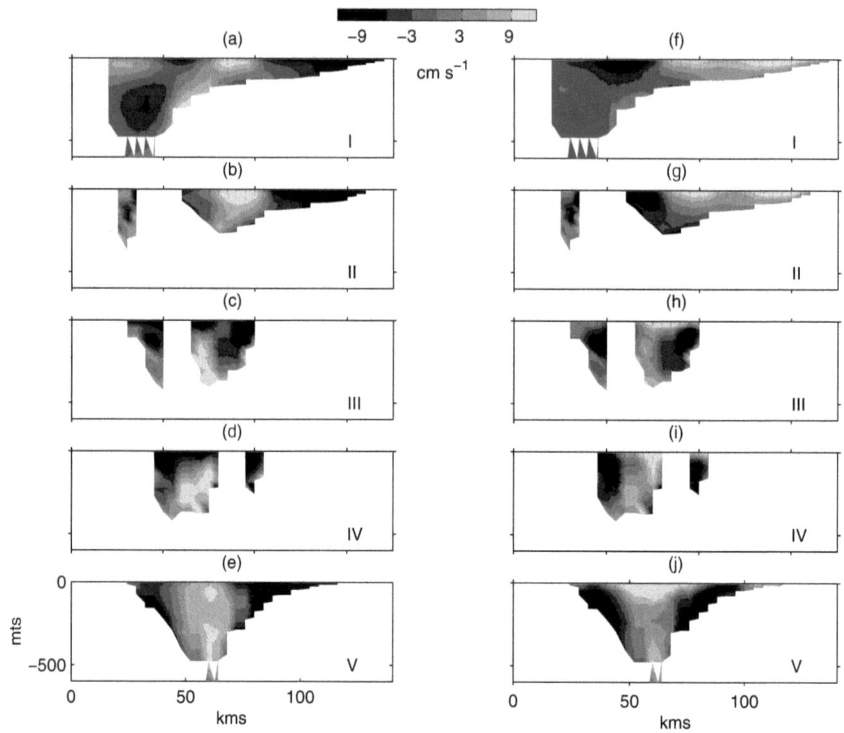

Figure 10. Snapshots of the along gulf residual currents for transversal sections I to V (see Figure 1 for their location) for February (left) and July (right) during spring tides.

similar pattern, and are not shown. A very rich vertical structure is revealed. In the upper layers, the February southward flow that enters the area over the mainland shelf can be tracked from section I (Figure 10a) to section V (Figure 10e); in July, the reverse surface flow in the same areas is apparent (Figure 10f-j). During February, from section V in the south (Figure 10e) to section I in the north (Figure 10a), there is a coherent near-bottom ingoing flow, which reaches section I (Figure 10a) on the eastern side of the section to feed the cyclonic gyre there. During July, this bottom current reaches only up to Tiburón basin (section III, Figure 10h), although traces of a very weak northward bottom current are still present in section I (Figure 10f).

The vertical structure of the gyres described in Figure 9 can be appreciated in Figure 10. Section I (Figures 10a and f) show the seasonally reversing gyre over Delfín basin; as observed [Lavín *et al.*, 1997; Carrillo *et al.*, 2002] it covers the entire water column it is tilted to the east, and not centered on the basin. On the right side of Section V (Figures 10e and j), which crosses San Pedro Basin, the anticyclonic gyre described before is quite apparent; it extends over the entire water column, and seems to feed the northward flow over the sills just north of the basin, which is clearly seen in Sections III (Figures 10c and h) and IV (Figures 10d and i). In February a two-layer system is present with inward bottom flow and outward surface flow (Figures 10c and d); in July (Figures 10h and i) northward flow occurs in the bottom and in the surface layers adjacent to San Esteban Island, while southward flow occurs only on the peninsula side.

The vertical structure of the along-gulf residual currents for the longitudinal sections (see Figure 1 for their location) through Tiburón Basin (Tiburón Section, TS hereafter) and through the Ballenas-Salsipuedes Channel (BS hereafter) are shown in Figure 11. Once again, the springs February (Figure 11a and b) and Springs July (Figure 11c and d) data are shown (note that the first few points, from the extreme left to the * symbol, just before San Lorenzo and San Esteban sills, are the same in the two sections). There is two-layer flow (going into the northern gulf in the bottom layers, going out in the top layers) in the two seasons over San Lorenzo and North Ballenas Channel sills as well as in the Salsipuedes-Ballenas Channel (Figures 11a and c). San Esteban sill shows two layer flow only during February (Figure 11b). The strong currents over the sills cause extensive mixing, as shown, for example, in Figure 6.

The temporal evolution for 30 days of the along-gulf residual flow over San Lorenzo Sill (the cross-point of Sections IV and BS) is shown in Figure 12, for February (Figure 12a) and July (Figure 12b). During February (Figure 12a), with winds from the NW, there is a two-layer flow system: outgoing flow is from the surface to about 220 m, and from there to the bottom is into the northern gulf. The strongest residual currents are found at the surface and at the bottom. The Sp-Np modulation of the two-layer system can clearly be appreciated, with stronger outward and inward currents during spring tides than during neap tides. Also, the interface between the two layers seems to rise $\sim$40 m during neap tides. During July (Figure 12b) there is a three-layer flow system, which switches to two layers during spring tides. There is a wind-driven in-flowing surface layer 10-20 m thick (which disappears during springs), then there is an outgoing layer down to 300-320 m, and lastly there is an ingoing bottom layer. The maximum of the outgoing flow is in the middle of the second layer, while the inflowing currents are maxima in the surface and at the bottom. This seasonally changing two and three-layer system of residual flow was proposed by Bray (1988), and first reproduced in a numerical model by Marinone (2003). Observations of inflowing bottom currents over San Lorenzo sill were first reported by Badan *et al.* (1991b) for a short period of time, and by Argote *et al.* (2003) for many months. The observations of Badan *et al.* (1991b) show the residual flow arresting and overcoming completely the

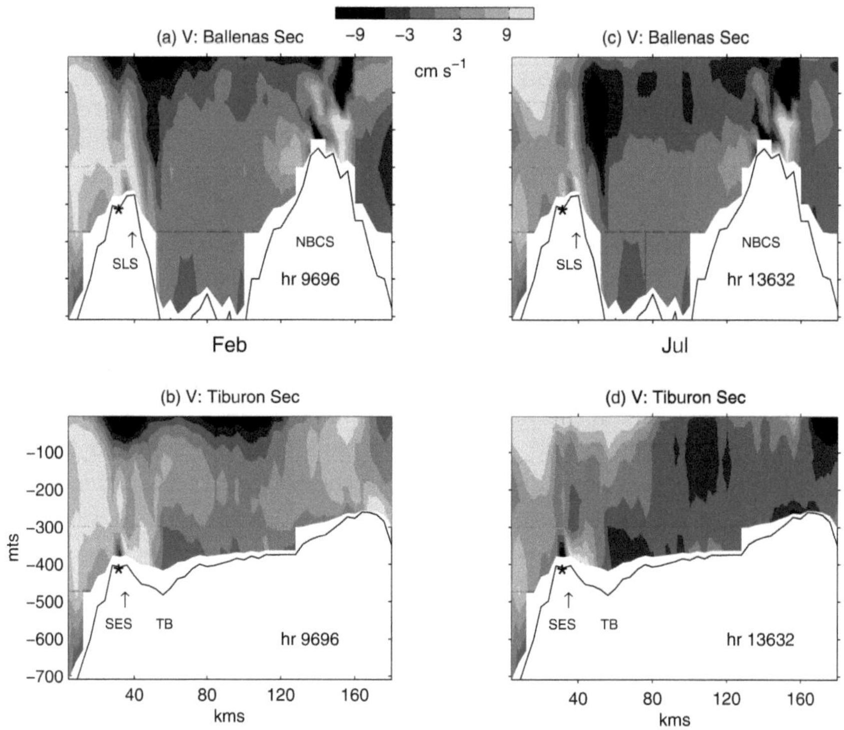

Figure 11. Snapshots of the along gulf residual currents for Ballenas (top) and Tiburón (bottom) longitudinal sections during February (left) and July (right). See Figure 1 for their location. SLS, NBCS, SES, and TB stands for San Lorenzo sill, North Ballenas channel sill, San Esteban sill, and Tiburón basin, respectively.

ebbing tidal currents, while in this model the tidal currents are stronger than the residual flow.

4. Conclusions

The three-dimensional numerical model of the dynamics and thermodynamics of the GC of Marinone (2003) is used to describe in some detail the distribution patterns of the tidal flow, the seasonal circulation and some vertical mixing parameters in the archipelago of the central gulf. The model is forced at the mouth of the gulf with the 7 principal diurnal and semidiurnal tidal harmonics of the surface elevation, and with the climatological annual and semiannual variations of the sea level height and of the temperature and salinity fields. The model was also forced at the surface with the climatological (annual) variation of wind and air-sea heat

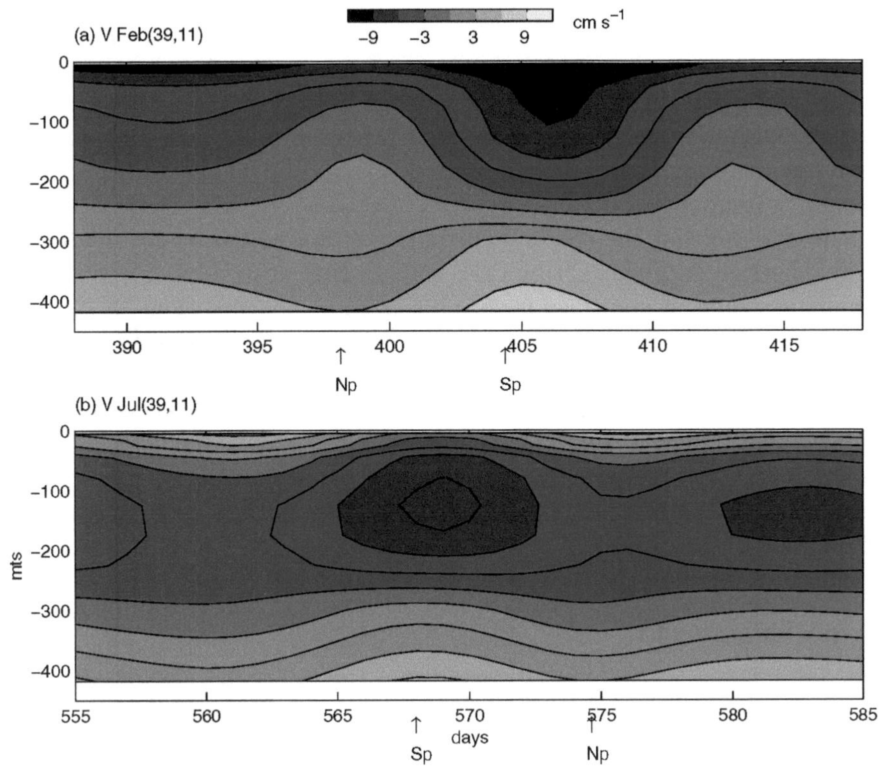

Figure 12. Time series of the along gulf residual velocity during February (top) and July (bottom) over San Lorenzo Sill (the cross-point of Sections IV and BS, see Figure 1).

and water fluxes; the latter were calculated from climatological (annual) meteorological variables and the modelled sea surface temperature.

The instantaneous barotropic currents are dominated by the tidal signal, with an rms of the speed, averaged over the area and over the tidal cycle, of $\sim$30 cm s^{-1} (Table I). These currents are strongest in some shallow areas and especially in constrictions like silled channels between the islands (San Esteban Sill, San Lorenzo Sill, and North Ballenas Sill) where they reach values close to 1 m s^{-1}. The instantaneous baroclinic (or internal) currents, which are produced by the barotropic tide as it passes over sills and basins, present extensive spatial varability (suggesting internal tides), and although they are generally less energetic than the barotropic flow (Table I), they imprint their variability on the total velocity field. The internal currents show much larger shears than the barotropic currents over topographic features, reflecting the topographic steering of the tidal currents. The areas where the largest total currents are found are the southern sills and over

San Pedro Basin, where a permanent anticyclonic gyre is produced.

The barotropic tidal flow causes strong mixing by friction against the bottom, and the internal tidal flow causes mixing in the interior of the fluid by rising Fr above the critical value of 2. In both cases the strongest mixing occurs over and close to the sills, where $Fr > 2$ even during neap tides. Also, mixing is tidally modulated, at the diurnal, semidiurnal and fortnightly frequencies.

It was found that the residual currents are modulated by the Sp-Np cycle, being stronger during spring tides than during neap tides. They are also strongly seasonal, the surface flow being in opposite directions in summer and winter (mostly following the seasonal wind changes) in areas where no drastic topographic features are present. However, over some basins (e.g. San Pedro Basin) the residual flow is in the same direction all year round, leading to permanent gyres. The residual currents show large vertical shear, forming a two or three-layer system during February and July, respectively. Close to the bottom over of San Esteban and San Lorenzo sills the residual currents are always up-gulf.

Acknowledgements

This research was financed by CONACyT, through grants 4300P-T and 35351-T, and by CICESE's regular budget.

References

Argote, M. L., A. Amador, M. F. Lavín, and J. Hunter. Tidal Dissipation and Stratification in the Gulf of California. *J. Geophys. Res.*, 100(6):16103–16118, 1995.

Argote, M. L., M. F. Lavín, and A. Amador. Barotropic Eulerian Residual Circulation in the Gulf of California Due to the M2 Tide and Wind Stress. *Atmósfera*, 11:173–197, 1998.

Argote, M. L., R. Romero-Centeno, F. Plaza, and A. Amador. Circulation and Subsurface Water Exchange Through the Central Archipelago of the Gulf of California. *J. Geophys. Res.*, submitted.

Backhaus, J. O. A Three-dimensional Model for the Simulation of Shelf Sea Dynamics. *Deutsche Hydrographisique Zeitschrift*, 38:165–187, 1985.

Badan-Dangon, A., D. J. Koblinsky, and T. Baumgartner. Spring and Summer in the Gulf of California: Observations of Surface Thermal Patterns. *Ocean. Acta*, 8:13–22, 1985.

Badan-Dangon, A., C. E. Dorman, M. A. Merrifield, and C. D. Winant. The Lower Atmosphere Over the Gulf of California. *J. Geophys. Res.*, 96:16877–16896, 1991.

Badan-Dangon, A., M. C. Hendershott, and M. F. Lavín. Underway Doppler Current Profiles in the Gulf of California. *Eos Trans. AGU*, 72(209):217–218, 1991.

Beier, E. A Numerical Investigation of the Annual Variability in the Gulf of California. *J. Phys. Oceanogr.*, 27:615–632, 1997.

Beier, E. Estudio de la Marea y la Circulación Estacional en el Golfo de California Mediante un Modelo de Dos Capas Heterogéneas. PhD thesis, CICESE, Ensenada, B.C., México. 1999.

Beier, E., and P. Ripa. Seasonal Gyres in the Northern Gulf of California. *J. Phys. Oceanogr.*, 29:302–311, 1999.

Berón-Vera, F. J., and P. Ripa. Three-dimensional Aspects of the Seasonal Heat Balance in the Gulf of California. *J. Geophys. Res.*, 105:11441–11457, 2000.

Berón-Vera, F. J., and P. Ripa. Seasonal Salinity Balance in the Gulf of California. *J. Geophys. Res.*, 107:15.1-15.15, 2002.

Bray, N. A. Thermohaline Circulation in the Gulf of California. *J. Geophys. Res.* 93:4993–5020, 1988.

Carbajal, N. Modeling of the Circulation in the Gulf of California. *PhD thesis, Institute fur Meereskunde, Hamburg.* 1993.

Carrillo, L. E., M. F. Lavín, and E. Palacios-Hernández. Seasonal Evolution of the Geostrophic Circulation in the Northern Gulf of California. *Estuarine, Coastal and Shelf Sci.*, 54:157–173, 2002.

Castro, R. Variabilidad Termohalina e Intercambios de Calor, Sal y Agua en la Entrada al Golfo de California. PhD thesis, UABC, Ensenada, México. 2001.

Castro, R., M. F. Lavín, and P. Ripa. Seasonal Heat Balance in the Gulf of California. *J. Geophys. Res.*, 99:3249–3261, 1994.

Collins, C. A., N. Garfield, A. S. Mascarenhas, M. G. Spearman, and T. A. Rago. Ocean Currents Across the Entrance to the Gulf of California. *J. Geophys. Res.*, 102:20927–20936, 1997.

Filonov, A., and M. Lavín. Internal Tides in the Northern Gulf of California. *J. Phys. Oceanogr.*, in press.

García, G., and S. G. Marinone. Tidal Dynamics and Energy Budget in the Gulf of California. *Ciencias Marinas*, 26:323–353, 2000.

Lavín, M. F., R. Durazo, E. Palacios, M. L. Argote, and L. Carrillo. Lagrangian Observations of the Circulation in the Northern Gulf of California. *J. Phys. Oceanogr.*, 27:2298–2305, 1997.

LeBlond, P. H., and L. A. Mysak. *Waves in the Ocean.* Elsevier Oceanographic Series, 1978.

López, M., and J. García. Moored Current and Temperature Observations in the Northern Gulf of California: A Strong Current Near the Bottom. *J. Geophys. Res.*, 108:3048–3065, 2003.

Marinone, S. G. Tidal Residual Currents in the Gulf of California: Is the M2 Tidal Constituent Sufficient to Induce Them?. *J. Geophys. Res.*, 102:8611–8623, 1997.

Marinone, S. G. A Three Dimensional Model of the Mean and Seasonal Circulation of the Gulf of California. *J. Geophys. Res.* submmited.

Marinone, S. G., S. Pond, and J. Fyfe. A Three-dimensional Model of Tide and Wind-induced Residual Currents in the Central Strait of Georgia, Canada. *Estuarine, Coastal and Shelf Sci.*, 43:157–182, 1996.

Orlanski, I. A Simple Boundary Condition on Unbounded Hyperbolic Flows. *J. Comput. Phys.*, 21:251–269, 1976.

Paden, C. A., M. R. Abbott, and C. D. Winant. Tidal and Atmospheric Forcing of the Upper Ocean in the Gulf of California 1. Sea Surface Temperature variability. *J. Geophys. Res.*, 96:18337–18359, 1991.

Palacios-Hernández, E., E. Beier, M. F. Lavín and P. Ripa. The Effect of the Seasonal Variation of Stratification on the Circulation on the Northern Gulf of California. *J. Phys. Oceanogr.*, 32:705–728, 2002.

Ripa, P. Seasonal Circulation in the Gulf of California. *Annales Geophys.*, 8:559–564, 1990.

Ripa, P. Towards a Physical Explanation of the Seasonal Dynamics and Thermodynamics of the Gulf of California. *J. Phys. Oceanogr.*, 27:597–614, 1997.

Simpson, J. H., A. J. Souza, and M. F. Lavín. Tidal Mixing in the Gulf of California. in *Mixing and Transport in the Environment.*, Edited by K. J. Beven, P. C. Chatwin and J. H. Millbank, 160–182 pp. John Wiley, New York. 1994.

Soto-Mardones, L., S. G. Marinone, and A. Parés-Sierra. Time and Spatial Variability of Sea Surface Temperature in the Gulf of California. *Ciencias Marinas*, 25:1–30, 1999.

Sverdrup, H. U. The Gulf of California: Preliminary Discussion of the Cruise of the "E. W. Scripps" in February and March. *Sixth Pacific Science Congress Proceedinds*, III:161–166, 1941.

Zimmerman, J. T. F. Vorticity Transfer by Tidal Currents Over an Irregular Topography. *J. Mar. Syst.*, 38:601–630, 1980.

A DESCRIPTION OF GEOSTROPHIC GYRES IN THE SOUTHERN GULF OF CALIFORNIA

J. M. FIGUEROA, S. G. MARINONE AND M. F. LAVÍN
Departamento de Oceanografía Física, CICESE
Ensenada, Baja California, México

Abstract. Historical hydrographic data are used to investigate the geostrophic circulation in the Southern Gulf of California, focusing on the description of evidences supporting the existence of gyres. It is found that the horizontal dimensions and sense of rotation of the gyres are variable, and that their position are not tied to basins and sills as has been previously proposed. In the vertical, the gyres reach at least 500 meters, and probably 1000 meters. A seasonal behavior cannot be ascertained due to data limitation.

Key words: geostrophic gyres, semienclosed basins, Gulf of California

1. Introduction

Geostrophic gyres are a common occurrence in the oceans. In closed and semienclosed basins, Emery and Csanady (1973) noted that a cyclonic circulation tends to occur more often than an anticyclonic one, but there are many exceptions. The Adriatic, for instance, has a global cyclonic circulation, broken into three recirculation cells in the northern, central and southern sub-basins; the first cell is influenced by the discharge of the Po river, while the latter two are controlled by the bathymetry (Poulain, 2001). In the Irish Sea, tidal mixing fronts induce a cyclonic gyre during the summer (Hill, 1993).

In the Gulf of California (GC hereafter) (Figure 1), the circulation in the northern region is dominated by a seasonally reversing gyre of a diameter comparable to the width of the gulf (Lavín *et al.*, 1997). Numerical models suggest that the gyre and its behavior are due to bathymetric control of the flow forced by winds and by the Pacific Ocean (Beier and Ripa, 1999). In the southern region of the Gulf of California (abbreviated SGC), the presence of geostrophic gyres was first observed with drifting buoys and hydrography in August 1978 by Emilsson and Alatorre (1997). Previous reports based on hydrographic data were made by Figueroa and Robles (1989). The presence of gyres and jets in the GC is evident in infrared and

O.U. Velasco Fuentes et al. (eds.), Nonlinear Processes in Geophysical Fluid Dynamics, 237-255.

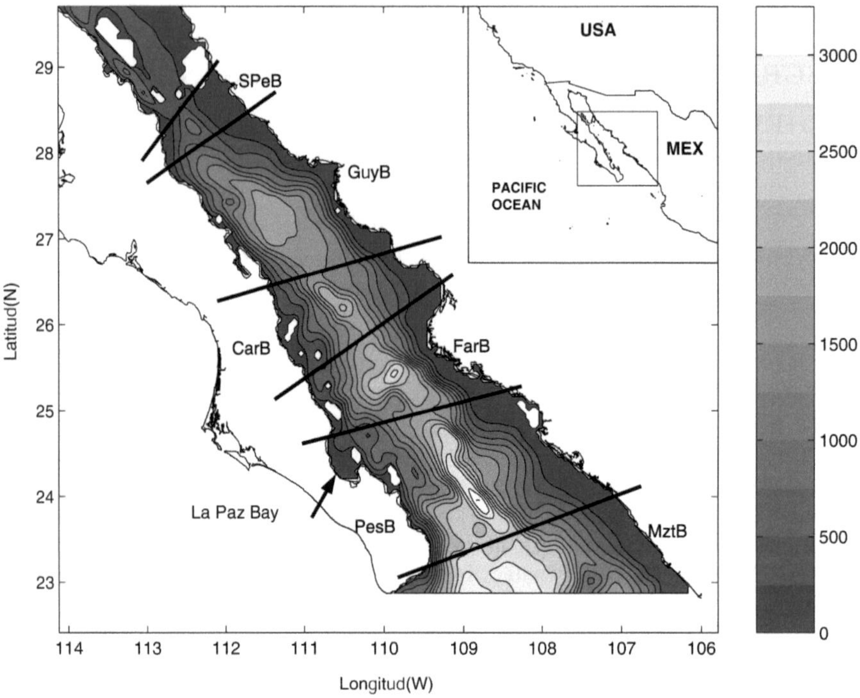

Figure 1. Study area and bathymetry (depth in meters). Heavy lines crossing the gulf indicates the sills separating the basins. The names of the basins are coded as follows: Mazatlán Basin (MztB), Pescadero Basin (PesB), Farallón Basin (FarB), del Carmen Basin (CarB), Guaymas Basin (GuyB) and San Pedro Mártir Basin (SPeB).

color satellite images (Badan-Dangon *et al.*, 1985; Gaxiola *et al.*, 1999). The images often suggest (e.g. Figure 9, of Badan-Dangon et al., 1985) the presence of a sequence of basin-wide counter-rotating gyres along the length of the SGC. There are also cool jets between the gyres, which seem to shoot off from coastal capes; these jets have been proposed as a possible mechanism in the transport of eggs and larvae of sardine (Hamman *et al.*, 1988). Recently Amador et al. (2003) report a northward-flowing coastal jet just off La Paz Bay which seems to be the western limb of a cyclonic gyre, very similar to that observed by Emilsson and Alatorre (1997). Although no rigorous analysis of these features has been carried out, it has been proposed that this sequence of gyres is a permanent feature of the circulation, and that their position coincides with the basins of the SGC (Fernández-Barajas *et al.*, 1994; their Figure 2).

In this paper, the GC hydrographic data bank is used to explore the dynamic topography of the GC in a search for structures suggesting the presence of geostrophic gyres like those reported in literature. Availability of data limits our investigation to describe -if they are present- the horizontal and vertical extension of gyres, and their possible relation to the position of bathymetric features like basins and sills. This work is a census of identifiable structures in the mass field through their signature in anomalies of dynamic topography.

2. Data and techniques

The data used here were selected from the hydrographic data bank of the GC, selection was done based on vertical resolution and horizontal coverage of the SGC. Cruises with not enough casts to allow a proper definition of structures in the size of SGC width where excluded, and so were those with a dense sampling net but with a vertical coverage less than 250 m. All casts in the strongly tidally energetic area around the archipelago (Argote *et al.*, 1995) where not included either. Temperature and salinity data in the older surveys were collected with reversing thermometers and sampling bottles, while CTDs have been used since the 1980s; for consistency, data every 10 meters were obtained from all profiles, by interpolation or by subsampling, respectively. Unfortunately, we can not assert anything about errors in data; there is no information about them in the data reports that we used to get them. Not having information about position and instrumental errors preclude us to use techniques like objective mapping to obtain field maps, and to make any correction for the presence of non geostrophic like structures. We selected 18 of 47 cruises in the gulf. The 18 cruises have 1320 casts but we use only 625 due to their vertical and horizontal coverage.

Dynamic topography maps were calculated using 500 db as a reference level (rl) in most cases. In a few cases 250 db had to be used as rl, and in one case it was possible to use 1000 db as rl. The average was removed from the dynamic height distributions; therefore the maps represent an anomaly of the dynamic topography. Maps drawing were done using MATLAB standard contouring programs with a quadratic level of interpolation, data used as a base to interpolate where carefully selected inside a polygon defined by usable casts, such polygon is shown in all of the maps enclosing the contour lines. The vertical distributions of geostrophic velocity were also obtained, but only selected sections will be shown.

The data coverage, both in time and in space (vertical as well as horizontal), is so poor that it precludes the use of harmonic analysis or of Empirical Orthogonal Functions (EOF) analysis, like that used by Carrillo *et al.* (2002) and Palacios *et al.* (2002) for the study of the circulation in the northern region of the GC. Therefore, the fields of dynamic height anomaly are described in a monthly basis, attempting to capture possible seasonal patterns. The best sampled months are January (1984), February (1957 and 1990), March (1939, 1983, 1984 and 1985), April (1956), October (1974, 1983 and 1994) and November (1961, 1972, 1984 and 1985). The surveys of April, June, and August 1957 do not allow the construction of maps of dynamic heights because the lines of stations across the gulf are too far apart; however, the same cross-sections were well sampled in the three surveys, therefore schematic diagrams of the inferred circulation are constructed. In order to investigate a possible relationship between the position of the circulation features and the basins of the gulf, the sills between the different basins are indicated, in each dynamic height anomaly map, by lines crossing the gulf. The basins shown in Figure 1 are marked in each map as follows: Mazatlán Basin (MztB), Pescadero Basin (PesB), Farallón Basin (FarB), del Carmen Basin (CarB), Guaymas Basin (GuyB) and San Pedro Mártir Basin (SPeB).

3. Results

Results are presented as dynamic topography anomalies. Closed contour lines are therefore indicative of the presence of anticyclones (if positive values, denoted by light shading) or cyclones (negative values, denoted by dark shading), while open/parallel lines denote the presence of jet-like structures. The identification of all structures mentioned in next section is based on this.

3.1. SURFACE CIRCULATION

January 1984. The dynamic topography (Figure 2a) shows two gyres. The diameter of the gyres is of the order of the local width of the gulf. The southernmost anticyclone is almost totally contained in Pescadero Basin; the cyclone that follows to the north is again almost totally contained in Farallón Basin. There seems to be evidence for an anticyclone in Guaymas Basin.

February 1957. Two gyres are found (not clearly defined)in Pescadero Basin, both on the peninsular side and with a radius of $\sim$ 25 km: an anticyclone on the southern part of the basin and a cyclone in the north, near the sill (Figure 2b). No gyres are found in Farallón Basin, only southward flow and a narrow northward flow in the mainland side. Centered in the sill between del Carmen and Guaymas basins, a 40 km diameter cyclonic gyre is located on the mainland side, while a southward jet-like flow is present close to the peninsula. Guaymas Basin once again contains an anticyclonic gyre in its center, with a 50 km radius.

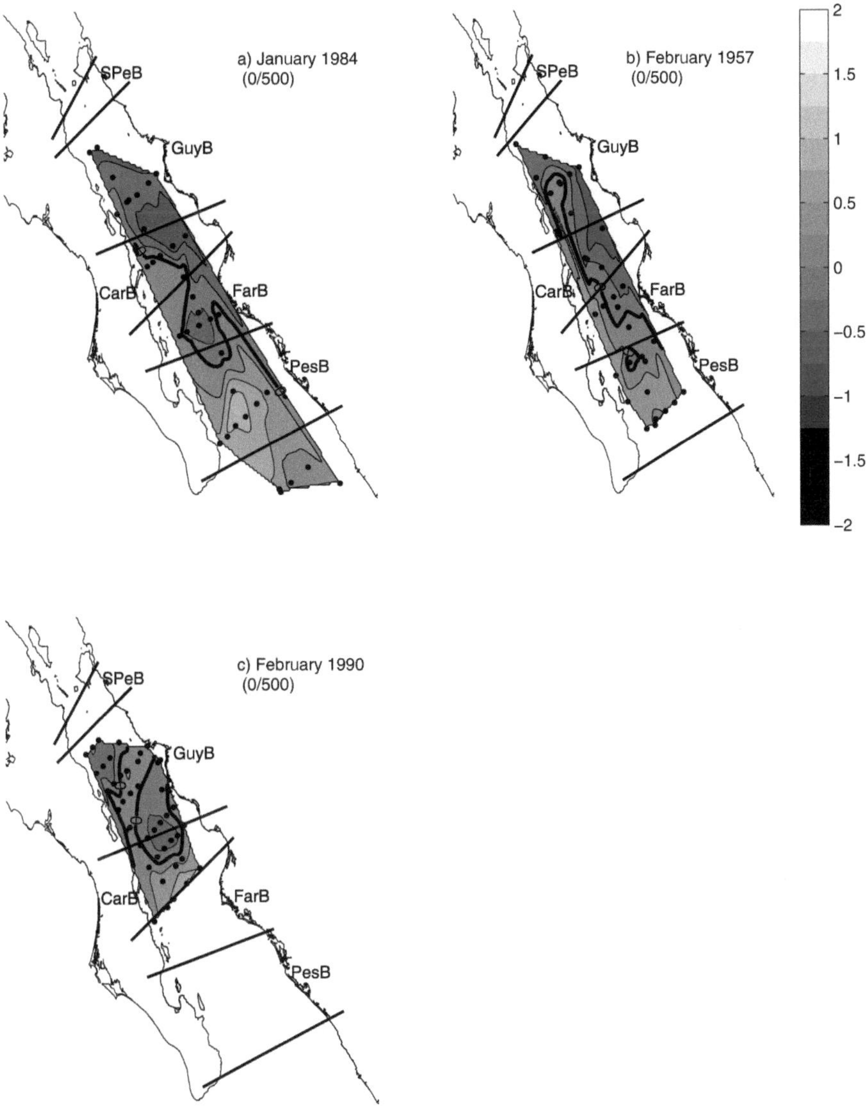

Figure 2. Dynamic topography anomaly (contours in gpm) for: (a) January 1984, (b) February 1957, and (c) February 1990. The reference level is indicated in each case and dots indicate cast positions.

February 1990. This survey sampled only to the north of del Carmen Basin (Figure 2c). A cyclonic gyre with a radius of $\sim$ 50 km straddles the Carmen-Guaymas sill.

March 1939. The separation of the cross-gulf hydrographic sections in this survey ($\sim$ 150 km; Figure 3a) may be too large to properly resolve gyres. However, an elongated cyclone gyre is suggested, in Pescadero Basin, close to the peninsula. In Farallón and del Carmen basins there is only weak southward flow. A cyclonic gyre of $\sim$ 50 km radius is located in the center of Guaymas Basin.

March 1983. The dynamic topography (Figure 3b) shows an alternating series of gyres. In Pescadero Basin, an anticyclonic gyre is suggested, although somewhat elongated, with minor semiaxis of $\sim$ 40 km. A cyclonic gyre is found over Farallón Basin, although a small section of it encroaches upon the sill to the north. The next structure can be interpreted either as a very elongated anticyclonic gyre ($\sim$ 130 km along-gulf) or as two smaller gyres ($\sim$ 50 km) of the same sign, covering all of del Carmen Basin and most of Guaymas Basin.

March 1984. A beautiful sequence of four alternating gyres is present in this survey (Figure 3c). The series starts at the mouth with an almost circular anticyclone with radius $>$ 50 km contained inside Pescadero Basin. The following cyclone is also circular and of the same approximate dimensions; it straddles the Pescadero-Farallón sill, although most of it is in Farallón Basin. This gyre is similar to that observed by Emilsson and Alatorre (1997). Over the Farallón-Carmen sill an anticyclone is found, similar in size to the previous two, but weaker. Finally, a situation similar to that shown in Figure 3b is found north of the Carmen-Guaymas sill: first a cyclone is found over the sill, and further north a feature suggests either another gyre or a continuation or elongation of the first.

March 1985. This survey covers only the Guaymas Basin (Figure 3d), and the data only allows calculations with rl at 250 db. A well-defined anticyclone is found at the northern end.

April 1956. Very few cross-sections can be constructed for this survey. The dynamic topography (Figure 4a) suggests a pair of elongated gyres, larger than those described previously. The first gyre is anticyclonic, and covers Pescadero and Farallón basins. The second gyre is cyclonic, and covers entirely del Carmen and Guaymas basins. Although these two gyres may be fictitious due to undersampling, at least it is possible to assert that the flow alternated sign between sections.

April 1957. The rl for this cruise is only 250 db, and the lines of stations are so far apart (Figure 4b) that a map of dynamic topography cannot be drawn. Therefore the surface velocity normal to the cross-sections is simply sketched. In the line across the mouth, there is outflow on the extremes of

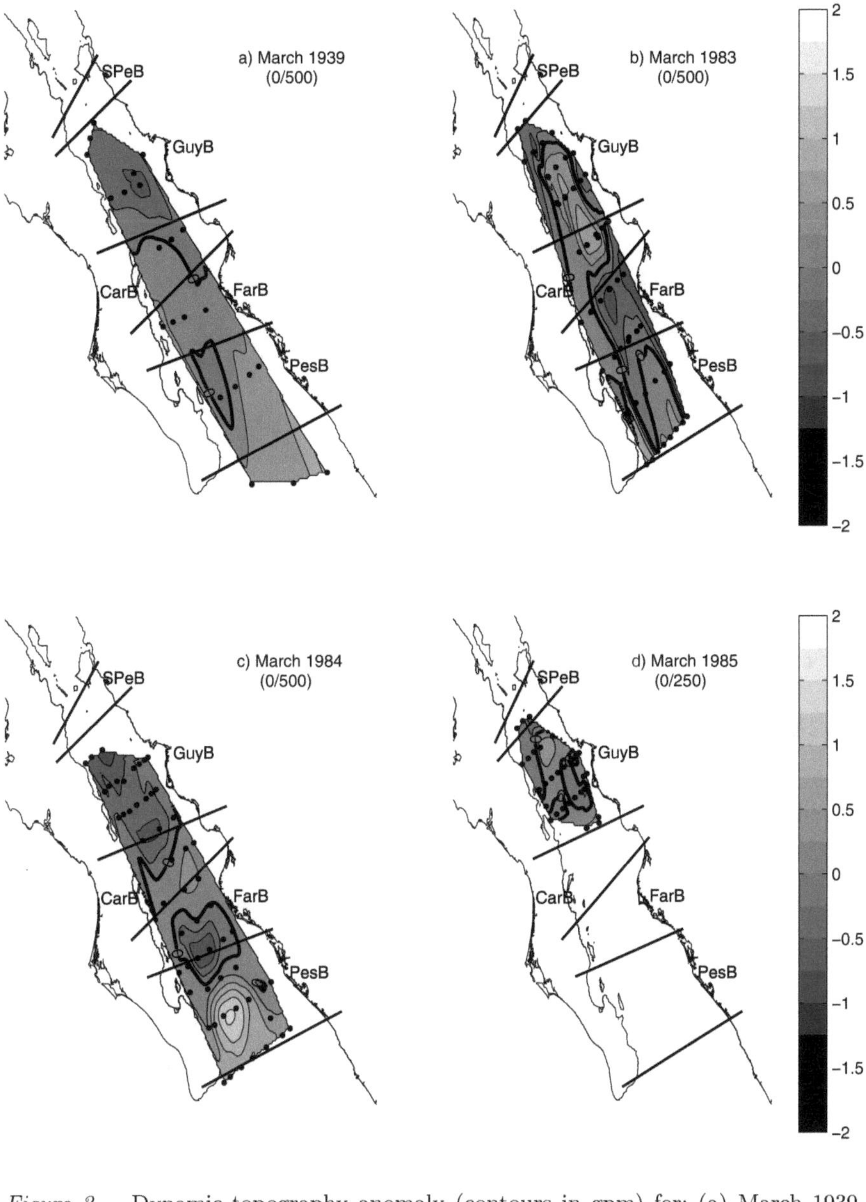

Figure 3. Dynamic topography anomaly (contours in gpm) for: (a) March 1939, (b) March 1983, (c) March 1984 and (d) March 1985.

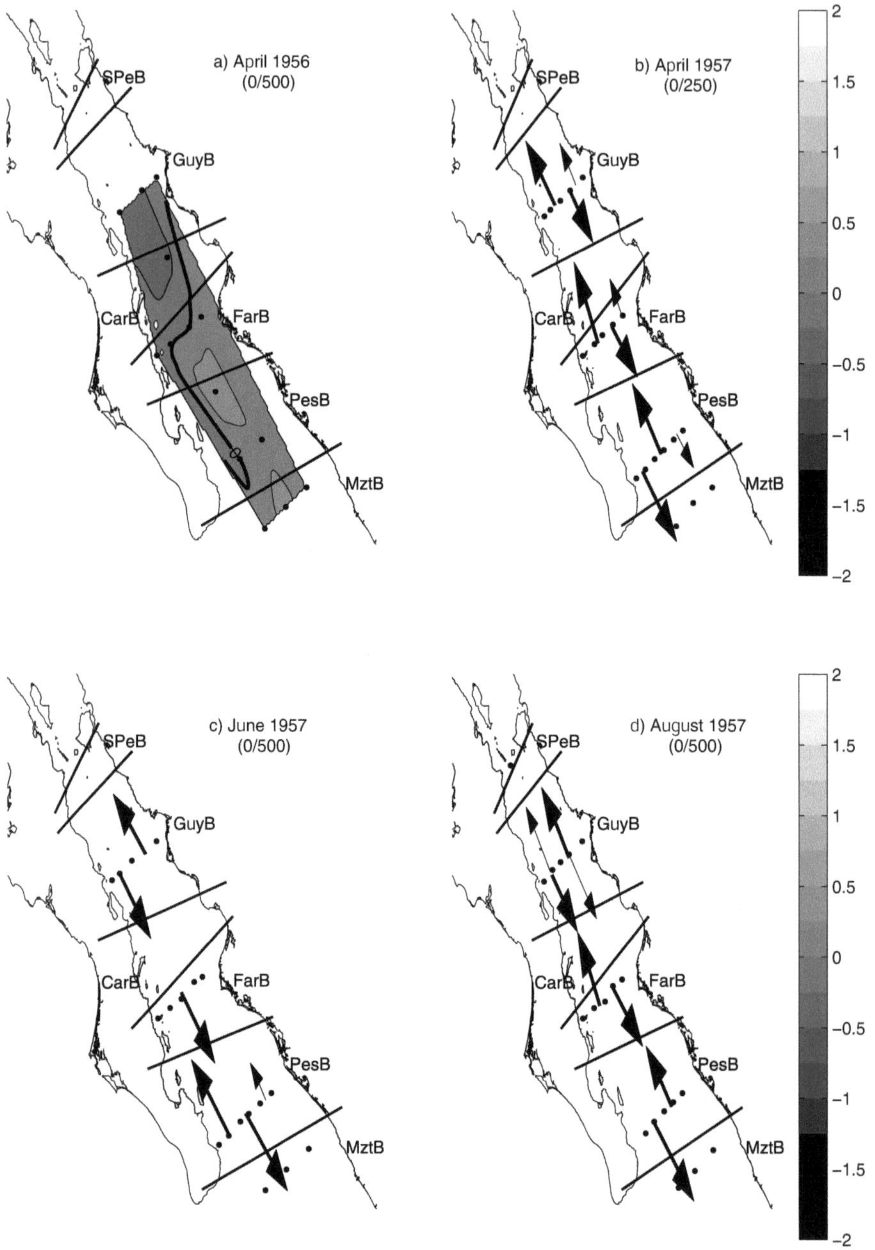

Figure 4. (a) Dynamic topography anomaly (contours in gpm) for April 1956. (b) Sketched surface geostrophic flow for April 1957, (c) Sketched surface geostrophic flow for June 1957, (d) Sketched surface geostrophic flow for August 1957: solid arrow represent the flow in the upper 100 m, and thin arrows represent the flow below that depth.

the line and inflow in the center; the outflow close to the peninsula is the strongest. In the other two lines (one close to the Farallón-Carmen sill, the other across Guaymas Basin) the flow pattern is the reverse to that in the mouth: strong inflow close to the peninsula, strong outflow in the central part and weak inflow on the eastern side.

June 1957. The same sparse sampling grid than in April 1957 was used, but deeper bottle casts permit the use of a rl at 500 db (Figure 4c). Across the mouth, the flow is opposite to that in April, with inflow close to the peninsula and off the mainland, and outflow in the central stations. Close to the Farallón-Carmen sill, the flow is mostly to the south. In Guaymas Basin, flow is again opposite to that in April, with outflow close to the peninsula and inflow on the eastern half of the section.

August 1957. The last of the 1957 cruises, with the same grid of stations as in April and June; 500 db is used as rl. The same pattern as in April is found in the sections across the mouth and near the Farallón-Carmen sill (Figure 4d): anticyclonic in the first, cyclonic in the latter. In Guaymas Basin the circulation pattern is quite complicated, with flow at depth (below $\sim$100 m, thin arrows) in the opposite direction to flow in the top 100 m (solid arrows).

October 1974. The sampling in this survey allows a good description of an anticyclonic gyre covering Guaymas Basin almost entirely (Figure 5a), and a few data suggest the presence of another one covering almost the rest of the basins.

October 1983. Reference level for this survey is 250 db (Figure 5b). Again Guaymas Basin is well sampled, and contains a well-defined anticyclone in the northernmost two-thirds. Further south, two side-by-side elongated gyres are suggested, anticyclonic on the peninsula side and cyclonic in the mainland side; the distance between the cross-sections is so large that the features may be spurious. The same survey is shown schematically in Figure 5c.

October 1994. Only Pescadero Basin was sampled, but quite densely (Figure 5d). No closed gyres are present, but the circulation in the NW area is cyclonic while that at the SE is anticyclonic. The extremely dense sampling across the mouth show a series of in- and out-flowing jets which are typical of the region, as reported by Castro *et al.* (2000).

November 1961. In this cruise a reasonably good sampling was made from Pescadero-Farallón sill to San Pedro Mártir Basin (Figure 6a). In the southernmost sampled area, anticyclonic circulation is present although without a closed gyre. The circulation of the rest of the sampled region is dominated by an elongated cyclonic pattern.

November 1972. The sampling was horizontally very good but unfortunately too shallow, so that a 250 db rl was used (Figure 6b). In Guaymas

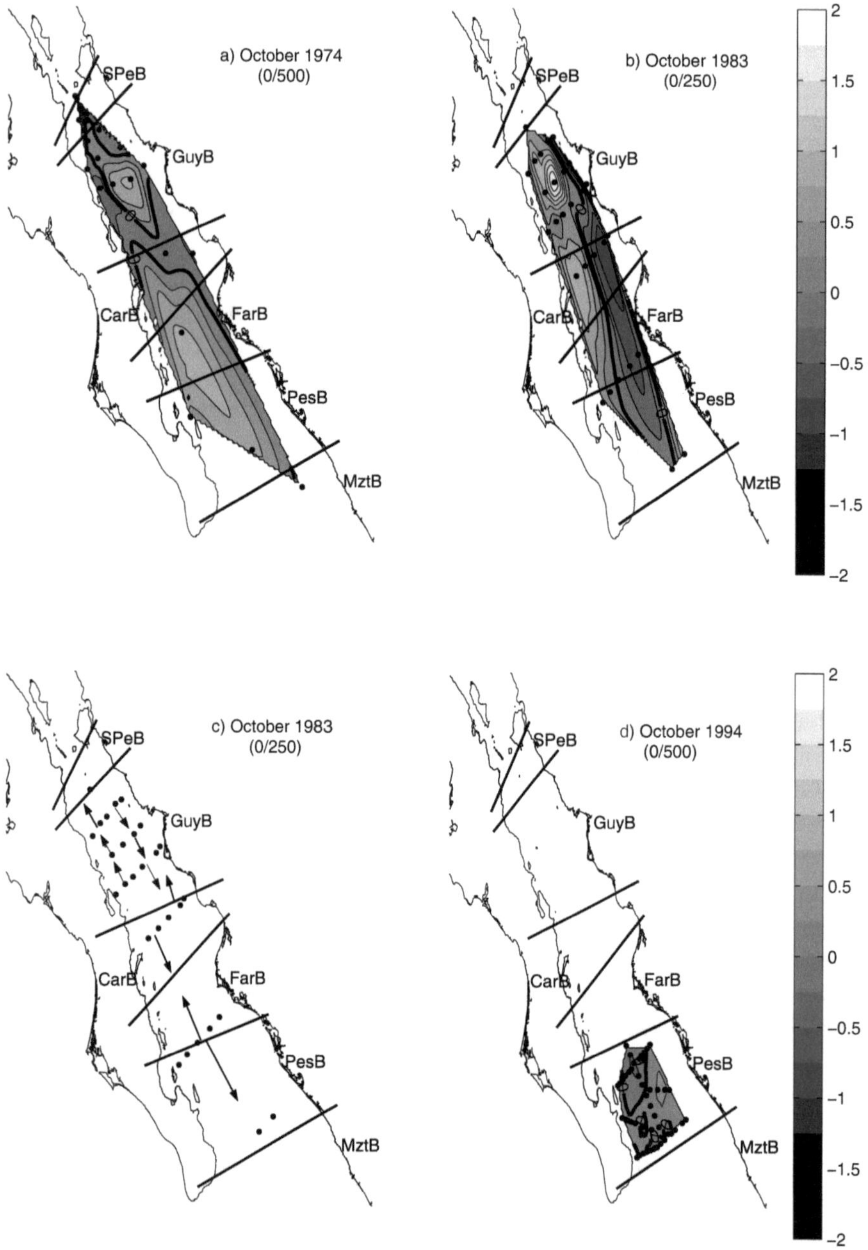

Figure 5. Same as previous figures, for: (a) October 1974, (b) October 1983, (c) October 1983, (d) October 1994.

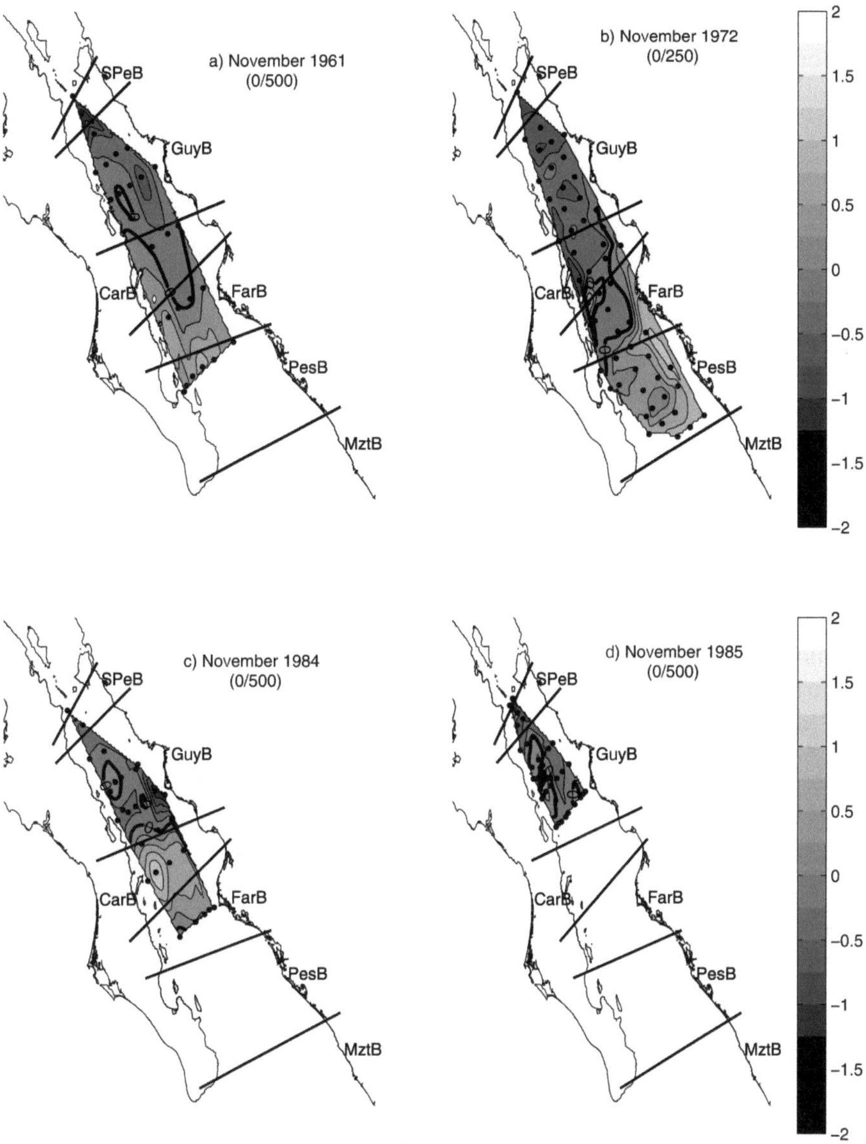

Figure 6. Same as previous figures, for: (a) November 1961, (b) November 1972, (c) November 1984, (d) November 1985.

Basin there was a very weak cyclone, but further south no gyres seem to be present. There are, however, clearly defined coastal jets on both sides of the gulf.

November 1984. The area north of Farallón Basin was very well sampled in this cruise (Figure 6c). An anticyclonic gyre over del Carmen Basin is the most salient feature. In Guaymas Basin there seems to be a cyclone-anticyclone pair, but it is not well defined.

November 1985. Guaymas Basin was thoroughly sampled in this survey (Figure 6d). A cyclone-anticyclone pair is found once again, with the anticyclone on the peninsula side.

3.2. VERTICAL STRUCTURE

The vertical structure of the geostrophic velocity field was calculated by the classical method, using the maximum common depth as a rl. Calculations were done for each line of stations for all the cruises, both along and across the gulf. In most cases it was observed that the surface structure extended coherently down to the rl (500 db in most cases) keeping its sign, suggesting that they may extend further down. The entire collection of geostrophic velocity cross-sections is not shown here; instead, selected sections of the cruise with the best horizontal and vertical coverage, March 1984, are used as illustration. The surface dynamic topography for this cruise is shown in Figure 3c (and repeated in Figure 7a), and it is described in Section 3.1.

Figure 7b shows the vertical structure of geostrophic velocity approximately across the center of the southernmost anticyclonic gyre shown in Figure 7a. Most of the flow is divided in two parts: (a) positive flow (to the north) from station 109 to midpoint between 111 and 112, with a clearly defined nucleus from the surface to the first 100 meters with a maximum velocity of 50 cm s^{-1}; (b) a negative-flow region (out of the gulf), which is wider but with a maximum magnitude of 40 cm s^{-1}. The nuclei of both regimes are in the first hundred meters. Flow direction is conserved in the entire sampled water column with exception of the area close to the peninsula west to station 111. Although a strong vertical shear is present, this velocity distribution shows that the anticyclonic structure observed at the surface is coherent all the way down to the rl in the rest of the section.

Figure 7c shows the geostrophic velocity structure normal to the line near the boundary between the anticyclone just described and the nearby cyclone (see Figure 7a). The flow is divided in three regimes: (a) southward flow close to the peninsular side, (b) northward flow in the central part and, (c) southward in the mainland side. Here, the direction of the flow is conserved in the entire sampled water column. The speed is lower than in the section described above (max. $\sim$10 cm s^{-1}). Figure 7d shows the geostrophic velocity normal to the line crossing the cyclonic gyre off La Paz

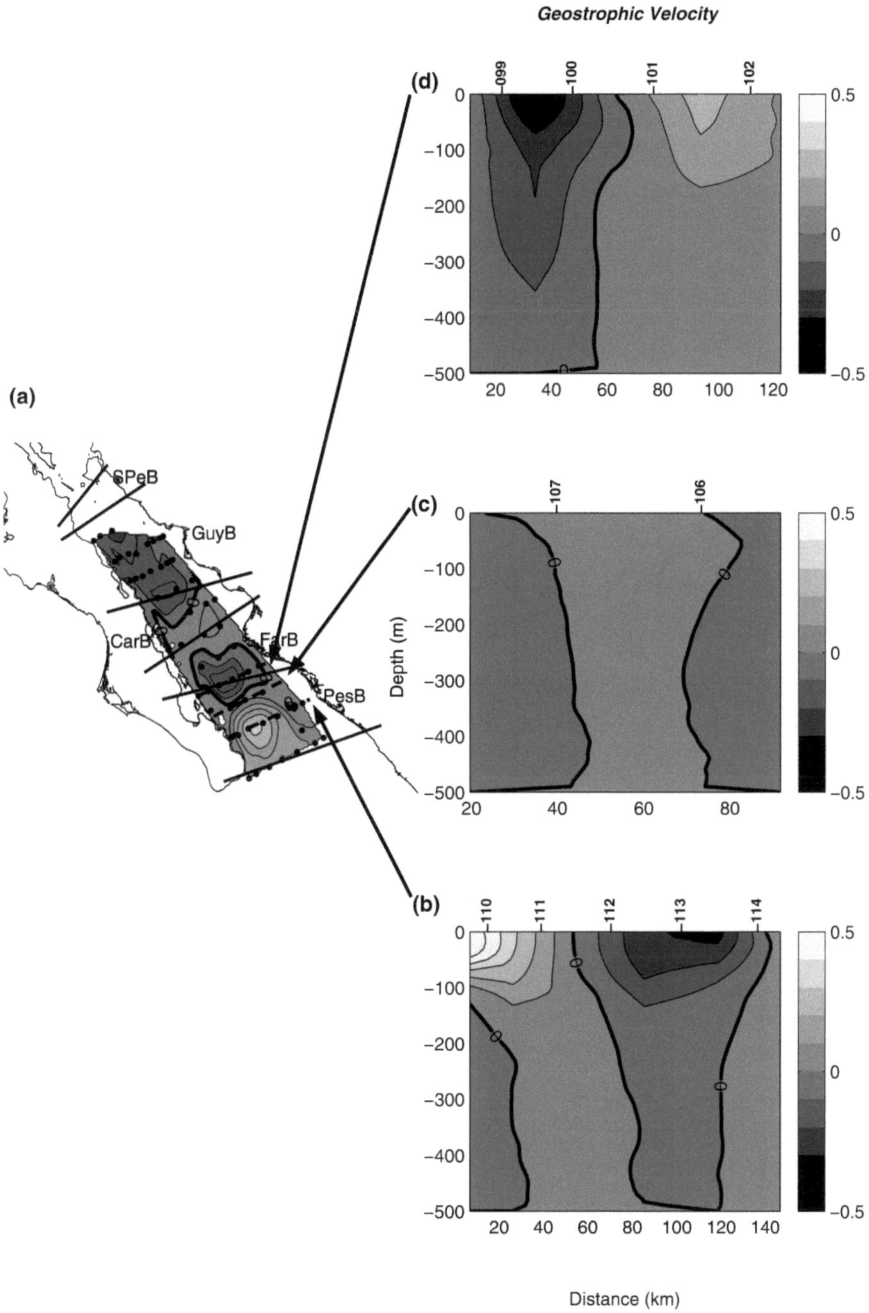

Figure 7. (a) Dynamic topography anomaly for March 1984 (same as Figure 3c). Arrows indicate the cross-sections of geostrophic velocity (0/500 m) calculations: (b) across an anticyclone, (c) border between two gyres with opposite sense of rotation, and (d) across a cyclone. Velocity contours in meters per second.

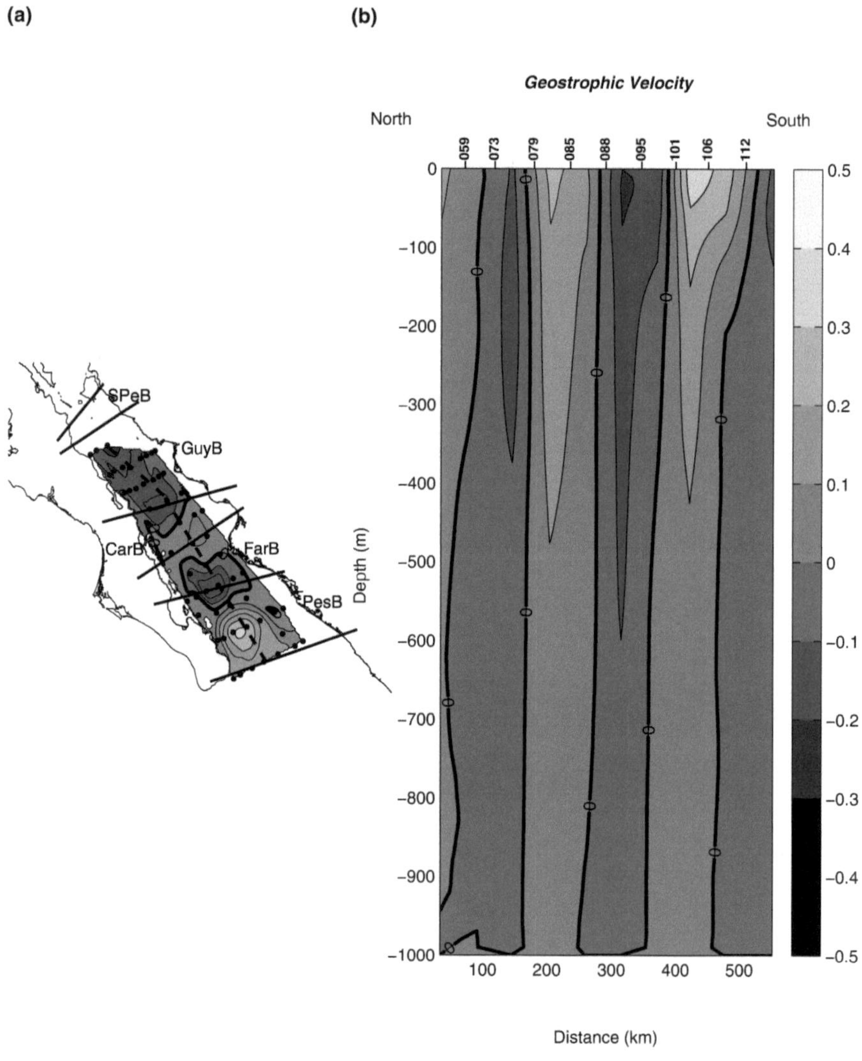

Figure 8. As in figure 7, but for an along-gulf section, and the reference level used is 1000 db.

Bay (see Figure 7a). The flow distribution is the opposite of that described for Figure 7b, and shows the cyclonic gyre extending down to the rl.

The vertical extent of the gyres was investigated further by calculating the geostrophic velocity normal to a line of stations along the gulf axis (Figure 8a), which allowed the use of a rl at 1000 db. A sequence of nuclei of alternating sign (positive toward the mainland) is evident. The position and sense of these structures coincides with the sequence of cyclone-anticyclone shown again in Figure 8a: maxima correspond to boundaries between gyres, and the zeros correspond to crest and valleys. The structures are coherent from the surface to 1000 db; the exception is only at the south, below station 112.

4. Discussion and Conclusions

In the search for patterns in this variegated collection, a schematic diagram (Figure 9) is used to summarize the above descriptions of the gyres. The gyres reported by Emilsson and Alatorre (1997) and by Amador *et al.* (2003) are included, as well as the gyre pattern proposed by Fernández-Barajas *et al.* (1994). The seasonal cycle is not well sampled in any of the basins, with especially poor coverage from April to September, which is half of the annual cycle; in addition there are no cruises in December and only one in January. The poor temporal and spatial coverage allow only sketched descriptive patterns, summarized as follows for each basin:

Guaymas Basin: In October, basin-wide anticyclonic gyres dominated the circulation, but in November, January, February and March gyres of either sign can be present, frequently in smaller pairs. Many gyres were seen to cross over the sill to the south. Ripa and Marinone (1989) found no evidence of the presence of an average or a seasonal signal in the vorticity of the geostrophic flow in Guaymas Basin. The detailed study of the structure of the mesoscale thermohaline structures made by Navarro-Olache (1989) did not show a basin-wide gyre either. Therefore, there is no support for the presence of a persistent gyre in the geostrophic circulation in Guaymas Basin, such as that proposed by Fernández-Barajas *et al.* (1994).

Del Carmen Basin: Gyres in this basin are often shared with the basins to the north (Guaymas Basin) or to the south (Farallón Basin). From November to April, a cyclone is often shared with Guaymas Basin; the sense of this gyre agrees with that proposed by Fernández-Barajas *et al.* (1994) for Guaymas Basin. Anticyclones are found over the sill to the south in October, January, February and March, which agrees with the northern edge of the anticyclonic gyre proposed by Fernández-Barajas *et al.* (1994) as covering del Carmen and Farallón basins (in Figure 9 it looks elongated because the axis have different scales: the combined area of del Carmen and Farallón basins is about the same as that of the Guaymas and Pescadero basins alone). However, this is the basin with the poorest sampling, with no data for several months.

Farallón Basin: Mainly anticyclonic circulation can be claimed in October and November, and mainly cyclonic in January and March. The cyclonic gyre reported by Emilsson and Alatorre (1997) during August 1978 (in gray in Figure 9) was partly in Farallón Basin and partly in Pescadero Basin;

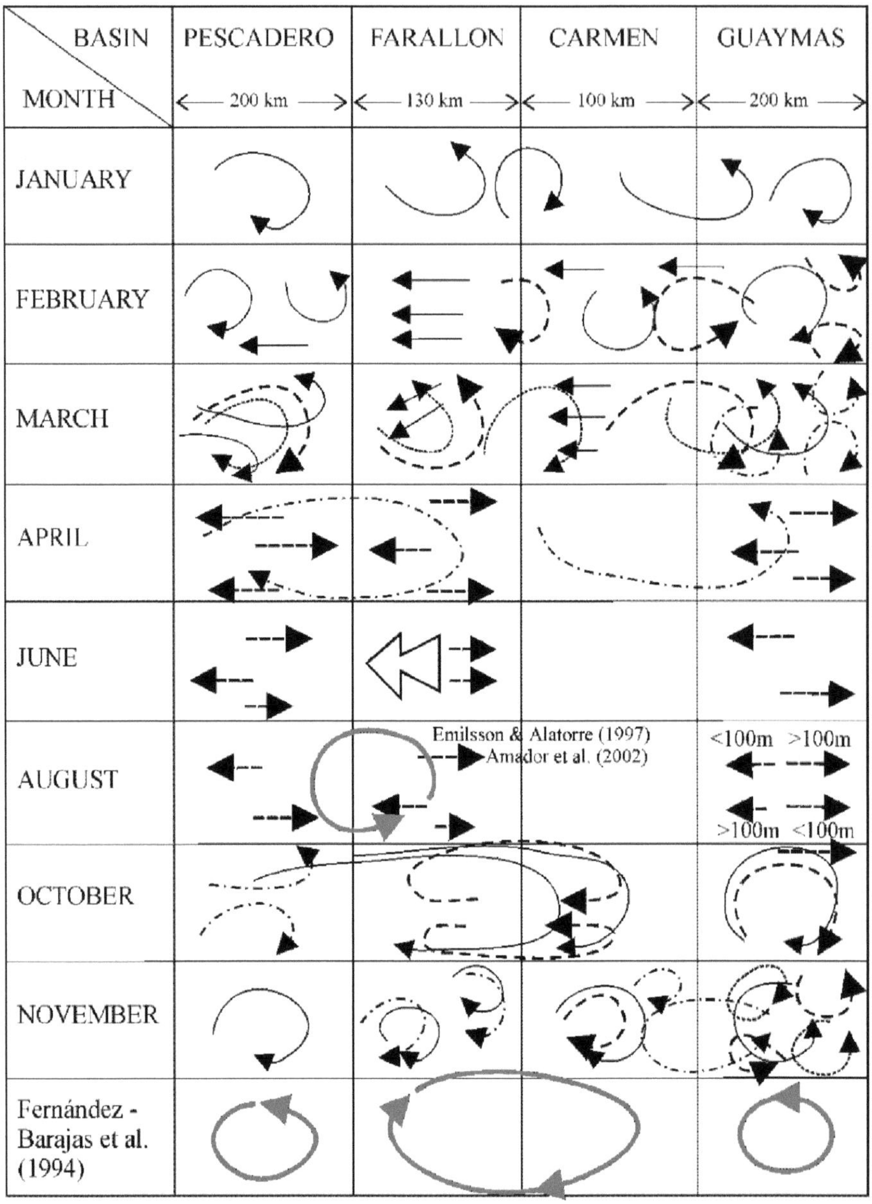

Figure 9. A monthly resume of the main features observed in the dynamic topography anomaly for each basin. Note that basin dimensions are not to scale, a rough estimation of their diameter is indicated. Gyres proposed by other authors are shown in gray. The different lines represent the different cruises for each of the months shown.

its sense of rotation is opposite to that proposed by Fernández-Barajas *et al.* (1994) for Farallón Basin, but agrees with that proposed by them for Pescadero Basin.

Pescadero Basin: According to Figure 9, circulation can be cyclonic or anticyclonic, with anticyclonic circulation most often seen from October to March, which is contrary to the mean pattern proposed for Pescadero Basin by Fernández-Barajas *et al.* (1994). Being at the entrance to the gulf, this basin is characterized by the presence of fronts between water masses and by gyres and meanders of different sizes, which are readily apparent in satellite infrared images (e.g. Amador *et al.*, 2003). There are, however, several studies (the most recent and trustworthy: Collins *et al.*, 1997; Castro *et al.*, 2000) indicating that the mean circulation pattern across the entrance to the gulf consists of inflow on the mainland side and outflow in the peninsula side; whether this circulation is evidence of a gyre such as that proposed by Fernández-Barajas et al. (1994), is not certain.

In summary: although the data set contains far less information than is necessary to unravel possible seasonal cycles of circulation in the SGC, the presence of geostrophic gyres, and along-shore jets is evident. Most of the gyres have a length scale of the order of the width of the gulf ($<$ 200 km), and they tend to alternate their sense of rotation along the gulf. Gyre couplets are often present either of smaller dimensions or stretched along the gulf (these latter may be spurious due to undersampling of the field). No definite pattern is observed in the position of the gyres or in their sense of rotation. In other words, the gyres do not appear more frequently or in a preferred sense of rotation in a given season. Nor do the gyres seem to be associated with bathymetric features: they can be found inside a single basin or on top of the sills, covering more than one basin. The best-defined gyres in the data set are surprisingly deep; they can be detected to depths of 1000 m; this is in agreement with the measurements of Collins *et al.* (1997), whose acoustically tracked dropsondes (Pegassus) measured speeds of $\sim$ 0.1 m s^{-1} at depths exceeding 1000 m in Pescadero Basin. The geostrophic currents calculated by Fernández-Barajas *et al.* (1994) for February 1992 also suggest gyres reaching to 1000 m; however, the quality of the latter data have been questioned by Navarro-Olache *et al.* (1997).

Coastal jets were detected along both coasts, as was observed in infrared images by Badan-Dangon *et al.* (1985). They are often associated with the presence of a gyre, but they often flow in the opposite sense to that of the closest branch of the gyre; this feature was observed with a ship-borne ADCP by Amador *et al.* (2003) off La Paz Bay. Jets are also present independently of gyres, and sometimes they cross the gulf; these across-gulf jets were studied in some detail by Navarro-Olache (1989) for Guaymas basin, but no relation was found between the jets and basin-wide geostrophic gyres.

This study shows that the circulation in the SGC is rich in deep (over 1000 m) geostrophic gyres and geostrophic coastal jets. However, simple circulation diagrams like that proposed by Fernández-Barajas *et al.* (1994), obtained from a single along gulf transect, are not supported by the available data. For example, the cyclonic gyre in Pescadero Basin reported by Emilsson and Alatorre (1997), Fernández-Barajas *et al.* (1994) and Amador *et al.* (2003) should not be considered a permanent feature. The main limitation of this study is the lack of quantitative arguments, which is imposed by the limitations in data coverage (in time and space). The data do not reveal a seasonal behavior, yet it may be present in some regions, but the scarcity of data preclude its detection. It is obvious that more data are necessary, with direct current measurements most desirable. It is recommended that future cruises in the SGC use station grids that can resolve the gyres and coastal jets.

Acknowledgements

We appreciate suggestions and comments from P. Ripa and two anonymous reviewers on an earlier version of this manuscript. We also thank the many contributors to the hydrographic data bank of the Gulf of California and José Domínguez for the elaboration of the last figure. This study was financed with CICESE's regular budget and by CONACyT (México), through contracts No. 4300P-T, 35251-T and 255-T9712.

References

Amador-Buenrostro, A., A. Trasviña-Castro, A. Muhlia-Melo, and M. L. Argote. Estructura de la Circulacion Sobre el Bajo Espiritu Santo y la Cuenca de Farallón en el Golfo de California. *Geofísica Internacional*, in press.

Argote, M. L., M. F. Lavín, and A. Amador. Barotropic Eulerian Residual Circulation in the Gulf of California Due to the M2 Tide and Wind Stress. *Atmósfera*, 11:173–197, 1998.

Badan-Dangon, A., D. J. Koblinsky, and T. Baumgartner. Spring and Summer in the Gulf of California: Observations of Surface Thermal Patterns. *Ocean. Acta*, 8:13-22, 1985.

Beier, E. and P. Ripa. Seasonal Gyres in the Northern Gulf of California. *J. Phys. Oceanogr.*, 29:302–311, 1999.

Carrillo, L. E., M. F. Lavín, and E. Palacios-Hernández. Seasonal Evolution of the Geostrophic Circulation in the Northern Gulf of California. *Estuarine, Coastal and Shelf Sci.*, 54:157–173, 2002.

Castro, R., A. S. Mascarenhas, R. Durazo, and C. A. Collins. Seasonal Variation of the Temperature and Salinity at the Entrance to the Gulf of California. *Ciencias Marinas*, 26:561–583, 2000.

Collins, C. A., N. Garfield, A. S. Mascarenhas, M. G. Spearman, and T. A. Rago. Ocean Currents Across the Entrance to the Gulf of California. *J. Geophys. Res.*, 102:20927–20936, 1997.

Emery, K. O., and G. T. Csanady. Surface Circulation of Lake and Nearly Land-locked Seas. *Proc. Natl. Acad. Sci.*, USA, 70:93–97, 1973.

Emilsson, I., and M. A. Alatorre. Evidencias de un Remolino Ciclónico de Mesoescala en la parte Sur del Golfo de California. in *Contribuciones a la Oceanografía Física en México*, Unión Geofísica Mexicana, Monografía, 3:113–139, 1997.

Fernández-Barajas, M. E., M. A. Monreal, and A. Molina-Cruz. Thermohaline Structure and Geostrophic Flow in the Gulf of California, during 1992. *Ciencias Marinas*, 20:267–286, 1994.

Figueroa, J. M. *Eddies in the Gulf of California*, Abstract of: The Oceanographic Society Inaugural Meeting, Monterrey, CA, USA., 1989.

Gaxiola-Castro, G., S. Alvarez-Borrego, M. F. Lavín, A. Zirino, and S. Nájera-Martinez. Spatial Variability of the Photosynthetic Parameters and Biomass of the Gulf of California *J. Plankton. Res.*, 21:231–245, 1999.

Hammon, M. C., T. R. Baumgartner, and A. Badan-Dangon. Coupling of the Pacific Sardine (*Sardinops Sagax Caeluleux) Life Cycle with the Gulf of California Pelagic Environment. CalCOFI Reps.*, 29:102–109, 1988.

Hill, A. E. Seasonal Gyres in Shelf Seas. *Annales Geophysicae*, 11:1130–1137, 1993.

Lavín, M. F., R. Durazo, E. Palacios, M. L. Argote, and L. Carrillo. Lagrangian Observations of the Circulation in the Northern Gulf of California. *J. Phys. Oceanogr.*, 27:2298–2305, 1997.

Navarro-Olache, L. F. Mesoestructuras Termohalinas en la Parte Central del Golfo de California. MSc thesis, CICESE, Ensenada, B. C. 1989.

Navarro-Olache, L. F., A. S. Mascarenhas, R. Durazo, and C. A. Collins. A Note on Upper Ocean Temperature and Salinity at the Entrance to the Gulf of California in August 1992. *Ciencias Marinas*, 23:273–282, 1997.

Palacios-Hernández, E., E. Beier, M. F. Lavín and P. Ripa. The Effect of the Seasonal Variation of Stratification on the Circulation on the Northern Gulf of California. *J. Phys. Oceanogr.*, 32:705–728, 2002.

Poulain, P-D. Adriatic Sea Surface Circulation Derived from Drifter Data Between 1990 and 1999. *J. Mar. Syst.*, 29:3–32, 2001.

Ripa, P., and S. G. Marinone. Seasonal Variability of Temperature, Salinity, Velocity and Sea Level in the Central Gulf of California, as Inferred from Historical Data. *Quarterly J. of the Royal Meteor. Soc.*, 115:887–913, 1989.

NONLINEAR INTERNAL WAVES NEAR MEXICO'S CENTRAL PACIFIC COAST

A.E. FILONOV
Departmento de Física
Universidad de Guadalajara
A.P. 4-040, Guadalajara, Jalisco, 44421, México

K.V. KONYAEV
Andreev Institute of Acoustics, Moscow, Russia

Abstract. Features of the generation and structure of internal tides near Mexico's Central Pacific Coast are discussed using: the hydrographic data of a field survey, and the temperature time series obtained by two anchored instruments. The shelf in the study area is extremely narrow; a steep continental slope starts near the coast. This kind of bathymetry is uncommon in studies of internal tide generation. The internal tide wave disintegrates near the coast; sets of nonlinear waves form as a result of the transformation of the baroclinic tide. Linear and non-linear theories results were used for the interpretation of the experimental measurements of the intense internal waves. The movement of the internal tide over the continental shelf is revised. The semidiurnal tidal ray patterns constructed given the sea-bottom profile and buoyancy frequency data show that all internal tide rays are directed from the segments of the slope with a critical ocean-ward tilt of the bottom.

Key words: internal waves, tides, continental shelf

1. Introduction

A series of theoretical works (Baines, 1982; Graig, 1987; Holloway, 1985, 1987), have shown that the generation of the internal tide is carried out mainly in the bank adjacent to the continental shelf which, as it rises from the deep towards the surface, forms an obstacle for the spread of the barotropic tidal waves. The slope generates intense baroclinic tidal waves, which move away toward the ocean interior and also toward the coast.

The shelf near Mexico's Central Pacific Coast is extremely narrow. The continental slope starts near the coastline (Figure 1). The bottom slope near the coast is as large as 20 to 30 m/km. Although the generation of internal tides at a shelf break is clearly a three-dimensional problem, it may

O.U. Velasco Fuentes et al. (eds.), Nonlinear Processes in Geophysical Fluid Dynamics, 257-274.

vary significantly in different sections along the coast. The following is a simple criterion to assess whether a particular section is likely to generate an internal tide, which only depends on the bottom slope ($\alpha = dz/dx$) and on the density stratification (N^2). The latter in turn determines the tilt angle (θ) of the characteristic ray path along which the internal tide propagates (La Fond, 1962; Baines, 1982; Graig, 1987; Holloway, 1987):

$$\theta(z) = \arctan((N(z)^2 - f_t^2)/(f_t^2 - f_i^2))^{1/2} \tag{1}$$

where f_t is the frequency of the wave; f_i is the inertial frequency. The energy transmission from the barotropic to the baroclinic tide is more effective when $\alpha/\theta \approx 1$. When $\alpha/\theta < 1$ (> 1) the transmission of energy results in inshore (offshore) propagation. A steep continental slope is usually separated from the shoreline by a relatively flat shelf; therefore, the bottom segments with critical tilts for the semidiurnal internal tide are found on the outer edge of the shelf. These segments experience a strong internal tide, which then moves coastward and oceanward (Baines, 1982; Huthnance, 1989; Pingree and New, 1991; Konyaev and Sabinin, 1992; Filonov and Trasviña, 2000).

Previous investigations (Filonov *et al.*, 1996a; 1996b; Filonov, 2000c; Konyaev and Filonov, 2002) showed that the main characteristics of the internal tides in the continental shelf near Mexico's Central Pacific Coast are clear. The internal tide includes a dominant semidiurnal tide, which is represented by a strong peak in the spectrum of temperature oscillations. The barotropic tide in the survey area has a mixed character with a dominant semidiurnal component. Studies based on the Baines model (1982) showed that here the energy flow from barotropic tides to the internal tide can vary significantly at different sections of the slope, depending on the arrival angle of the barotropic wave to the slope. The maximum energy flux, in case of perpendicular arrival, is estimated to be 763 $\mathrm{Jm^{-1}s^{-1}}$ on a meter of the continental shelf edge, and the extent of the initial internal disturbance is of 4.9 m (Filonov *et al.*, 1996a).

This paper presents new information of the space-time structure of the internal tidal waves on the western coast of Central Mexico. The character of non-linear transformation of waves extending on a shelf on the coast side and waves reflected from a continental slope extending along the open ocean side is discussed.

2. Measurements and methodology

The measurements of internal waves were performed at a Water Ecological Polygon, located at the coast of Jalisco and Colima (Figure 1). The field

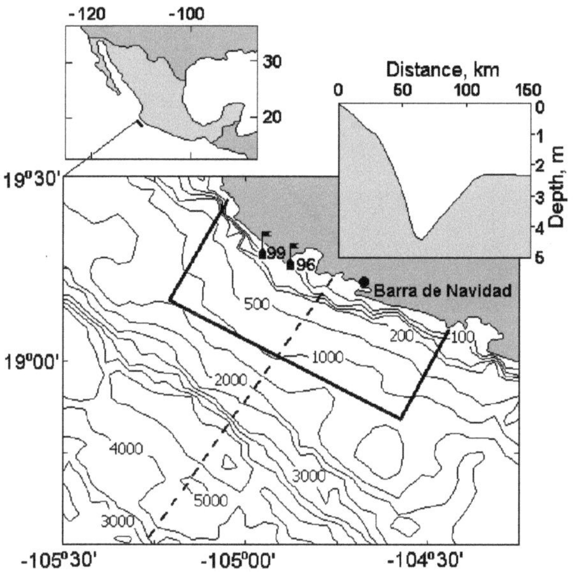

Figure 1. Study area. The rectangular line shows the area of Water Ecological Survey Polygon. Flags show the buoys positions in 1996 and 1999, the dashed line points to the bottom line measured. In the right top corner a depth profile of the coast line is shown.

work was carried out from a small research vessel "BIP-V", that belongs to the Coastal Ecological Center of the University of Guadalajara.

The main data sets to be analyzed are time series of temperature and salinity, collected by two moorings (see locations in Figure 1). The first of them operated in a bottom depth of 70 m (2 km from the coast), from September 22nd to October 30th, 1996, and was instrumented at depth 38 m (below the surface). The bouy had a conductivity-temperature-depth (CTD) meter SBE-16 (made by Sea-Bird Electronics). The instrument has a measuring precision of 0.01 °C for temperature, 0.001 S/m for conductivity, and 0.25 % of the entire scale range for pressure. The sampling rate was 1 minute.

The second mooring operated in a bottom depth of 52 m (1 km from the coast) and operated 28 days in April 1999. This anchored buoy was equipped with two digital thermographs a TDS-85 and a BoxCar Pro 4.0. The first of them, at a depth of 38 m below the surface, was operated almost one month and sampled every 4 min with a temperature-measuring accuracy of ±0.2 °C. The second one, at a depth of 33 m, was sampled every minute for only two weeks with an accuracy ±0.32 °C.

The temperature and salinity fields were surveyed for ten hours on May 18, 1999, over an area of 30 x 50 Km (Figure 2). The survey site extends from 70 to 1000 m depth. The survey includes a total of 61 vertical

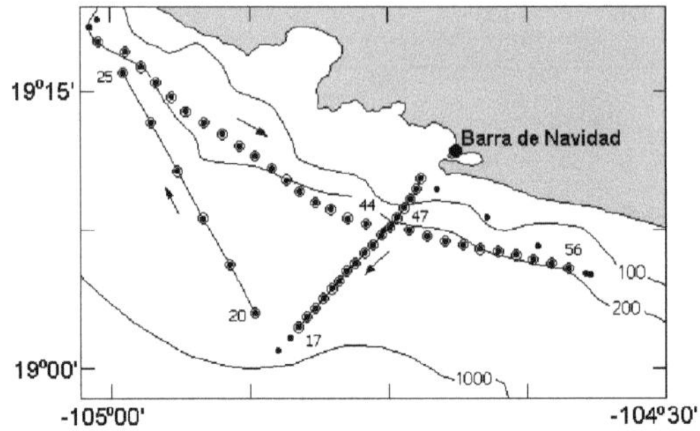

Figure 2. Pattern of vessel route and distribution of measurement sites (numbered consecutively) made in May 18, 1999.

temperature and salinity profiles with a timing of 10 minutes and a spacing of 1 to 3 Km. The density was calculated from the temperature and salinity by the equation of state for seawater. To avoid random pulsations, the mean density profile was smoothed with depth and this was used to calculate the buoyancy frequency and orbit slope of the semidiurnal internal tide.

The spatial characteristics of the internal tide can be measured with a CTD, if done with close enough profiles and in a very fast way. At least 3-4 profiles should be obtained in a wavelength distance. For this purpose, we have used a measurement method (Filonov *et al.*, 1996b) that can be summarized as follows. At full vessel speed of about 10 km/h the CTD profiler placed in a streamlined box moves near the surface. For profiling purposes, the vessel, while keeping its speed unchanged, makes a couple of circles to allow the profiler sink, which it does at a speed of as much as 1 m/s and to a depth defined by the length of the towing cable. Then, the vessel moves straigth to the next profiling site. The survey positioning is recorded with the help of a Global Positioning System (GPS) receiver. The sample rate for depth, temperature, and salinity was twice per second (with a depth step of 0.5 m when profiling); then, the measured data were smoothed using a cosine filter with a half width of 2 m over depth and recorded with a step of 1 second.

2.1. VERTICAL DEVIATIONS OF THE WATER LAYERS ON THE TEMPERATURE PROFILES DATA

On average the temperature decreases monotonically with depth, and the vertical temperature gradient varies moderately in the below-thermocline layer under consideration. Therefore, the temperature profiles can conveniently be used to calculate the vertical deviations of water layers. Going from measured temperature profiles to vertical deviations, which are largely determined by internal waves, requires a number of intermediate operations (Konyaev and Filonov, 2002).

First, a few small inversions on the measured profiles should be removed. Since the inversions create ambiguity in calculating the deviations from the mean profile, their presence in the data is unacceptable. They are removed through averaging two inversions-free profiles obtained by moving along the profile from top to bottom and vice versa with hysteresis jumps in the inversion layers. The averaged profile smoothly aligns the inversion areas and does not differ from the initial profile at the remaining levels. Then, for all individual profiles, the mean temperature profile $T_0(z)$ is calculated and taken as its unperturbed value. Due to the absence of inversions, the profile $T_0(z)$ uniquely defines the inverse dependence (profile) $z(T_0)$.

The vertical deviations of water layers $\triangle z$ are calculated as functions of mean temperature T_0 by comparing the individual $z(T)$ and mean $z(T_0)$ profiles: $\triangle z(T_0) = z(T) - z(T_0)$, where each deviation $\triangle z(T_0)$ refers to mean profile depth $z(T_0$; thus, is transformed into $\triangle z(z)$. In the near-surface layer, the individual and mean profiles begin with different temperatures; this makes it impossible to calculate vertical deviations in this layer for a number of profiles. The thickness of this uncertainty layer is almost equal to the thickness of the upper quasi-homogeneous layer (in this case, it is slightly larger than 20 m).

2.2. SYNTHETIC APERTURE METHOD

The survey of temperature or density usually covered two horizontal coordinates, depth and time. However, since the vertical profiles were performed at different points in space and time they are not used to detect internal waves. The spectrum turns out to be represented in the true rather than the observed frequency coordinates (Doppler shifts for both the frequency and wave vector are missing). Radar scanning and radio-astronomy use the synthetic aperture method. The method groups all signals depending on their position in time and space. The method allows to realize the potential resolution on data arbitrarily distributed in space and time (Konyaev, 1990, 2000).

In the synthetic aperture method the four-dimensional Fourier Transformation is reduced to summation of field measured values following the course of the sensor (Konyaev, 2000):

$$c_r(\overrightarrow{K}, k_z) = p\sum_z\sum_n T(n,z)\exp(-i2\pi(\overrightarrow{K}\bigotimes\overrightarrow{X}(n) + k_z z)) \qquad (2)$$

where c_r –is the amplitude spectrum of the temperature field.

$n = 1, ..., N$ – is the value number of the trajectory line.

$\overrightarrow{X} = \{t(n), x(n), y(n)\}$ – is the vector formed by $t(n)$ and the horizontal spatial coordinates $x(n)$ and $y(n)$ of the parametric form of the sensor trajectory.

z – depth.

$T(n, z)$ – temperature values on the trajectory points.

$\overrightarrow{K} = \{f, k_x, k_y\}$ – vector formed by frequency f and the two horizontal components of the wave vector k_x and k_y.

$\bigotimes$ – scalar multiplication of vectors $\overrightarrow{K} \bigotimes \overrightarrow{X}(n) = f \cdot t(n) + k_x \cdot x(n) + k_y \cdot y(n)$.

k_z – vertical component of wave vector.

p – normalizer which depends on the volume of the spatial and temporal measurements along axes x,y,z and t.

The duration of this survey was shorter than the period of the semidiurnal tide, which makes the frequency resolution low. The absence of intense internal waves at other frequencies (in particular, a diurnal internal tide) is important for a reliable treatment of the spatial spectrum. Judging from temperature oscillations at the anchored buoy, semidiurnal oscillations prevail here (Figure 5).

3. Temporary variations

Despite the unusual bottom relief, the semidiurnal internal tide in the upper oceanic layer along the seacoast of Mexico is very strong. Measurements performed along a line 2 km away from the coast in October 1996, have shown that internal waves cause the thermocline to sink till 25-30 m. This sinking shows a sharp forward front, with time gradients reaching 9 °C/h for temperature and up to 1.5 psu/h for salinity (Figure 3).

The short waves present the same amplitude as the original tidal wave and, therefore, a great slant. In these waves, the horizontal orbital currents can reach tens of centimeters in a second, that is, their magnitudes approximate the phase velocities. In them, the vertical orbital movements frequently are near 5-10 mm/s. Usually, the orbital movements of the short internal waves change the roughness of the ocean surface, forming alternate stripes of flat and rough surface in the ocean (La Fond, 1962; Konyaev and Sabinin, 1992).

The thermocline's solitary downwelling has a far-from-sinusoidal shape, with the front part steeper than the back part. This is because as the internal tide moves from the edge of the continental shelf towards the coast its group speed diminishes. In order for the internal tide to preserve its horizontal energy flux both its energy density and wave amplitude increase, producing non-linear deformation and its disintegration into shorter wave groups.

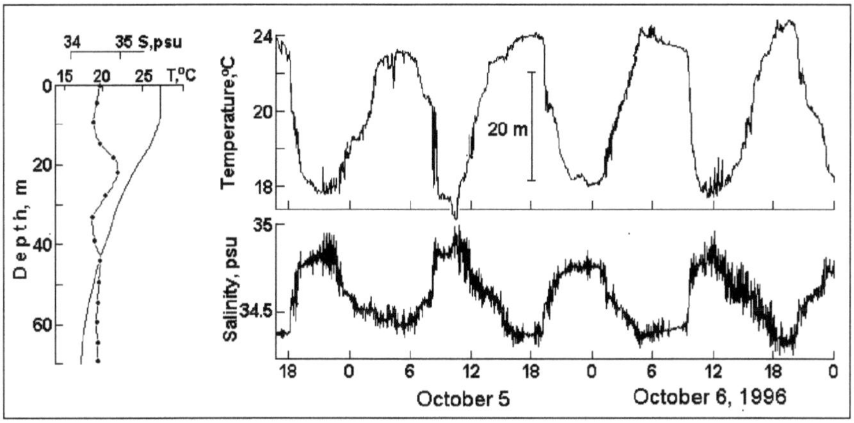

Figure 3. (a) Temperature and salinity variations at the buoy anchored 2 km out of the shore and at a local depth of 70 m. Measurements were carried out on October 16th, 1996. (b) Temperature and salinity oscillations at a level of 35 m.

Measurements performed on the mooring 1 km away from the coast (in 1999) have oscillations which are very different from the ones obtained 2 km from the coast in 1996. This suggests that there is a large change of the internal oscillations between the two measuring positions (as internal waves travel to the coast, they disintegrate into irregular groups of waves; some of which look like Korteweg and de Vries (KdeV) solitons).

In Figure 4, the oscillations observed present, in their majority, a thermocline shape with sharp troughs and smooth crests. The oscillations were significantly asymmetric and presented the shape of solitons. Linear and non-linear models were used for the interpretation of the above mentioned examples of the experimental measurements of the intense internal waves (Figure 4a, b shows these waves with numbers), as used by Osborne and Burch (1980) and Holloway (1987). Parameters for linear internal waves were found using the numerical solution to the boundary value problem (Krauss, 1966) for vertical profiles of the Brunt-Väisälä frequency, obtained using data from the sounding made near the mooring:

$$W_{zz} + k_h^2 \frac{N^2 - \omega^2}{\omega^2 - f^2} W = 0 \tag{3}$$

where $W = 0$ at $z = 0$ and $z = -H$; ω is the internal wave frequency; f is the Coriolis parameter; k_h is the horizontal wave number; N^2 is the Brunt-Väisälä frequency; $\rho = \rho(z)$ is density.

For each wave shown in figure 4b the eigenfunctions W were calculated and the wave numbers and velocities of phase C_0 for the first mode waves were found. Then these parameters were compared with the parameters of

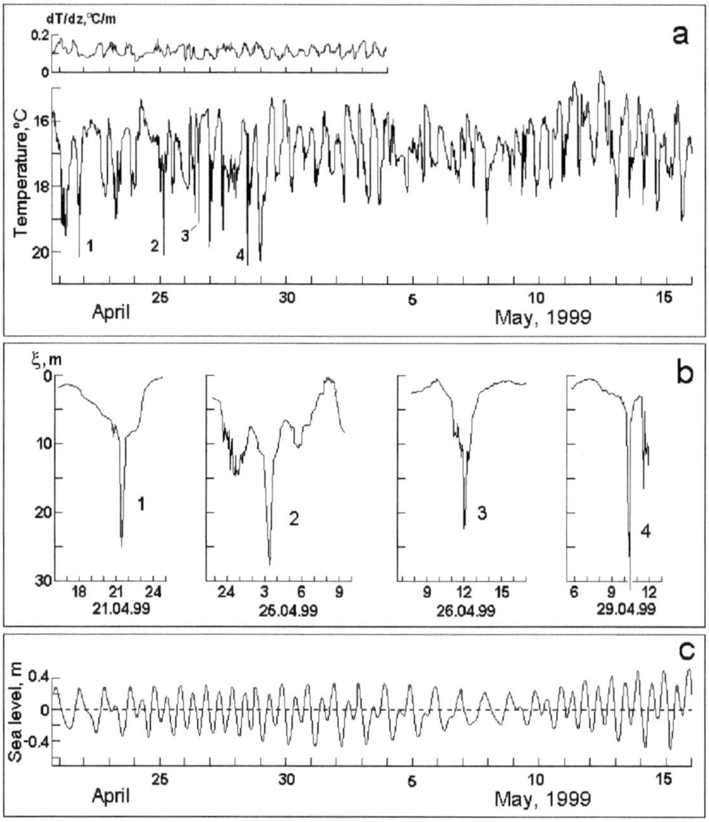

Figure 4. (a) Temperature at a level of 38 m below the surface and 1 km away from the coast at a depth of 51 m. Intense internal waves shown with numbers. In the figure's top part vertical temperature gradient variations are given. They were calculated on the data measurements by two termographs one located at a depth of 38 m and the other at a depth of 33 m. (b) The profiles of the four most intense internal waves are shown in figure 4a. (c) Sea level oscillations in the port of Manzanillo (50 Km southward from the anchor).

the non-linear waves in the context of the KdV model. We calculated the non-linearity α and dispersion β parameters. The KdV equation is written as (Osborne and Burch, 1980; Holloway, 1987; Ostrovskiy and Stepanyants, 1989):

$$\xi_t + C_0\xi_x + \alpha\xi\xi_x + \beta\xi_{xxx} = 0 \tag{4}$$

where ξ is vertical displacement; C_0 is the linear waves phase speed; t is time, and x is the horizontal coordinate. Parameters α and β were estimated for the continuos stratification case using the eigenfunction $W(z)$ of the boundary value problem (3), by numerically integrating the formulae:

$$\alpha = \frac{3}{2} C_0 \frac{\int_{-H}^{0} \rho W_z^3 dz}{\int_{-H}^{0} \rho W_z^2 dz}; \quad \beta = \frac{C_0 \int_{-H}^{0} \rho W_z^3 dz}{2 \int_{-H}^{0} \rho W_z^2 dz}. \tag{5}$$

Based on these estimates of α and β and on the measured wave heights ξ_0, the speed of KdV solitons C_s and its characteristic horizontal length L_s was calculated with the following formulae:

$$C_s = C_0 + \frac{\alpha \xi_0}{3}, \quad L_s = \left[\frac{12\beta}{\alpha \xi_0}\right]^{1/2}. \tag{6}$$

Since a particular solution to equation (4) gives the form of a soliton of the type:

$$\xi(z) = \xi_0 \mathrm{sech}^2[(x - C_s t)/L_s], \tag{7}$$

then the period $\tau_0 = 2\pi/(k_h C_0)$ was determined to be the time duration of the wave, on the level $sech^2(1) = 0.42$ of its height, measured from the trough (see page 82 in the monograph by Konyaev and Sabinin, 1992). This wave period was used for both internal wave models. The height of the waves was defined from the vertical displacements estimated by equation:

$$\Delta\xi_z(t) = \xi_z(t_1) - \xi_z(t_0) = [T(t_1) - T(t_0)]/(\overline{dT/dz}), \tag{8}$$

where $\xi_z(t_1)$, $\xi_z(t_0)$, are vertical displacements at level z at times t_1 and t_0,which originate temperature fluctuations $T(t_1)$, $T(t_0)$. $(\overline{dT/dz})$ is the average vertical temperature gradient during $(t_1 - t_0)$, which was estimated using data from two thermographs, a TDS-85 and a BoxCar, which were displaced on a 5 m vertical line. The results of the calculations made with equations (3) to (8) are presented in table I.

The mean height of the solitary waves is found to be between 16 and 28 m. They last from 17 to 32 min, with a phase speed ranging from 0.29 to 0.36 m/s and a characteristic length (L_s) that goes from 72 to 204 m. From equation (7) it follows that the speed of the soliton and its length scale depend on its height, when the hydrology is constant. Therefore, the bigger the soliton the faster it moves. At the same time the shape of the soliton readjusts itself such that with an increase in height the soliton becomes more compressed. Summarising, the non-linear transformation of the baroclinic tide in the continental shelf generates a wide spectrum of soliton-like waves. Because of their different heights they quickly disperse in space, forming the specific scheme of fluctuations recorded by the instruments (Figures 4a, b). Figure 5 shows the power spectrum of the temperature and sea level oscillations illustrated in Figures 4a, c. A strong peak in the

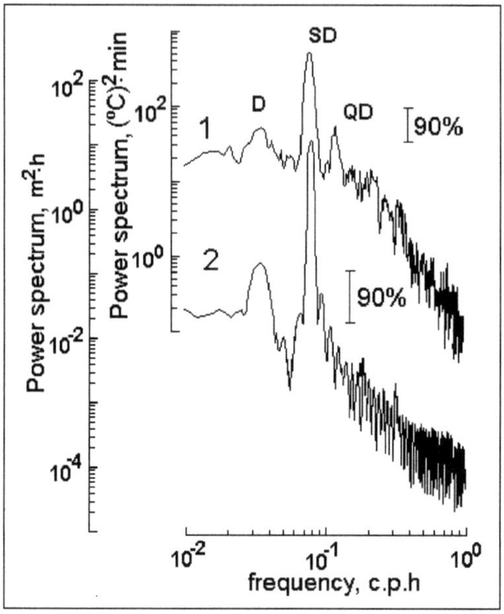

Figure 5. Power spectrum temperature (1) and sea level (2) oscillations illustrated in figures Fig. 4a,b. Vertical line shows the 90 % confidence interval.

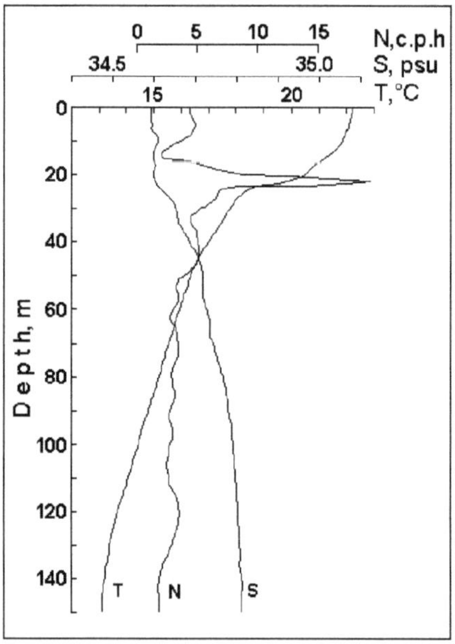

Figure 6. Survey-area-mean variation of the temperature, salinity and buoyancy frequency with depth.

TABLE I. Parameters of the internal wave, based on the linear and non-linear models.

Averaging period	21-22 h, 21.04.1999	02-05 h, 25.04.1999	11-13h, 26.04.1999	10-11 h, 29.04.1999
dT/dz [$°C/m$]	0.132	0.117	0.141	0.119
No. of wave	1	2	3	4
Measured parameter of the wave				
ζ_0	16.2	23.1	17.3	27.8
τ_0	21.3	30.7	32.3	17.1
Parameters calculated from models				
Linear wave model				
C_0 [m/s]	0.25	0.34	0.36	0.23
L_0 [m]	324.6	622.5	697.4	230.9
KdV soliton model				
$\alpha \cdot 10^2$ [l/s]	-0.345	-0.115	-0.101	-0.421
β [m^2]	48.9	61.7	60.3	50.7
C_s [m/s]	0.32	0.35	0.36	0.29
L_s [m]	102.5	166.9	203.5	72.1

spectrum of temperature oscillations occurs only at a semidiurnal frequency. The maximum buoyancy frequency is reached at a level of 20 m (Figure 6).

4. Spatial spectra

The horizontal spatial spectrum of the vertical oscillations was calculated for a semidiurnal frequency in 10 m-thick layers shifted sequentially by 5 m and were averaged over these layers in certain depth interval. The horizontal spectra varied significantly with each layer, and this dependence slows down with increasing layer depth.

The survey duration and area were close to the period and length of the wave under study, respectively. Normally, the spectral window shape depends appreciably on the phase of oscillations. This dependence can be cancelled by averaging the spectral estimate for different initial phase implementations (averaging over levels). For a wave length of 500 m the phase changes by 50 degrees within 70 m, this, as seen below, is the observed shift in the vertical.

The mean horizontal spatial spectrum for the depths between 50 and 120 m has a single mayor peak. As it is evident from comparison with a test sine wave, the height of this peak is consistent with oscillations of a depth-mean amplitude of 11 m (see Figure 7a; to make the patterns of the small peaks clearly distinguishable against the background of the major peak, the amplitude spectra rather than its square is plotted). The shape of the peak is not much different from that of the spectral window averaged over the phase oscillations; thus, these oscillations involve a single prevailing wave that moves westward and has a wavelength of 62 km.

The vertical structure of the wave is determined from the section of three-dimensional spatial spectrum for the same layer. The section goes through both the vertical wave-number axis and the horizontal spatial-peak. The peak in this section is at the vertical wave number +2 cycle/km, which corresponds to an oblique wave with a vertical length of 500 m, an upward phase motion, and a group down motion (see Figure 7b).

The average of the horizontal spectrum from 20 up to 75 m, shows one more peak which matches the wave of length 20 km, and moves southward. Its vertical structure corresponds to an oblique wave of vertical length about 400 m, downward phase velocity, and upward group velocity (Figure 7c, d).

5. Ray patterns of the internal tide

The pattern of the internal tide emerging along the continental slope can be described by rays, or characteristics, that are formed in areas with a critical bottom tilt (i.e. which coincides with the tilt of tidal orbits). The variation of the tidal orbits with depth governs the internal-tide ray path. The tidal wave group moves along the ray, the tidal mid-ray oscillation amplitude is a maximum, and the oscillation phase is the same along the whole ray. The phase motion of the wave is directed across the ray and ray trajectory coincides with a line of equal phase (Baines, 1982; Holloway and Merrifield, 1999).

The internal-tide intensity, however, depends smoothly on the bottom tilt angle, so that energy in the baroclinic tide can be generated in all areas of a sloping bottom. Some observational and numerical modelling data indicate that the ray approach can be used for an approximate description of internal tide generation along a continental slope. The ray structure is clearly tracked over a couple of cycles of ray reflection from the bottom and surface of the ocean (Pingree and New, 1991; Holloway and Merrifield, 1999). Then, the rays are diffused, and if the oceanic depth varies slightly, the notion of the vertical mode structure of a tidal wave becomes adequate.

The ray paths are constructed from measured data of the bottom and buoyancy frequency profiles (Figure 5). The buoyancy frequency is cal-

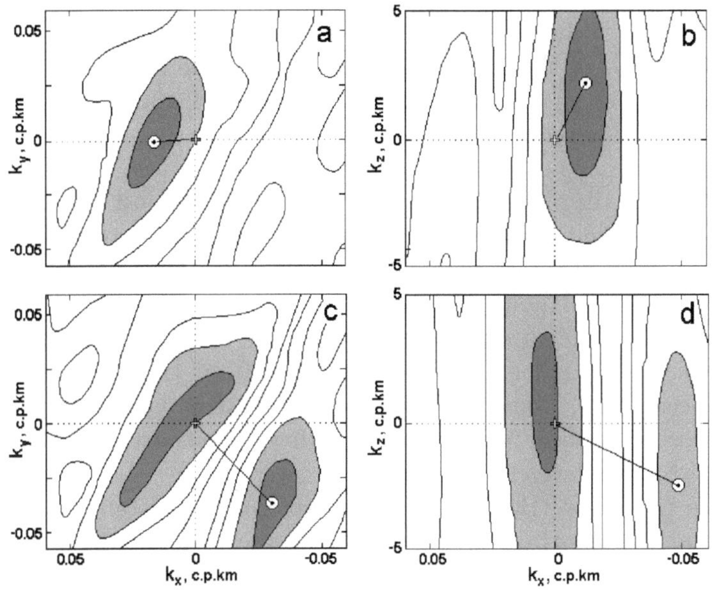

Figure 7. Horizontal spectrum of vertical deviations (a, c) and the vertical section of the corresponding three-dimensional spatial spectrum that passes through the main peak of the horizontal spectrum (b, d). The basic wave extending downward, was detected on horizons 50-120 m. The short wave extending upwards, is detected on measurements on horizons 20-75 m. The circle denotes spectral peaks, the crosses correspond to the origin of coordinates, the lines stand for the wave vectors, the contours are depicted every 0.2 of the maximum value, and the shaded area denotes the peak areas surface.

culated from the mean-measured profile of density in the upper 120 m layer of the ocean. Below, the buoyancy frequency is assumed to decrease exponentially with depth to 0.5 cycle/h in the 5000-m layer (Filonov *et al.*, 1996a).

The wave vector tilt angle θ (1) with respect to the horizontal depends on the tidal frequency $f_t = 0.0805$ cycle/h; the inertial frequency $f_i = 0.0271$ cycle/h; and the buoyancy frequency $N(z)$, which varies with depth. The ray path at each depth is perpendicular to the wave vector (La Fond, 1962; Konyaev and Sabinin, 1992).

Normally the critical bottom tilt is found on the near-discontinuity between the flat shelf and the steep continental slope. Here, three internal-tide rays are formed (Baines, 1982). The first ray is upward (coastward) and generates a short tidal coastward wave on the shelf due to the reflection from the ocean bottom and surface. The second ray is upward and oceanward and reaches the surface not too far away. The third ray is downward and oceanward and appears at the surface far from the continental slope after being reflected from the bottom. Far away from the continental slope and over large depths, the rays going oceanward form a long tidal first-mode wave. The length of this wave is determined by the distance between the neighboring points at the ocean surface reached by the rays (Prinsenberg and Rattray, 1975; Baines, 1982).

Along the coast near Barra de Navidad, the bottom tilt is somewhat greater than the tilt of the semidiurnal tidal ray for almost the whole slope. The bottom profile used in the calculations includes an area of critical tilt at a level of 970 m (Figure 8a,b). If this area includes three rays as before, the first ray (upward, going along the slope and coastward) does not reach the surface, it reaches a steeper slope segment lying shallower inshore, at a depth of 190 m and a distance of 11 km from the coast, where it is reflected oceanward. Thus, two outgoing rays (rather than one, as is usual) reach the ocean surface not too far away from the coast, and the distance between them is relatively small (in this case, 35 km). The ray oscillation phases are the same (Baines, 1982); therefore, the distance between the rays governs the length of the internal tidal wave. Within the survey area, the corresponding short wave goes upward to the ocean surface and, on being reflected, goes downward. The third ray, outgoing downward from the area, appears at the surface approximately 150 km away from the coast after being reflected from the bottom.

The bottom profile used has at least one additional noticeable area with a critical tilt at a level of 190 m (Figure 8c, d). For another bottom profile not to far away from the first, areas with critical tilts are found at different levels. This makes the overall situation more complicated, but the above features of the internal tide seem to remain unchanged.

Internal waves are formed on both sites mentioned above; then they move toward open sea. This does not mean, however, that near the coast there are no sites with a critical inclination to generate internal waves which propagate toward the coast (where the waves undergo nonlinear disintegration, as it is suggested in Figure 4).

6. Discussion

Here we discuss a plausible scenario, given the limited observations made of the research area, as well as the prevailing theories which have been successful in describing other measurements (Prinsenberg and Rattray, 1975; Baines, 1982; Holloway, 1985; Graig, 1987; Filonov, 2000c), of a coherent view of the internal wave near Mexico's Central Pacific Coast.

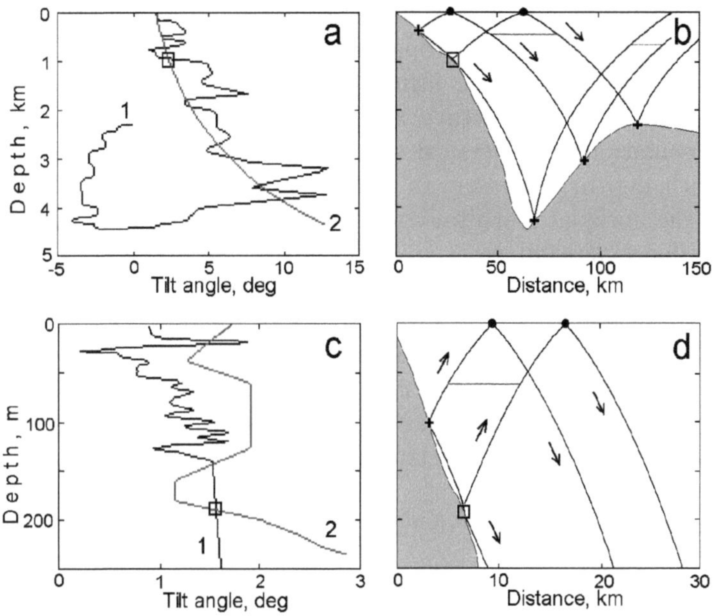

Figure 8. (a) Change with depth of (1) the bottom tilt and (2) the ray slope for the diurnal internal tide and (b) the ray pattern for the area with a critical bottom tilt at a depth of 970 m. The same for depth 190 m (c,d). The squares denote the points with a critical bottom tilt (a,c) and the place of generation in the continental slope (b,d). The circles denote the points of ray reflections from the ocean surface. The asterisks denote the points of ray reflections from the slope and from the bottom. The arrow shows the direction of group wave propagation along the rays (b,d). Negative tilts of the bottom are located on the external side of the Central American trough (a).

6.1. INTERNAL WAVES ON A SHELF.

The internal tides near Mexico's Central Pacific Coast travel from its generation point in the continental slope towards the coast with an almost semi-diurnal peridiocity, shaped as waves that strongly deformated the thermocline and the fields of other physical and chemical water properties in the whole continental shelf region. In the continental shelf the internal tides are non-linear oscillations with a steeper front part. The front part sometimes can be almost vertical (Figure 3); a common temperature decrease is 8-10 °C/h and up to 1.5 psu/h for salinity. At the continental shelf, these waves can reach amplitudes of 20-25 m for the vertical fluctuations of the thermocline, with a 2 to 8 km longitude. While, at the continental slope of the ocean, the wave's amplitudes are not bigger than 5 m, but their have longitude varies between 25 and 30 km (Filonov, 2000a, 2000c).

In the coastal zone the frontal part of the internal tides are sharper. This contributes to their disintegration in groups of soliton-like waves, that are shorter, of larger amplitude and hence slope, which evidently disipate their energy entirely as they propagate towards the coast.

The internal tides in the continental shelf near Mexico's Central Pacific Coast present a modal structure, always with the presence of only the first mode of oscillation. The vertical shift phase for the water layers does not change with depth; therefore, in the upper part of the thermocline the phases of the horizontal orbital currents must differ by 180° (see Figures 4 and 7 in Filonov, 2000c).

The internal tides in the continental shelf conserve the main characteristics of the barotropic waves: inequality of the daily and semi-monthly ranges (clearly on a monthly basis), therefore, a multidirectional study of the parameter's variability could provide a statistical forecast of the variations of oceanological fields (up to a phase) on the continental shelf zone where they have not been transformed into short wave groups.

6.2. INTERNAL WAVES ON AN OCEANIC SLOPE

On the basis of the survey's temperature and salinity data, the internal tide is described in two different ways. Spectral analysis in space gives parameters of the dominant internal tide; here, the wave turned out to be unexpectedly short in length and to propagate ocean-ward. In constructing the ray patterns of the internal tide, it is shown that, along the Pacific coast of Mexico, al three internal-tide rays generated in areas with critical bottom tilts can go out oceanward. Two of these rays go to the upper oceanic layers not too far away from the coast and over the continental slope, which is matched by a short near-surface tidal wave propagating oceanward.

In the survey area, the major wave propagates downward after being reflected from the ocean surface. This major wave should correspond to an oblique wave going upward to the surface. This oblique wave is not seen in the vertical spectrum of oscillations, since the survey area is evidently small in size.

The principal wave propagates almost westward at an acute angle to the continental slope line, although the propagation would apparently be expected to be southward or southwestward and perpendicular to the slope. This behaviour of the wave can be explained by the influence of a current with a horizontal velocity shear. The third description (using rays) gives a working hypothesis for explaining the emergence of the short tidal wave propagating oceanward.

The processes of nonlinear transformation of an internal tide with soliton formation are active and observed consistently on the shelf when the internal tide propagates coastward (Konyaev and Sabinin, 1992). These phenomena are observed more rarely in areas where the continental shelf's depth is larger; for example, groups of solitons outgoing from the coast and observed at the ocean surface along the coast of Kamchatka (Konyaev and Sabinin, 1998). As in our case, the shelf near Kamchatka also has a small width and the pycnocline is located close to the ocean surface.

Certain soliton groups can appear far away from the continental slope, in an area of the open ocean where the internal-tide ray rises to the surface after being reflected from the bottom (Pingree and Mardell, 1985). The same phenomenon can occur in the area under consideration, approximately 150 km away from the coast, especially as the external side of the deep-ocean tray (at which the first reflection of rays occurs) is tilted against the rays, so that the rays come closer to one another on being reflected (Figure 8b).

Apparently, the bulk of the internal tide energy goes oceanward; however, part of this energy goes coastward. As in the normal case, these waves are also subjected to a nonlinear transformation, which has already been observed (Filonov *et al.*, 1996a; Filonov, 2000c).

Acknowledgements

The authors are grateful to the "BIP-V" crew and their co-workers Irina Tereshchenko, Cesar Monzn and Daniel Kosonoy. This work was supported by CONACYT through projects: 1449-PT, 35553-T and the Russian Foundation for Basic Research, project no. 01-05-64289.

References

Baines, P.G. Internal Tide Generation Models. *Deep Sea Res.*, 29:307–338, 1982.

Filonov, A.E., C.O.Monzn and I.E.Tereshchenko. On the Conditions of Internal Wave Generation Along the West Coast of Mexico. *Ciencias Mar.*, 23:255–272, 1996a.

Filonov, A.E., C.O.Monzn and I.E.Tereshchenko. A Technique for Fast Conductivity-Temperature-Depth Oceanographic Surveys. *Geofís. Int.*, 35:415–420, 1996b.

Filonov, A.E. Spatial Structure of the Temperature and Salinity Fields in the Presence of Internal Waves on the Continental Shelf of the States of Jalisco and Colima México. *Cienc. Mar.*, 26:1–21, 2000a.

Filonov, A. E. Thermal Structure and Intense Internal Waves on the Narrow Continental Shelf of the Black Sea. In the Special Issue *J. Mar. Sys.*, 24:27–40, 2000b.

Filonov, A.E. Internal Tide and Tsunami Waves in the Continental Shelf of the Mexican Western Coast. Oceanography of the Eastern Pacific, CICESE-UGM, México, 1:31–45, 2000c.

Filonov, A.E. and A.Trasviña. Internal Waves on the Continental Shelf of the Gulf of Tehuantepec, México. *Est. Coast. Shelf Sci.*, 50:531–548, 2000.

Graig, P.D. Solution for Internal Tide Generation Over Coastal Topography. *J. Mar. Res.*, 45:83–105, 1987.

Holloway, P.E. A Comparison of Semidiurnal Internal Tides from Different Bathymetric Locations on the Australian North West Shelf. *J. Phys. Oceanogr.*, 15:240–251, 1985.

Holloway, P.E. Internal Hydraulic Jumps and Solitons at a Shelf Break Region on the Australian North West Shelf. *J. Geophys. Res.*, 92:5405–5416, 1987.

Holloway, P.E. and M.A. Merrifield. Internal Tide Generation by Seamounts, Ridges, and Island. *J. Geophys. Res.*, 104:25937–25951, 1999.
Huthnance, J.M. Internal Tides and Waves Near the Continental Shelf Edge. *Geophys. Astrophys. Fluid Dyn.*, 48:81–106, 1989.
Konyaev, K.V. Spectral Analysis of Physical Oceanographic Data. A. BALKEMA, ROTTERDAM, 210 p., 1990.
Konyaev, K.V. Internal Tide at the Critical Latitude. *Izvestiya RAN. Atmospheric and Oceanic Physics*, 36:363–375, 2000.
Konyaev, K.V. and K.D.Sabinin. *Waves Inside the Ocean.* Sankt-Petersburg. Hydrometeoizdat, Sankt-Petersburg, (in Russian), 272 p., 1992.
Konyaev, K.V. and K.D.Sabinin. Intence Internal Waves in the Vicinity of the Kamchatka Pacific Coast. *Okeanologia*, (Moscow), 38:31–36, 1998.
Konyaev, K.V. and A.E. Filonov. Internal Tide Along the Pacific Coast of México. *Izvestiya RAN. Atmospheric and Oceanic Physics*, 38:259–266, 2002.
Krauss, W. *Interne Wellen.* Gebrder Borntrager, Berlin, 248 p., 1966.
La Fond, E.C. *Internal Waves, in The Sea.* M. N. Hill, editor. Wiley, Inter-Science Series, New York, 731–756, 1962.
Osborne, A.R., and T.L. Burch. Internal Solitons in the Andaman Sea. *Science*, 208:451–460, 1980.
Ostrovsky, L.A. and, Yu.A. Stepanyants. Do Internal Solitons exist in the Ocean? *Rev. Geophys.*, 27:293–310, 1989.
Prinsenberg, S.J. and M.Jr. Rattray. Effects of Continental Slope and Variable Brunt-Väisälä Frequency in the Coastal Generation of Internal Tides. *Deep Sea Res.*, 22:251–263, 1975.
Pingree, R.D. and G.T. Mardell. Solitary Internal Waves in the Seltic Sea. *Progr. Oceanogr.*, 14:431–442, 1985.
Pingree, R.D. and A.L. New. Abissal Penetration and Bottom Reflection of Internal Tidal Energy in the Bay of Biscay. *J. Phys. Oceanogr.*, 21:28–39, 1991.

CANEK: MEASURING TRANSPORT IN THE YUCATAN CHANNEL

J. OCHOA, A. BADAN, J. SHEINBAUM & J. CANDELA
Departamento de Oceanografía Física, CICESE
Ensenada, Baja California, México

In memory of Pedro Ripa 1946–2001

Abstract. The Yucatan Channel is one of the key restrictions of the North Atlantic surface circulation, and also a privileged location to understand the circulation within the Gulf of Mexico and the Caribbean Sea. From September 1999 to June 2000 a set of eight instrumented moorings measured currents and temperatures across the Yucatan Channel. The main result of such measurements, published elsewhere, is that the mean transport amounts to only 23 Sv instead of the 28-30 Sv believed until recently to be a robust value. This result implies the need of an overall review of the other Gulf Stream's sources. Here we show that the second period of measurements with essentially the same array, from July 2000 to May 2001 is in good agreement with the first period's mean transport. The correlation functions of both velocity components, in the subinertial band, show 500 m in the vertical and 70 km in the horizontal as the characteristic scales. An array of only fourteen currentmeters, slightly optimized in position, scattered through the section allows an adequate estimation of transport and fluctuations, with a standard error of 0.3 Sv. But an array of even more time series from only three moorings, each consisting of a near-surface, upward-looking ADCP plus two single point currentmeters distributed below, degrades the skill down to 2.5 Sv; this is because each ADCP measures highly correlated series, and the other meters are too far apart, thus producing a poor coverage.

Key words: Transport, Yucatan

1. Introduction

The Gulf Stream is the most studied current in the world's oceans; it is mainly fed by the Loop Current which occupies the eastern portion of the Gulf of Mexico. The transport at 26° 45'N between West Palm Beach, Florida and Settlement Point, Bahamas has been monitored via voltage measurements of a calibrated cable for the past 20 years (Larsen, 1992)

O.U. Velasco Fuentes et al. (eds.), Nonlinear Processes in Geophysical Fluid Dynamics, 275-286.

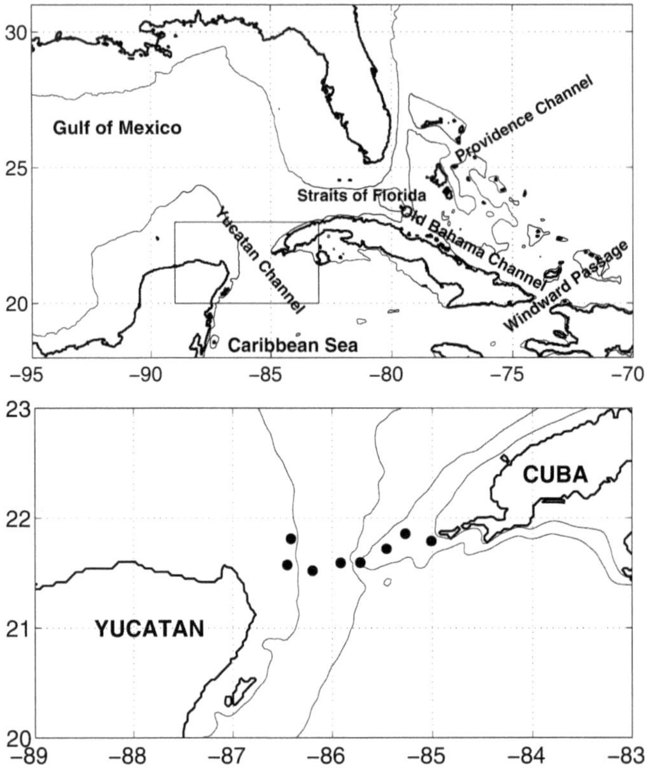

Figure 1. Maps with geographic names and layout of moorings during the second measuring period (*i.e.* the large dots between Cuba and Yucatan). The 500 m isoline is shown in both maps and the 2000 m in the lower map.

and shows a mean transport of 34.2 Sv of a dominant northward flow. The transport of the Loop Current is the outflow from the Gulf of Mexico, which is dominantly eastward between Key West, Florida and La Habana, Cuba at 81° 45'W, just 200 km south of the cable. Given the Gulf's closed geometry this outflow must be the same as the transport through the Yucatan Channel and only two other plausibly passages can add to the Gulf's outflow before the cable: the Old Bahamas Channel which adds water flowing northwestwardly where the Gulf's outflow turns northward, and the Providence Channel which also feeds waters flowing northwestwardly at 25° 45'N (see Figure 1). Recent and previous estimates of the transport outflowing from the Gulf of Mexico fluctuate between 28 and 30 (see, for example: Mooers and Maul, 1998; Schmitz, 1996). Therefore the transports through these channels or other sources to the well known mean transport between West Palm Beach and Settlement Point must be reviewed. Also, all the flows into and out of the Caribbean Sea should be consistent with the 23 Sv transport through the Yucatan Channel.

The following section shows the general features of our measurements and how the transport in the Yucatan Channel was computed. Sheinbaum *et al.* (2002) report the first calculations; here we show that the rest of the data confirm those results. The contributions included here are: i) ancillary information about the spatial structure of the flow in the channel (*i.e.* correlation functions), ii) the transport calculations of the second period of measurements, as well as the confidence intervals for the mean transports, and iii) an analysis of the limitations inherent to inferring transports from a three mooring array, instead of the full coverage across the section. A final section summarizes the results. It is worth noticing that the name channel or strait does not apply in the dynamical sense to the boundary between the Gulf of Mexico and the Caribbean Sea; we have called Yucatan Channel and sometimes the reader will find the name Yucatan Straits, but is neither long nor narrow enough; the first internal Rossby radii of deformation are 44, 25, 15, and 13 km, all computed with respect the sill depth (2040 m), whereas there are 200 km from Yucatan to Cuba.

2. Measured Transports

An analysis of the structure and variability of the currents in the Yucatan Channel during the first measuring period is found in Sheinbaum *et al.* (2002). Figure 1 shows the distribution and percentage of data recovered during a second period, from July 2000 to May 2001, with very much the same spatial coverage. The distribution of currentmeters was based on the availability of meters, a desire for a uniform coverage, and the alignment with hydrographic features such as the extremes of salinity and oxygen or isotherms. The focus here is on subinertial motions, whence the original series have been low pass filtered; a 32 h period is the half power cut off of the filter.

In order to estimate transport through the section an interpolation and extrapolation method is required, and we used an objective mapping technique as described in Sheinbaum *et al.* (2002). The spatial scales of the mapping are consistent with straightforward estimates of correlation functions. The important feature of these functions, as shown in Figure 2, is that they allow choosing the type of correlation as a function of the lag, and the corresponding parameters. The statistics deviate from homogeneity mainly because the westernmost measurements, which have been excluded in computing the correlation functions depicted in Figure 2. The across-channel velocity component, denoted by u, has a monotonous decaying correlation in both directions which, when fitted to exponentials, gives the

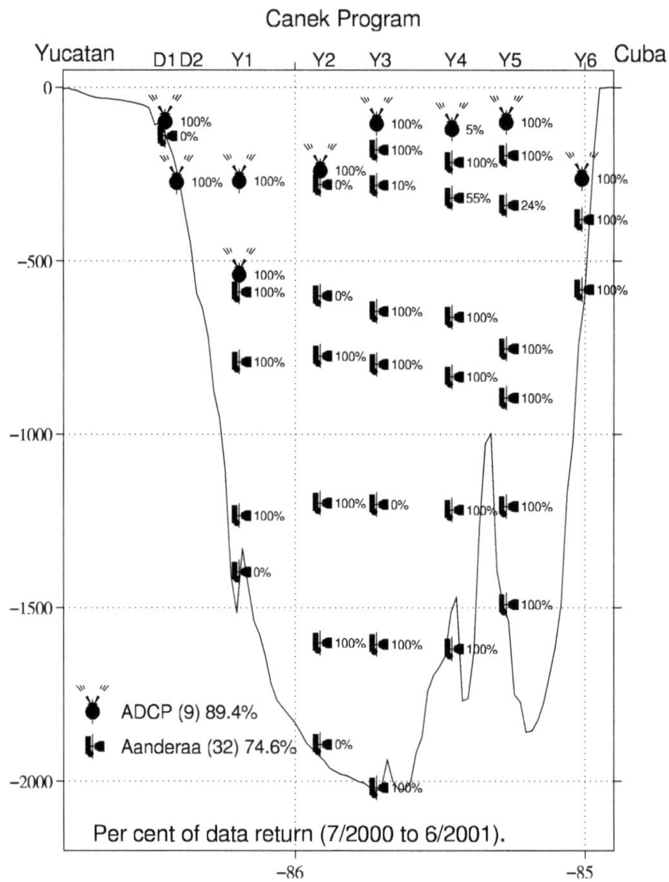

Figure 2. Layout of instruments and percentage of data return during the second measurement period.

scales of 500 m in the vertical, and 70 km in the horizontal. Eventhough the correlation points from the mooring that showed the largest deviations are not included, it is clear that the statistics are unlikely to be homogeneous as several correlation estimates differ more than their uncertainty from simple plausible correlation functions. Also, the intensity of the velocity variability (Figure 4b) shows considerable structure within the section. The velocity component along the channel, denoted by v, shows similar behavior in the correlation as function of the vertical lag as the across-channel component (u); its scale is the same when fitted to an exponential. A substantial difference of v relative to u is its correlation as a function of the horizontal lag; hence we fitted a function of the form

$$\rho(\Delta x) \equiv < v(x)v(x+\Delta x) > = cos(2\pi\Delta x/L_c)e^{-2\pi\Delta x/L_e} , \qquad (1)$$

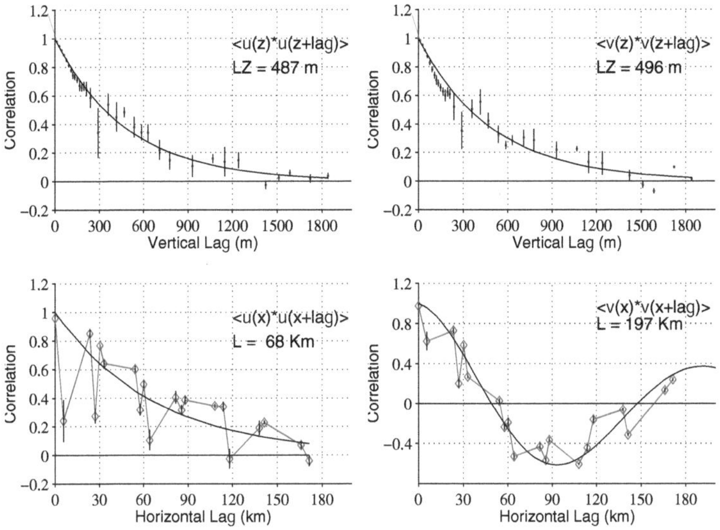

Figure 3. Correlations of the velocity components across (u) and along (v) the channel as functions of the vertical separation (at same horizontal position; *i.e.* same mooring) and horizontal separation (within 40 m depth separation). Data from the easternmost mooring is not included. Fitting of an exponential function or the same multiplied by a cosine provide the L and LZ parameters shown in the figure (see text).

which gives as parameters L_c= 140 km and L_e=200 km. Fitting with a single scale parameter by forcing $L_c = L_e$ gives 200 km. This last function is very much the same as the fitted curve shown in Figure 3 with the first zero crossing near 50 km. The fitting of a simple exponential function produces $L = 70$ km, as with u. An extensive analysis of the velocity fluctuations and of the structure of the flow is given in Abascal *et al.* (2003).

This analysis supports the functions and scales used in the objective mapping: 70 km in the horizontal, and 400 m in the vertical, because the mapping is insensitive to the use of the simple exponential or Equation 1 as correlation function. Although the correlation functions of Figure 3 are incomplete representation of the statistics, since these are inhomogeneous, they indicate how smooth instantaneous maps should be; variability with scales smaller than the characteristic values is not intense.

Once equipped with an interpolation/extrapolation scheme in space, the maps and estimates of transport as a function of time follow directly. The mean distributions of the currents in separate frequency bands are shown in Figure 4. These maps are very much in agreement with those of the first deployment; the importance is that the year to year variability was small during the periods of measurements.

The time series of transports through the section are plotted in Figure

4, in this case for both measurment periods. Purposely exaggerating the amount of significant digits, the mean transports were 23.75 and 22.03 Sv for the first and second periods, and the overall mean is 22.83. Since the currentmeters also measured temperatures, the separation of transport above and below a given isotherm is possible. The temperature field was mapped using objective mapping with the same simple exponential correlation functions and the same scales as those for velocities. Transport calculations are very insensitive to the interpolation scheme, as long as it is somehow reasonable.

In order to compute confidence limits on the mean transport we tried three calculations; all based in estimating the equivalent degrees of freedom (N_{EQ}) for the available series. Two of these follow Priestley's (1989) formula:

$$N_{EQ} = N \; [\sum_{k=-(N-1)}^{N-1} (1 - \frac{k}{N}) \rho(k\Delta t) \;]^{-1} \; , \tag{2}$$

where the correlations ($i.e.$ $\rho(k\Delta t) \equiv < \Delta T(t) \Delta T(t + k\Delta t) > / < \Delta T^2 >$, where ΔT is the transport anomaly) were computed with the transport time series itself, either directly or by fitting a first order autoregressive processes. Figure 5 shows the transport time series. The third estimation of the equivalent degrees of freedom follows Richman *et al.* (1977) who used the squared correlations instead of the correlations in Equation 2. Their formula is:

$$N_{EQ} = N \; [\sum_{k=0}^{N-1} \rho^2(k\Delta t) \;]^{-1} \tag{3}$$

Table I lists the confidence intervals for each measurement period. Richman *et al.* (1977) argue, given that subsequent values of the sampled correlation are highly dependent of each other, that this formula should give more pessimistic and realistic values compared with Priestley's version. Nonetheless the pessimistic confidence limit happens in this case with the autoregresive estimate. Regardless of the version or period used the 90% confidence limit is below 1.1 Sv, whence the means of each period are statistically different, since they differ by 1.7 Sv.

A result not shown in Table I is that by adding the two measurment periods the 90% confidence limit is 0.5 or 0.8 Sv, depending on the estimate being used. By avoiding the pessimistic interval, the mean is between 22.3 and 23.3 with 90% confidence. The pessimistic confidence limit, from the autoregressive estimate of the correlation, is 0.8 Sv, in which case the mean is between 22.0 and 23.6, with 90% confidence. A mean transport close to 23 Sv in Yucatan Channel, well below the historical 28 Sv, is remarkable.

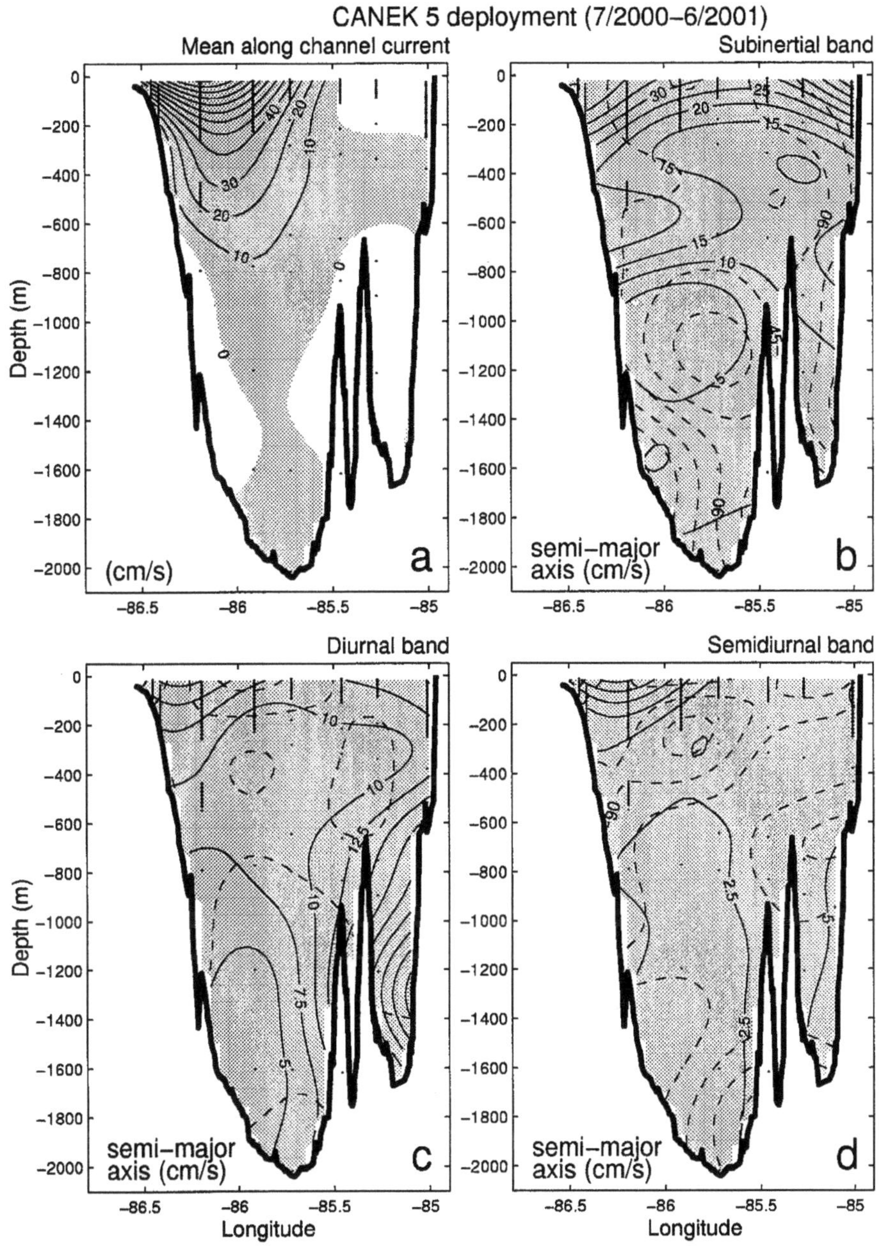

Figure 4. Mean distributions of currents in separate frequency bands. (a) The mean along-channel velocity component. Shading indicates flow into the Gulf of Mexico. (b), (c), and (d) The variability in the subinertial, diurnal and semidiurnal bands in terms of the ellipse amplitude (continuous contours) and orientation (broken contours).

TABLE I. Calculations for each measurement period of the equivalent degrees of freedom (N_{EQ}), the decorrelation time scale (*i.e.* the time to achieve one degree of freedom), and the 90% confidence limit in Sv for the mean transport.

	Period	Priestley raw	Priestley AR(1)	Richman
N_{EQ}	1st	50	21	30
	2st	33	30	51
Total measuring time/N_{EQ} in days	1st	5.5	13.0	5.0
	2st	9.7	11.0	3.4
90% confidence limit in Sv	1st	0.7	1.1	0.7
	2st	0.9	0.9	0.5

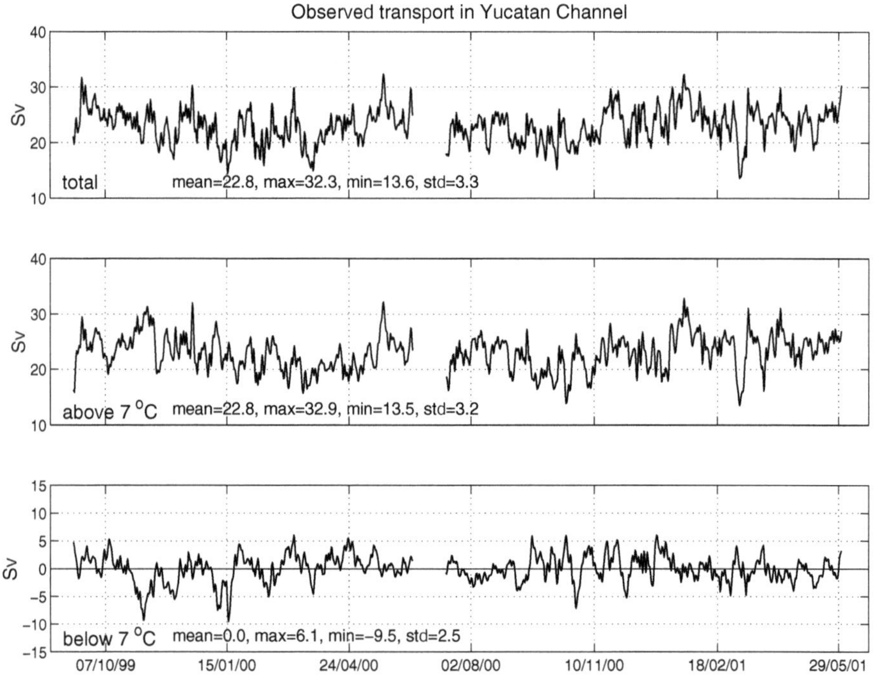

Figure 5. Time series of transport, a) total, b) above 7 °C, and c) below the same temperature. Both measurement periods are included.

3. Inferring Transport with Fewer Measurements

An important inquiry is the issue of optimizing the number of instruments, meaning the use of fewer, to measure the transport. Since the correlations are considerable for distances less than 70 km in the horizontal, and 500 m in the vertical, an array of only fourteen single point meters but scattered through the section proves to be sufficient. There is nothing special about fourteen; it is just an example with less than half the number of meters used; there were thirty-two single point meters in the second measurment period, of which twenty-three had 100% data recovery (see Figure 2). With the use of the objective maps as a true field and positioning fourteen meters at some grid points of the map, we compute via a straightforward least square fit the coefficients α_k, $k = 1, 2, .., M$ in:

$$T(t) = \sum_{k=1}^{M} \alpha_k v_k(t) \ + \ \epsilon(t), \tag{4}$$

where $T = T(t)$ is the time series of the known transport, and $v_k = v_k(t)$ is one (*i.e.* the kth chosen time series) of the fourteen predictors. Then, we optimize by repositioning each meter in neighboring grid points, and testing if the fit improves. All this is done with the first measuring period, producing a standard error of the fitted transport of 0.23 Sv and an insignificant bias. A measure of the skill of those positions and coefficients is extracted through their use during the second measurement period. The result is a standard error of 0.37 Sv and -0.05 Sv of bias, both of which are well below the 90% confidence level uncertainty for one year of observations. Inverting the roles of the measurement periods produces a skill with a standard error of 0.33 Sv and a bias of -0.04 Sv. Figure 6a shows the positioning of the fourteen meters, which might seem promising given the number of meters, but the typical deployment is heavily restricted on the number of moorings, and not so much on that of individual meters.

Following a similar procedure, now considering moorings, a restriction on the sensors being vertically aligned must be included. For illustration purposes we test the use of only three moorings, as Figure 6b shows, each having an ADCP looking upward to sample the surface layer, and two single point meters at deeper levels. There is also nothing special about choosing only three moorings; the results are examples of the limitations. Lets assume that each ADCP produces thirteen time series corresponding to depths from 270 to 30 m in 20 m steps. Then, the amount of predictors has increased to forty-five; thirty-nine near surface series due to the three ADCPs, plus six deep meters. Following the same procedure as with the exercise of fourteen meters proves to be nonsense. The straightforward use of Equation 4 with the forty-five predictors produces a transport fit in the first measurement period with a root mean square (rms) difference of 0.4 Sv and negligible bias. The skill as tested using the second measurement period has a rms of 1.9 Sv and the bias is 0.5 Sv. If we interchange the periods, the fitting produces a rms of 0.5 Sv also with negligible bias, and the test produces a rms of 1.2 Sv and bias of -0.8 Sv. But these numbers do not really show that this situation is nonsense, what shows the failure is the coefficients themselves. In the situation of fourteen predictors all coefficients are positive, they vary from 4 to 30 km^2 (their sum is close to the 230 km^2 transversal area of the section; the fourteen coefficients add to 234.2 and 225.4 km^2) and the rms of their difference is 1.8 km^2. But in the case of the forty-five predictors the coefficients vary from -1.4×10^{12} to 1.6×10^{12} km^2 and the rms of their difference is 3×10 km^2; too large absolute values of both signs and a huge difference between the first and second periods.

These ill-results should not be surprising given that: i) the number of effective degrees of freedom is very close to the number of predictors in use (see N_{EQ} in Table I), and ii) the predictors out of each ADCP are highly correlated among them Therefore the straightforward least square solution of Equation 4 becomes an ill-posed problem, resulting in such a disparate coefficient values. For this type of situations a *regularization* procedure is the usual choice. What this means is the addition of constrains on the allowed coefficients, such as avoiding large differences among them. Without going into a general formulation, a simple constraint which is physically acceptable is to ask for the same coefficient for all the series provided by one ADCP. This is equivalent to use the average of each ADCP's series, thus producing only three predictors per mooring (*i.e.* there will be one coefficient per ADCP as well as per each meter). This solves the problem of the coefficient's wild behavior and this total of nine predictors from the three moorings array produces a skill of 2.5 Sv.

Table II summarizes the three examples of inferring transport with a limited number of meters. By construction $< \delta T^2 >$ is always smaller than $< \delta\tilde{T}^2 >$; we call $\delta T = \epsilon(t)$ the least-square residual of Equation 4 and $\delta\tilde{T} = \delta\tilde{T}(t)$ the residual when using the coefficients computed in the alternate measuring period. The difference in the two sets of coefficients provides a measure of the model inconsistency (*i.e.* of a specific use of Equation 4). As the two sets of coefficients converge to each other, the use of either one becomes consistent and its credibility increases.

4. Summary

The two periods of measurement in Yucatan Channel offer mean transports which are statistically significant different from each other, both are

TABLE II. Summary the degree of fit and the skill value of the three examples of using a limited amount of meters to infer transport.

Number of coefficients or predictors	Fit $\sqrt{<\delta T^2>}$	Test $\sqrt{<\delta\tilde{T}^2>}$ Skill	Test $<\delta\tilde{T}>$ Bias	Rms of $\delta\alpha$ $\sqrt{<\delta\alpha^2>}$ from two fits	Mean of α's $<\alpha>$
14	0.23	0.33	-0.04	1.8	16.7
	0.19	0.38	-0.05		16.1
45	0.4	1.9	0.5	3×10^{11}	5.0
	0.5	1.2	-0.8		4.7
9	2.0	2.6	-0.2	7.5	16.3
	2.2	2.5	-0.2		14.1

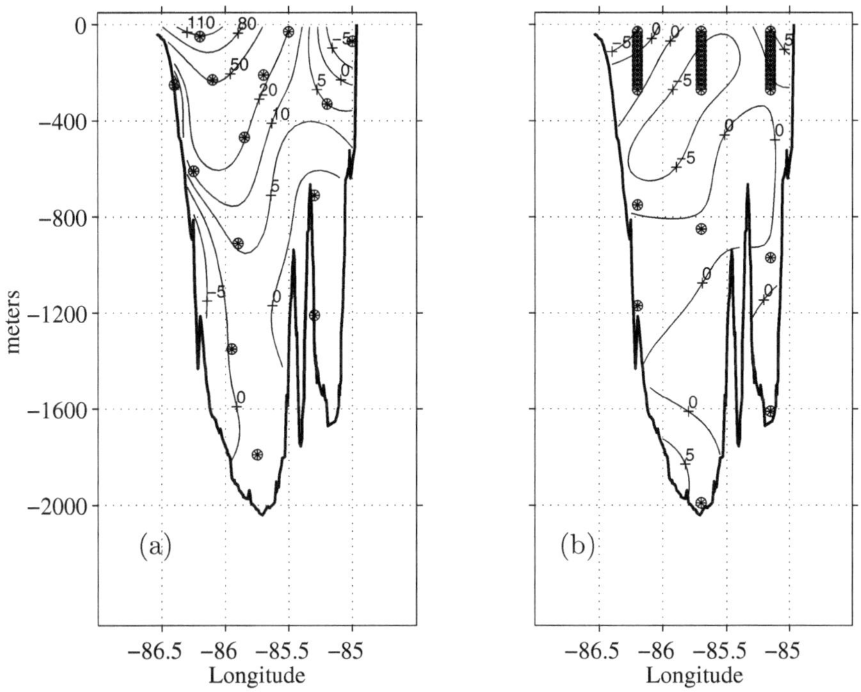

Figure 6. Maps with positions slightly optimized for the purpose of inferring transport. In frame (a) there are fourteen positions and in frame (b) three moorings which may give either forty-five predictors, or nine, depending if each ADCP data is considered as 13 predictors or just one (see text). The maps show, in (a) the mean of the along-channel velocity considering both measurement periods, while in (b) their difference, both in cm/s.

consistently well below the 28 Sv historical mean. Adding both periods, results in a conservative mean estimate between 22.0 to 23.6 Sv with a 90% confidence level. The transport time series via the cable in the Florida Current do not show that the periods of measurements in Yucatan were in any way anomalous. The full record of cable measurements does not show any significant trend, neither do the available measurements in Yucatan Channel. Thus there is reason to believe that the measurement periods in Yucatan are representative for longer periods.

The correlation among the measured time series, in subinertial frequencies, shows a characteristic scale of 500 m in the vertical and 70 km in the horizontal.

The exercises of inferring transports with fewer instruments show that three moorings will produce very limited estimates with errors on the order of 2.5 Sv when the standard deviation of the actual transport is 3.3 Sv. Thus the value of the measurements in both periods is high; its repetition is costly. Even an array optimized solely to infer transport with acceptable accuracy, requires more than three moorings.

Acknowledgements

We thank Oscar Velasco for his helpful review. The authors are grateful to the crew of the '*R/V Justo Sierra*', to the technical staff of the Physical Oceanography Department of CICESE, and to our colleagues from the Instituto de Oceanología, Cuba, and the Universidad Nacional Autónoma de México, who made this program possible. Financial support was provided by IAI, CONACYT, CICESE, and by the DeepStar Consortium.

References

Abascal, A., J. Sheinbaum, J. Candela, J. Ochoa and A. Badan. Analysis of flow variability in the Yucatan Channel. Submitted to *J. of Geophys. Res.*, 2003.

Mooers, C. N. K. and G. A. Maul. Intra-Americas Sea Circulation. In *The Sea*, edited by K. H. Brink and A. R. Robinson, pages 183–208, Wiley, New York, 1998.

Priestley, M. B. Spectral Analysis of Time Series. Academic Press. San Diego, California, 890 pp., 1981.

Larsen, J. G. Transport and heat flux of the Florida Current at 27°N derived from cross-stream voltage and profiling data: theory and observations. *Phil. Trans. R. Soc. London Ser. A*, 338:169–236, 1992.

Richman, J. G., C. Wunsch, and N.G. Hogg. Space and time scales of mesoscale motion in the western North Atlantic. *Rev. Geophys.*, 15:385–420, 1977.

Schmitz, W. J. On the World Ocean Circulation: Volume I. Technical Report WHOI-96-03, Woods Hole Oceanographic Institution, 1996.

Sheinbaum J., J. Candela, A. Badan and J. Ochoa. Flow structure and transport in the Yucatan Channel. *Geophys. Res. Letts.*, 29(3):10.1029/2001GL013990. 2002.

DIAGNOSTIC FORCE BALANCE AND ITS LIMITS *

JAMES C. MCWILLIAMS
Department of Atmospheric and Oceanic Sciences &
Institute of Geophysics and Planetary Physics
UCLA, Los Angeles, CA 90095-1565, U.S.A.

Abstract.

A perspective is presented on the premises, mathematical structure, regimes of validity, historical experience, evolutionary singularities, and unbalanced instabilities for the reduced fluid-dynamical system, the Balance Equations. The Balance Equations are an asymptotically consistent (but non-unique) set of approximations for rotating, stably stratified flows, built around the diagnostic force balances of hydrostacy in the vertical and gradient-wind balance in the horizontal divergence. It is widely agreed that the vast majority of the energy in the general circulations of the ocean and atmosphere is in balanced motions. There is a conundrum about large-scale energy dissipation in the ocean and atmosphere. Planetary forcing energizes large-scale balanced motions, including balanced mesoscale instabilities of the directly forced flows. Balanced flows are asymptotically characterized by an inverse energy cascade toward larger scales, but there are few and relatively inefficient dissipation mechanisms available at large and mesoscales (e.g., bottom drag and radiative cooling). Thus, the dynamical routes to dissipation at small scales remain uncertain. Unbalanced instabilities—especially centrifugal and anticyclonic, ageostrophic instabilities—may have an important role in furthering the route to dissipation. Finally, a brief analysis is made of Pedro Ripa's contributions to the mathematical characterization of balanced dynamics and its possible unbalanced instabilities.

Key words: rotating, stratified fluid dynamics; geostrophic balance; gradient wind balance; hydrostatic balance; instability; Balance Equations

1. Introduction

The majority of the kinetic and potential energy of the peristent oceanic currents occurs on relatively large spatial scales, $\sim$ 10s km horizontally and 100s m vertically or larger. Its flow intensity is weak enough so that the relevant Rossby and Froude numbers,

$$Ro = \frac{V}{fL} \quad \text{and} \quad Fr = \frac{V}{NH} \tag{1}$$

* This essay is written to honor the career of Pedro Ripa.

O.U. Velasco Fuentes et al. (eds.), Nonlinear Processes in Geophysical Fluid Dynamics, 287-303.

are not very large[1] In the asymptotic rotating, stratified regime (i.e. Ro, $Fr \rightarrow 0$), there are two primary paradigms for the fluid dynamics away from boundaries:

- *geostrophic currents*, whose nonlinear dynamics is called geostrophic turbulence and which has an inverse energy cascade, transferring energy to larger scales both vertically and horizontally, and a vanishing energy dissipation rate as $Re = VL/\nu \rightarrow \infty$ (Charney, 1971)[2];
- *inertia-gravity waves* (IGW), whose weak-wave turbulence has an efficient forward energy cascade to wave breaking at small scales that then instigates a Kolmogorov turbulent cascade with a finite energy dissipation rate at any Re.

These two regimes can, and in nature do, coexist in space and time. They are structurally distinguished by different characteristic aspect ratios, H/L (i.e. $\sim f/N \ll 1$ for geostrophic currents and ~ 1 for IGW), velocity anisotropy (i.e. small vertical velocity w for geostrophic currents), and evolutionary time scales (i.e. slow $\sim L/V$ and fast $\sim f^{-1}, N^{-1}$, respectively). These regimes are dynamically segregated by the process of geostrophic adjustment (i.e. IGW radiation leaving behind geostrophic currents) and by relatively weak advective coupling, compared to the advective influences within each regime.

However, there are many situations in the ocean where Ro and Fr are not asymptotically small, e.g. with stronger flow, smaller scale, weaker stratification, or nearer the Equator where f is small. What happens then to these paradigms? More specifically, if the energy residing in the general circulation is primarily geostrophic, and its forward energy cascade to dissipation is therefore inhibited, then which routes of energy transfer escape this constraint to accomplish the necessary dissipation?

2. Balanced Dynamics

Advective evolutionary rates are slower than IGW rates whenever $Ro, Fr < 1$. The concept of the slow manifold is the subset of all possible solutions to the fundamental fluid equations that evolve only on the slow rates of advection, V/L, or potential-vorticity differences (e.g. Rossby waves with $\sigma \sim \beta L$, where σ is the wave frequency and $\beta = df/dy$), thereby excluding all the faster acoustic wave and IGW solution behaviors (Leith, 1980; Lorenz, 1980). Balanced dynamics denotes the processes controlling the evolution

---

[1] V is a characteristic horizontal velocity, f is Coriolis frequency (Earth's rotation), N is buoyancy frequency (stable density stratification), and (L, H) are (horizontal, vertical) length scales.

[2] ν is the molecular viscosity.

on the slow manifold, and Balance Equations (BE) are a Partial Differential Equation (PDE) system that manifests balanced dynamics.

The essential ingredient of balanced dynamics is quasi-static, or so-called diagnostic, force balance in the momentum equations, with a small net acceleration. In the vertical direction (i.e. parallel to gravity, $-g\hat{\mathbf{z}}$, and to the dynamically significant component of Earth's rotation vector, $\hat{\mathbf{z}}f$), hydrostatic balance between the pressure gradient and buoyancy force,

$$\frac{\partial \phi}{\partial z} = -g\rho, \tag{2}$$

is asymptotically accurate to $\mathcal{O}(Ro^2 \cdot (H/L)^2)$ for geostrophic currents[3] (Gent and McWilliams, 1983; McWilliams, 1985). In the horizontal direction, geostrophic balance between the pressure gradient and Coriolis force is accurate to $\mathcal{O}(Ro)$, but more fundamentally, gradient-wind balance is an accurate approximation to the divergence of the horizontal momentum equation to $\mathcal{O}(Ro^2)$. Gradient-wind balance is defined by

$$\nabla_\perp^2 \phi = \hat{\mathbf{z}} \cdot \boldsymbol{\nabla}_\perp \times f\mathbf{v}_\perp + 2J_\perp[u,v], \tag{3}$$

whose three terms arise from the pressure gradient, Coriolis, and trajectory-normal centrifugal forces, respectively[4]. This relation is probably more familiar in its integrated form for axisymmetric flow, $\mathbf{v}_\perp = V(r,z)\hat{\theta}$, viz.

$$\frac{\partial \phi}{\partial r} = fV + \frac{V^2}{r}. \tag{4}$$

Since the turn of the nineteenth century, it has been recognized that synoptic and mesoscale winds and currents typically satisfy such balance relations. Furthermore they are essential ingredients for proper initialization of numerical weather forecasts and other types of data assimilation in models (Daley, 1991).

These relations comprise an under-determined dynamical system; e.g. geostrophic, hydrostatic balance provides three equations for four fields: ϕ, ρ, u, v. The earliest resolution of the under-determinacy (Charney, 1947) was the Quasigeostrophic Equations (QG), which adds either a prognostic vorticity or potential-vorticity equation with leading-order asymptotic validity as $Ro \to 0$. Since the middle of the twentieth century, many different BE proposals have been made (e.g. by Bolin, Thompson, Monin, Lorenz, Charney, Salmon, Ripa, Allen, Holm, McIntyre, etc.) for a well-posed PDE

[3] This assumes that $Ro \sim Fr$, as is conventional in scaling analysis of baroclinic, geostrophic currents.

[4] In (3), $\phi = p/\rho_o$ is the geopotential function, $\mathbf{v}_\perp = (u,v)$ and $\boldsymbol{\nabla}_\perp$ are the horizontal velocity and gradient operator, and $J_\perp$ is the horizontal Jacobian operator.

system, consistent with a diagnostic force balance (thus lacking IGW behavior) and with more accurate solutions than QG. Furthermore, where BE solutions have been obtained[5], they have often been found to closely track solutions from more fundamental fluid equations (e.g. incompressible Boussinesq Equations or hydrostatic Primitive Equations) with properly balanced initial conditions, even when Ro is not especially small. In these instances, the solution behavior appears to lie on the slow manifold even for the more fundamental models that also admit non-slow solutions.

With hindsight it seems clear that there is not a uniquely best choice for the BE. For successful performance as described immediately above, the important criteria for a good BE are that it be consistent with hydrostatic, gradient-wind balance (i.e. its errors are at least $\mathcal{O}(Ro^2)$) and that it yield good integration behavior without premature singularities (occurring at too small a Ro; see Sec. 3) or spurious high-frequency modes (not present in the fundamental equations). Many BE models meet these criteria, although many others have failed to do so. Among the more successful models, the distinctions seem, for now at least, to be more a matter of mathematical aesthetics than of fundamental physical validity.

In the face of this non-uniqueness, my own preferences for the BE are guided by the treble principles of minimality, i.e. maximal truncation of the fundamental PDE system consistent with asymptotically second-order accuracy; rotational velocity dominance (see below); and conservation of at least some important integral properties, but not necessarily all of those present in the fundamental fluid equations. The BE selection principle of minimality was first enunciated for energy conservation (Lornez, 1960), at the expense of exact parcel conservation of potential vorticity. To achieve the latter some degree of minimality must be lost (Charney, 1962), except when the truncation is made in an isentropic coordinate frame (Gent and McWilliams, 1984), where exact energy conservation is lost. To achieve both energy and potential vorticity conservation often conflicts with minimality, if not also with good integration behavior (Allen, 1991; Holm, 1996). A persistent theme in formulating BE models is making the approximations within Hamiltonian's Principle itself, which assures the full suite of conservation properties (Salmon, 1983, 1988), but not necessarily accuracy or minimality. Non-conservation need not be harmful if the error magnitude is small and non-accumulating under integration.

Another widespread theme is the choice between basing the BE on either a geostrophic or a rotational approximation to the horizontal velocity at

[5] BE proposals have been much more common than nontrivial, time-dependent BE solutions. This is because of the computational computational challenge in most BE to solve a nonlinear, implicit PDE system to evaluate the time derivatives for use in time integration; e.g. see (8).

leading order, i.e.

$$\mathbf{v}_\perp \;\approx\; (1/f)\hat{\mathbf{z}} \times \boldsymbol{\nabla}_\perp \phi \quad \text{or} \quad \mathbf{v}_\perp \;\approx\; \hat{\mathbf{z}} \times \boldsymbol{\nabla}_\perp \psi, \tag{5}$$

with an accompanying assumption that either the ageostrophic or horizontally irrotational velocity component is small. The well-known Semigeostrophic Equations model, based on the so-called geostrophic momentum approximation (Hoskins, 1975), follows the former path, but it sometimes has unsatisfactory, inaccurate solution behaviors, at least in part because it is not fully consistent with gradient-wind balance; however, this geostrophic path has been extended more successfully (Allen and Holm, 1996; Allen *et al.*, 2002). But the arena is still open for new BE proposals, not to mention debates about old ones.

Now to be more specific, consider the particular BE model that is minimal, has rotational flow dominance, and adiabatically conserves all material properties, including potential vorticity — sometimes called the Isentropic Balance Equations (IBE). With a Helmholtz decomposition of the horizontal velocity,

$$\mathbf{v}_\perp \;=\; \hat{\mathbf{z}} \times \boldsymbol{\nabla}_\perp \psi + \boldsymbol{\nabla}_\perp \chi, \tag{6}$$

the vertical component of vorticity is

$$\zeta \;=\; \nabla_\perp^2 \psi, \tag{7}$$

and the absolute vorticity is $\mathcal{A} \;=\; f + \zeta$. In terms of the variables (ψ, χ, Φ), the conservative[6] IBE are defined as follows (McWilliams *et al.*, 1998):

$$\begin{aligned}
\frac{\partial \mathcal{A}}{\partial t} + J_\perp[\psi, \mathcal{A}] + \boldsymbol{\nabla}_\perp \cdot (\mathcal{A} \boldsymbol{\nabla}_\perp \chi) \;&=\; 0 \\
\nabla_\perp^2 \Phi - \boldsymbol{\nabla}_\perp \cdot (f \boldsymbol{\nabla}_\perp \psi) - 2 J_\perp \left[\frac{\partial \psi}{\partial x}, \frac{\partial \psi}{\partial y} \right] \;&=\; 0 \\
\frac{\partial \mathcal{S}}{\partial t} + J_\perp\left[\psi, \mathcal{S} \right] + \boldsymbol{\nabla}_\perp \cdot (\mathcal{S} \boldsymbol{\nabla}_\perp \chi) \;&=\; 0,
\end{aligned} \tag{8}$$

where all derivatives are defined in isentropic coordinates (x, y, Z, t), Z is proportional to the potential density ρ_θ,

$$\Phi \;=\; \phi + g z \rho_\theta / \rho_o \tag{9}$$

is the Montgomery potential, and

$$\mathcal{S} \;=\; \frac{\partial^2 \Phi}{\partial Z^2} \tag{10}$$

[6] Of course, in any complex evolution, non-conservative (diffusive and dissipative) terms must be added to (8).

is an inverse measure of the local stratification. The equations in (8), respectively, are the vertical vorticity equation (i.e. the curl of the horizontal momentum equation), gradient-wind balance (cf. (3)), and conservation of mass and internal energy[7]. These equations conserve the hydrostatic approximation to Ertel's potential vorticity,

$$\Pi = \frac{\mathcal{A}}{\mathcal{S}}, \tag{11}$$

along horizontal trajectories at constant Z. Eq. (8) comprises a temporally first-order PDE system, in spite of the appearance of two time-derivative terms, because the gradient-wind balance constrains their independence; thus, it excludes both IGW solutions or any spurious modes so that it truly is a model for the slow manifold[8]. This system can be physically justified under any circumstance where $\chi \ll \psi$ (or equivalently $w \ll (H/L)\cdot \mathbf{v}_\perp$). This includes not only the regime of modest (Ro, Fr) values (with second-order asymptotic validity) but also of strong vortices (Shapiro and Montgomery, 1993) and fronts (Gent *et al.*, 1994)[9].

The various BE have been found to be useful from many perspectives. They can provide an accurate model for slow manifold behavior, whether for use in diagnostic analysis (e.g. forecast initialization) or as a potentially cheaper computational alternative to the standard use of Primitive Equations in general circulation models. They provide a parsimonious framework for dynamical interpretation by excluding extraneous IGW processes where apt. And they have the potential for defining a "process gap" between slow-manifold and IGW unresolved processes for parameterization of their effects in large-scale models (though this has not yet been exploited).

3. Evolutionary Limits of Balanced Dynamics

In consternation of the BE successes, it now seems clear that the slow manifold that they are intended to approximate often does not exist in the precise sense of an invariant inertial manifold. Solutions of the fundamental fluid equations that initially are slowly evolving do not remain entirely slow

[7] This assumes a simple equation of state. The oceanographically more realistic generalization is straightforward: parcel conservation of both potential temperature θ and salinity S, and evaluation of $\rho_\theta(\theta, S, p)$ using the equation of state for seawater for use in Φ.

[8] For comparison, the Boussinesq and Primitive Equation PDE systems are temporally third order.

[9] BE proposals have also been made for planetary-scale flows, ranging from Linear Balance (Lornez, 1960) to Global Balance (Gent and McWilliams, 1983) to Planetary Geostrophic Balance (Phillips, 1963; Vallis, 1996). Their focus is a more accurate expression of the Coriolis force than in QG.

for an appreciable time interval, even as $Ro \to 0$ with $Re < \infty$, although the discrepant behavior is vanishingly small, e.g. $\sim Ro^{-1/2} e^{-c/Ro}$ with $c > 0$. For this reason the slow manifold is sometimes called a fuzzy manifold. The cleanest demonstrations of this behavior have been made for systems with reduced degrees of freedom or particularly simple flow configurations (Vautard and Legras, 1986; Lorenz and Krishnamurthy, 1987; Camassa, 1995; Vanneste and Yavneh, 2002). Nevertheless, BE solutions are typically entirely slow, at least for small enough Ro, so that they do generate a precisely defined slow manifold that often well approximates the true fluid behavior. Of course, this operational definition of the slow manifold differs among the different BE models.

Mathematicians are interested in proving existence and uniqueness for PDE systems such as the BE. Sometimes physicists view this as a sterile inquiry, because they are convinced either that solutions exist, having plausibly calculated them, or that well formulated physical principles should not lead to non-existence or non-uniqueness. The family of BE proposals provide an interesting middle ground since it is typical that there exist parameter regimes and flow configurations for which solutions do not exist[10]; e.g. a sufficiently strong pressure maximum cannot have an axisymmetric flow that satisfies (4) for $f > 0$. For the particular BE model (8), an exact and concise characterization of the limits of time integrability has been found (McWilliams *et al.*, 1998)[11]. This PDE system changes type from degenerate-elliptic to hyperbolic under any of three conditions:

1. Change of sign of vertical stratification, $\mathcal{S}$;
2. Change of sign of absolute vorticity, $\mathcal{A}$;
3. Change of sign of $\mathcal{A} \pm \mathcal{S}t$,

where

$$\mathcal{S}t^2 \;=\; (u_x - v_y)^2 + (v_x + u_y)^2 \;=\; \zeta^2 - 4J_\perp[\psi_x, \psi_y] \qquad (12)$$

is the variance of the rotational-flow component of the horizontal strain rate (in isentropic coordinates). When these conditions are encountered, (8) cannot be integrated forward in time. None of these conditions occurs in the QG limit, since $\mathcal{A}$, $\mathcal{A} \pm \mathcal{S}t \to f + \mathcal{O}(Ro)$ and $\mathcal{S} \to \mathcal{S}_o + \mathcal{O}(Ro)$, where $\mathcal{S}_o(Z) > 0$ is the stable, resting-state stratification. Note the greater susceptibility of anticyclonic regions (i.e. with $\zeta/f < 0$) in the second and

[10] This statement is not universally true, although it does seem true for the more accurate balance models; e.g. the QG model always has solutions, but they are inaccurate except when Ro is rather small.

[11] This result was first found for the analogous BE model in (x, y, z, t) coordinates (i.e. as first proposed by Lorenz, 1960), where the integrability conditions are only approximately expressible in a similar form (Yavneh *et al.*, 1997).

third conditions[12]. Furthermore, note the greater susceptibility in the third condition, since $(\mathcal{A}/f - |\mathcal{S}t/f|) \leq \mathcal{A}/f$ for moderate values of Ro. The first and second conditions also are related to the potential vorticity (11). Since potential vorticity and buoyancy are conserved on parcels, except for mixing effects, there is an evolutionary inhibition for an unforced flow to spontaneously develop a violation of the first and second conditions. There is no such constraint with respect to the third condition, and this is another sense in which there may be a greater susceptibility to the third condition. Thus, for Ro and Fr large enough, especially in anticyclonic regions, the dynamical evolution must fail to remain balanced at least to some degree.

4. Unbalanced Instabilities

So what happens when BE solutions cease to exist? An interesting conjecture is that a change of type of the approximate PDE system corresponds to the onset of a type of new instability in the more fundamental system, presuming that a meaningful definition of instability can be adduced for the particular circumstances (e.g. for steady-state flows). This conjecture is physically interesting only if the accuracy of the approximate system remains high up to near the point of type change (which, e.g. is not true for QG). This conjecture invites us to assign instabilities to each of the integrability conditions above. It is easily done for the first—gravitational or convective instability (Chandrasekhar, 1961))—and second—centrifugal or inertial instability (Ooyama, 1966; Hoskins, 1974)— conditions, consistent with the conjecture. But this assignment is not easily made for the third condition.

It is, of course, difficult if not impossible to establish a universal categorization of instability types. In the context of strongly rotating, stratified flows, however, we can come close to doing so. The only linear instabilities of a steady current in QG dynamics are of the inflection-point type, requiring a change in sign of the mean potential vorticity gradient due to horizontal and/or vertical curvature in the flow (i.e. barotropic and/or baroclinic instabilities). Inflection-point instability continues to occur at larger Ro and Fr values, and it can be expected to remain well balanced over some range in these parameters beyond their asymptotic limit. Thus, its onset conditions have nothing to do with the limits of balance. In contrast, gravitational and centrifugal instabilities occur only for the Ro values of $\mathcal{O}(1)$ indicated above, and the growing linear modes for each of these exhibit significant

[12] There is often a greater susceptibility in the first condition as well when a balanced, anticyclonic flow implies a locally weaker stratification, e.g. in the center of an anticyclonic vortex with an interior velocity maximum.

departures from balanced dynamics and at finite amplitude lead to highly dissipative turbulent energy cascades.

After investigating the normal-mode stability of several types of 3D stratified shear flows—barotropic elliptical flow (McWilliams and Yavneh, 1998), Taylor-Couette flow (Molemaker *et al.*, 2000; Yavneh *et al.*, 2001), Eady's (1949) baroclinic flow (Molemaker *et al.*, 2002), and a barotropic boundary current (unpublished)—we have come to the inductive conclusion that there is another generic type of instability which occurs for all $Ro \neq 0$ in anticyclonic flows, which we therefore call anticyclonic, ageostrophic instability (AAI)[13].

The unstable growth rates for an anticyclonic Taylor-Couette flow (Figure 1) show a steep decay as $Ro \to 0$; in an asymptotic approximation, this rate can be shown (Yavneh *et al.*, 2001) to be predominantly exponential,

$$\sigma \approx f\sqrt{Ro}\, e^{-c/Ro}, \tag{13}$$

for c a constant, which depends on the radial eigenmode number. At the other limit of $\mathcal{A} \to 0$, σ increases to become comparable to the growth rate for centrifugal instability that occurs for this flow even in the absence of stable stratification (Chandrasekhar, 1961). There is no comparable instability in a cyclonic Taylor-Couette flow.

In Figure 1 the growth rate is plotted as a function of $\mathcal{A}/f - |\mathcal{S}t/f|$, the quantity in the third condition for loss of BE integrability. Although there is indeed a steep decrease in σ in the neighborhood of its change of sign—seemingly steeper than exponential in its shape—there is not an abrupt onset of instability there. Thus, with respect to the less familiar third condition, the conjecture above is only loosely confirmed. A better interpretation, perhaps, is that this anticyclonic, ageostrophic instability implies a progressive leakage into unbalanced motions with Ro, similar to the progression of the fuzziness of the slow manifold.

A uniform vertical shear flow in a uniformly rotating and stratified fluid is an anticyclonic flow in the sense that its Ertel potential vorticity (11) is smaller than the resting state counterpart, $f/\mathcal{S}_o$. The unstable growth rate (Figure 2) is nearly independent of Ro for the inflection-point mode first derived by Eady (1949). There is another unstable mode for $Ro \neq 0$ within the centrifugally and gravitationally stable regime, whose growth rate $\sigma(Ro)$ is qualitatively similar to that for Taylor-Couette flow: it vanishes as $Ro \to 0$ and becomes comparable to the traditional mode's rate as $Ro \to 1$, $\mathcal{A} \to 0$. Stone (1966, 1970) referred to the second mode as a shortwave instability, but we prefer to identify it as another example of anticyclonic, ageostrophic

[13] We believe a variety of previously discovered "frontal" instabilities for single-layer flows can also reasonably be associated with AAI (Griffiths *et al.*, 1982; Paldor and Ghil, 1997).

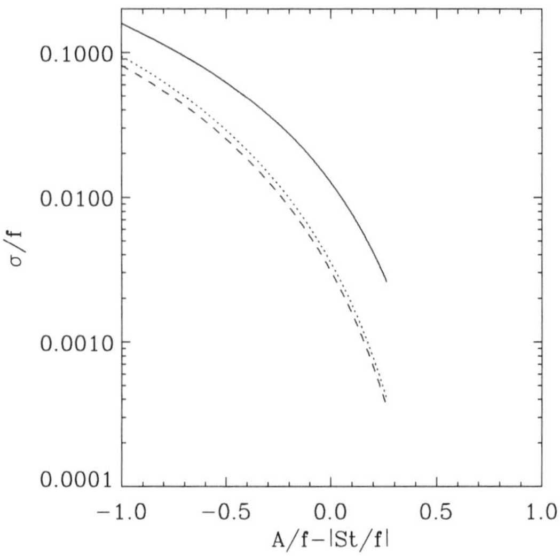

Figure 1. Growth rates for anticyclonic, ageostrophic instability in a Taylor-Couette flow, optimized over vertical and azimuthal wavenumbers (Molemaker *et al.*, 2000; Yavneh *et al.*, 2001). The upper curve is associated with the gravest radial mode and the lower family of curves with higher radial modes. The abscissa monitors the third condition for loss of BE integrability; its range is from where $\mathcal{A} = 0$ on the left to where $Ro = 0$ (i.e. QG) on the right.

instability. We can fit a closest balanced solution to these modes using a total energy norm as the misfit measure (Molemaker *et al.*, 2002). The result of this fitting (Figure 3) shows that the traditional mode remains well balanced for all Ro and has vanishing unbalance as $Ro \to 0$, while the anticyclonic, ageostrophic mode is primarily unbalanced for all Ro even as its growth rate vanishes as $Ro \to 0$. Its eigenmode structure (Figure 4) shows that the greatest degree of unbalance occurs in the vicinity of an inertia critical level—at the z level where $\omega - k\overline{U}(z) \pm f \approx 0$—which sharpens as Ro decreases[14].

Between the two alternative paradigms mentioned in the Introduction, the one pertaining to unbalanced motions is IGW. Thus the question arises whether it is appropriate to associate anticyclonic, ageostrophic instability with IGWs. The eigenvalue problem for a shear instability requires that all eigenvalues be either real or in complex conjugate pairs. Therefore, for an instability to occur (i.e. with nonzero growth rate σ, equal to the imaginary part of the complex eigenfrequency), it must do so through the coalescence

[14] Here ω is the eigenfrequency, k is the streamwise wavenumber, and $\overline{U}$ is the baroclinic mean flow.

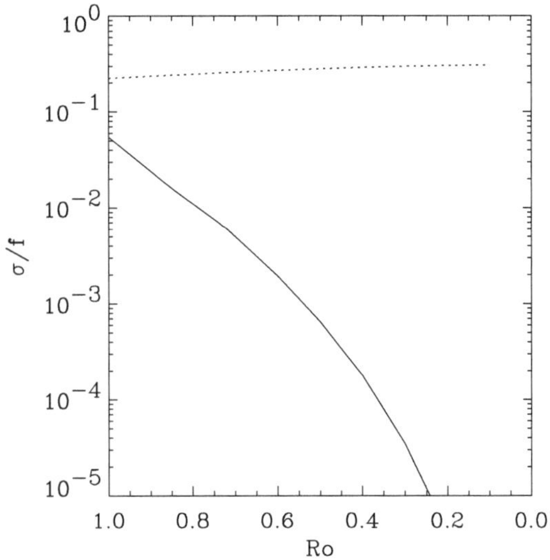

Figure 2. Instability growth rates for a uniform baroclinic shear flow, optimized over horizontal wavenumbers (Molemaker *et al.*, 2002). The upper curve corresponds to the traditional mode that occurs as $Ro \to 0$ (Eady, 1949), and the lower curve corresponds to the anticyclonic, ageostrophic mode. The abscissa Ro ranges from where $\mathcal{A} = 0$ on the left to where $Ro = 0$ (i.e. QG) on the right.

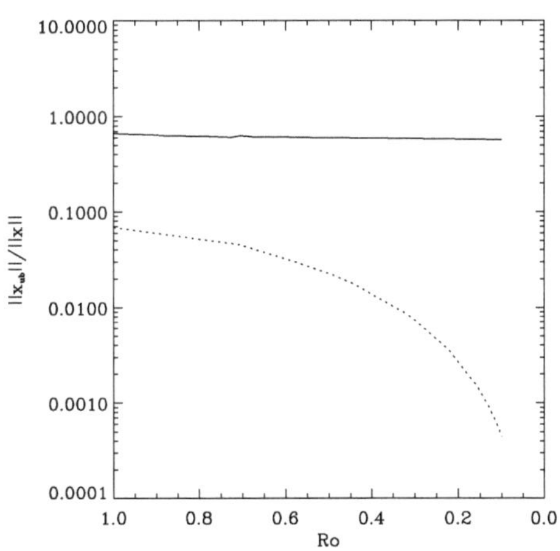

Figure 3. Ratio of unbalanced energy to total energy for the unstable eigenmodes in a uniform baroclinic shear flow. The dotted curve corresponds to the traditional mode that occurs as $Ro \to 0$ (Eady, 1949) and the solid curve to the anticyclonic, ageostrophic mode.

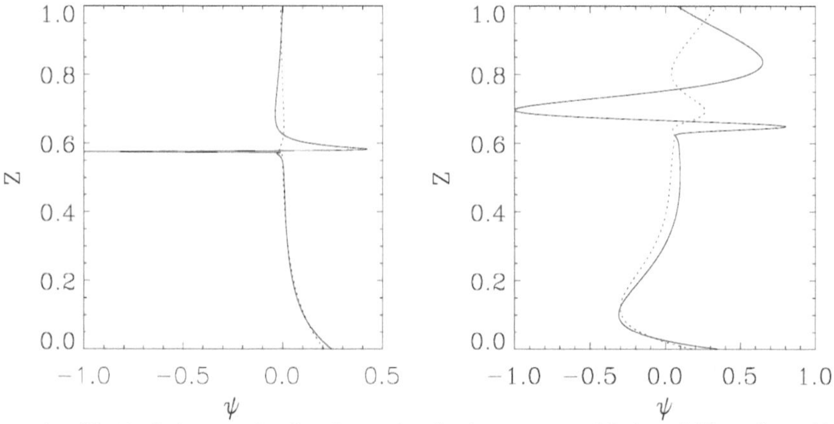

Figure 4. Vertical eigenmodes for the anticyclonic, ageostrophic instability of a uniform baroclinic shear flow at $Ro = 0.7$ (left) and 1.0 (right). The solid curve is the complete eigenmode for the Boussinesq Equations and the dotted curves are BE fits to it. The vertical axis is non-dimensionalized by the domain height. Note the near-occurrence of an inertial critical layer in each case.

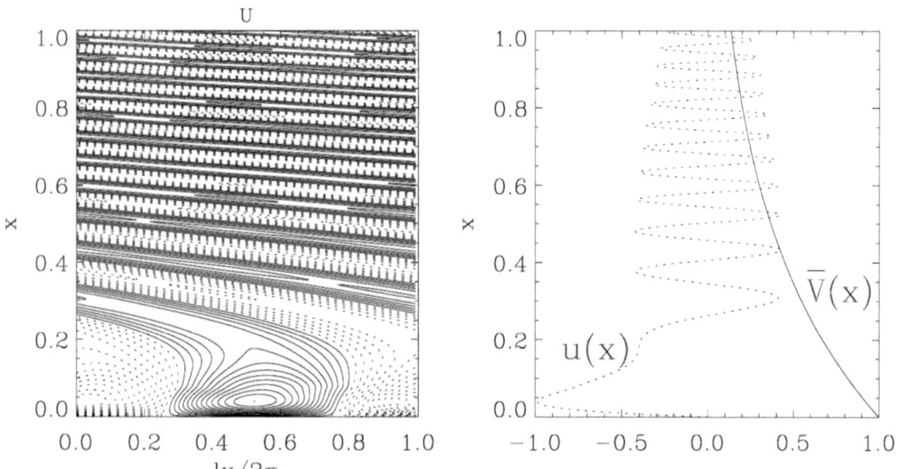

Figure 5. (Left) Horizontal eigenmode for the boundary-normal velocity u in an anticyclonic, ageostrophic instability of a barotropic boundary current. (Right) Offshore profiles of the eigenmode u and the normalized mean shear flow $\overline{V}(x)$. x is the non-dimensional distance normal to the boundary: $x = 0$ corresponds to the boundary location, and only a portion of the unbounded domain in x is plotted.

of two real eigenfrequencies at the margin of instability. It is reasonable for an unbalanced instability to identify at least one of these coalescing frequencies as of the IGW type, albeit possibly much altered from its resting-state character by the influence of the mean shear. In a bounded domain filled by the mean shear flow, it may be difficult to recognize the IGW character of the eigenmode (e.g. Figure 4). In a different instability problem for a barotropic boundary current—with an exponentially decaying mean-shear profile away from the boundary, hence no inflection point—the anticyclonic, instability mode does exhibit a weakly decaying oscillatory structure, closely matching the IGW dispersion relation, in the far-field interior (Figure 5).

5. Ripa's Contributions

Pedro Ripa exhibited a sustained interest in the formulation of approximate models for rotating, stratified flows and in the stability of its steady flows. His preferred framework for model formulation is a few, vertically stacked, hydrostatic fluid layers. This is a strong approximation in the permitted flow structures, compared to a 3D continuously stratified fluid, but it continues a long tradition in geophysical fluid dynamics, starting with the widely used Shallow-Water Equations (SWE) for a single layer of uniform density ρ_o, variable thickness h, and vertically uniform $\mathbf{v}_\perp$. The SWE are not an accurate representation of most natural flows—offshore tsunami waves are an exception—but their solutions exhibit many dynamically analogous behaviors in a PDE system with one fewer spatial dimension.

The usual generalization of the SWE is to N uniform-density layers (i.e. ρ_{oi}, $i = 1 \ldots N$, decreasing upward, plus $(h_i, \mathbf{v}_{\perp i}(\mathbf{x}_\perp))$, coupled through hydrostatic pressure forces associated with displacements of the bounding interfaces. Ripa (1993, 1995) devised a variant of this commonly used model (which he called Homogeneous Layered Primitive Equation Model, HLPEM) to include lateral density variations within each layer $\rho_i(\mathbf{x}_\perp)$, constrained to maintain gravitational stability and with a Galerkin-like, vertically truncated coupling to the $(h_i, \mathbf{v}_{\perp i})$. He called this new model the Inhomogeneous Layered Primitive Equation Model, ILPEM. Its advantage compared to HLPEM is in providing a modest increase in the vertical degrees of freedom for a given N.

Ripa (1996, 1999) proposed a BE for the ILPEM that he called ILQGM. As in QG, $\mathbf{v}_\perp = (1/f)\hat{\mathbf{z}} \times \boldsymbol{\nabla}_\perp \phi$. This model is therefore inconsistent with (3) and has approximation errors of $\mathcal{O}(Ro)$, which limits its accuracy in strong flows. The ILQGM is expressible in a Hamiltonian formulation, though it does not contain an exact form of Hamilton's Principle and does not preserve all the conservation laws of the fundamental fluid equations (e.g. its plausibly defined $\prod$ is not conserved on parcels moving with $\mathbf{v}_\perp$); he did not consider this latter property as inherently disadvantageous, nor do I. In the absence of explicit solutions of ILQGM, I can see no reason why its integration behavior would not be good and thus satisfactorily embody a slow inertial manifold (as does QG in 3D).

Ripa (1983, 1989, 1991) developed a theorem for the sufficient conditions for stability of a symmetric steady flow (e.g. an axisymmetric $\overline{V}_i(r)$) in N homogeneous hydrostatic layers, a HLPEM. This is discussed more fully elsewhere in this volume (Shepherd, 2003), but I will roughly paraphrase these conditions here as

1. $\overline{V}_i$ has no inflection points;
2. The flow is everywhere "subcritical": $||\overline{V}_i - \omega r|| < C_i$ for all i, ω, r,

where $C_i \sim \sqrt{\Delta\rho_{oi} h_i / \rho_o}$ is an internal gravity wave speed associated with layer i, ω is interpretable as a generic frequency, and $|| \cdot ||$ symbolizes a norm. The meaning of subcritical here is that the appropriately defined Froude number is small. The first condition is a familiar one, e.g. in balanced instabilities (Sec. 4). Because of C_i, the second condition obviously implicates IGW processes, hence the possibility of unbalanced instability modes, and it is suggestive of precluding a "resonance" (i.e. coalescence of real eigenfrequencies to allow a nonzero growth rate σ; Sec. 4) between IGW propagation and Doppler shifting by the mean shear flow. These features are consistent with the unbalanced instabilities presented above. But there is nothing in the second condition that obviously implicates the greater susceptibility of anticyclonic flows to instability. A more fundamental difficulty, recognized by Ripa as the ultraviolet problem, is that since $\Delta\rho_{oi}, h_i \rightarrow 0$ as $N \rightarrow \infty$, hence $C_i \rightarrow 0$, it is not obvious that the second condition can ever be satisfied in a continuously stratified flow. If not, then Ripa's theorem is not useful in assessing 3D flow stability except as a constraint on the larger vertical scales of motion. Nevertheless, the second condition in Ripa's theorem presages the possibility, and as $N \rightarrow \infty$ possibly the ubiquity, of unbalanced instabilities at finite *Ro* values.

6. Summary

Based on several decades of atmospheric and oceanic experience, there is emerging understanding of the slow manifold as an accurate approximation for most large- and mesoscale motions and of the BE as a non-unique embodiment of this behavior. However, the slow manifold is a fuzzy and leaky one, and the leakage of energy provides a route to dissipation at small scales that escapes the inhibition for downscale energy cascades within balanced dynamics. At small *Ro* the rate of leakage may be small and thus potentially negligible for the energy budget of the general circulation. But this cannot be a confident conclusion, particularly for the ocean with its nearly adiabatic interior dynamics, since it would imply that the dissipation is accomplished primarily near the solid boundaries and sustained by efficient boundaryward energy fluxes throughout the interior (Müller *et al.*, 2002). The mechanism of leakage in the interior may be indicated by unbalanced instabilities of balanced steady flows, especially anticyclonic, ageostrophic and centrifugal instabilities since they can occur in stable stratification at moderate *Ro* values. Whether these instabilities provide a significant energy leakage at finite amplitude is not yet known. It is also uncertain whether balanced turbulence contains the seeds of its own destruction by a cascade-induced enhancement of *Ro* values at smaller scales that could act to amplify the unbalanced instability and leakage rates[15].

Pedro Ripa grappled with some of the key dynamical issues addressed here, with great clarity of insight and expression and mathematical precision, albeit without being granted the time to see their fundamental resolutions. Would that he could have continued with us in these quests.

Acknowledgements

This essay is based substantially on my longstanding collaborations with Peter Gent, Jeroen Molemaker (who prepared the figures), and Irad Yavneh, all of whose contributions and comradeship I gratefully acknowledge. I also appreciate financial support by the National Science Foundation and the Office of Naval Research.

References

Allen, J. Balance Equations Based on Momentum Equations with Global Invariants of Potential Enstrophy and Energy. *J. Phys. Oceanogr.*, 21:265-276, 1991.

Allen, J., and D. Holm. Extended-geostrophic Hamiltonian Models for Rotating Shallow Water Motion. *Physica D*, 98:229-248, 1996.

Allen, J., D. Holm, and P. Newberger. Toward an Extended-geostrophic Euler-Poincaré Model for Mesoscale Oceanographic Flow. In J. Norbury and I. Roulstone, editors, *Large-Scale Atmosphere-Ocean Dynamics, Volume 1: Analytical Methods and Numerical Models*, 101-125. Cambridge University Press, 2002.

Camassa, R. On the geometry of an atmospheric slow manifold. *Physica D*, 84:357-397, 1995.

Chandrasekhar, S. *Hydrodynamic and Hydromagnetic Stability.* Oxford University Press, 1961.

Charney, J. The Dynamics of Long Waves in a Baroclinic Westerly Current. *J. Met.*, 4:135-163, 1947.

Charney, J. Integration of the Primitive and Balance Equations. *Proc. Int. Symp. Numerical Weather Prediction*, pages 131-152. Meteor. Soc. Japan, Tokyo, 1962.

[15] There is some indication that this can occur (Yavneh *et al.*, 1997).

Charney, J. Geostrophic Turbulence. *J. Atmos. Sci.*, 28:1087-1095, 1971.

Daley, R. *Atmospheric Data Analysis.* Cambridge University Press, 1991.

Eady, E. Long Waves and Cyclone Waves. *Tellus*, 1:33-52, 1949.

Gent, P., and J. McWilliams. Regimes of Validity for Balanced Models. *Dyn. Atmos. Oceans*, 7:167-183, 1983.

Gent, P., and J. McWilliams. Balanced Models in Isentropic Coordinates and the Shallow-water Equations. *Tellus*, 36A:166-171, 1984.

Gent, P., J. McWilliams, and C. Snyder. A Note on a Scaling Analysis of Curved Fronts: The Formal Validity of the Balance Equations and Semigeostrophy. *J. Atmos. Sci.*, 51:160-163, 1994.

Griffiths, R., P. Killworth, and M. Stern. Ageostrophic Instability of Ocean Currents. *J. Fluid Mech.*, 117:343-377, 1982.

Holm, D. Hamiltonian Balance Equations. *Physica D*, 98:379-414, 1996.

Hoskins, B. The role of Potential Vorticity in Symmetric Stability and Instability. *Q. J. Royal Met. Soc.*, 100:480-482, 1974.

Hoskins, B. The Geostrophic Momentum Approximation and the Semigeostrophic Equations. *J. Atmos. Sci.*, 32:233-242, 1975.

Leith, C. Nonlinear Normal Mode Initialization and the Generation of Gravity Waves. *J. Atmos. Sci.*, 37:958-968, 1980.

Lorenz, E. Energy and Numerical Weather Prediction. *Tellus*, 12:364-373, 1960.

Lorenz, E. Attractor Sets and Quasi-geostrophic Equilibrium. *J. Atmos. Sci.*, 37:1547-1557, 1980.

Lorenz, E., and V. Krishnamurthy. On the nonexistence of a slow manifold. *J. Atmos. Sci.*, 44:2940-2950, 1987.

McWilliams, J. A Note on a Uniformly Valid Model Spanning the Regimes of Geostrophic and Isotropic, Stratified Turbulence: Balanced Turbulence. *J. Atmos. Sci.*, 42:1773-1774, 1985.

McWilliams, J., I. Yavneh, M. Cullen, and P. Gent The Breakdown of Large-scale Flows in Rotating, Stratified Fluids. *Phys. Fluids*, 10:3178-3184, 1998.

McWilliams, J., and I. Yavneh. Fluctuation Growth and Instability Associated with a Singularity of the Balance Equations. *Phys. Fluids*, 10:2587-2596, 1998.

Molemaker, J., J. McWilliams, and I. Yavneh. Instability and Equilibration of Centrifugally Stable Stratified Taylor-Couette Flow. *Phys. Rev. Lett.*, 86:5270-5273, 2000.

Molemaker, J., J. McWilliams, and I. Yavneh. Ageostrophic Baroclinic Instability and Loss of Balance. *J. Phys. Ocean.*, submitted, 2002.

Müller, P., J. McWilliams, and J. Molemaker. Routes to Dissipation in the Ocean: The 2D/3D Turbulence Conundrum. In H. Baumert, J. Simpson, and J. Sundermann, editors, *Marine Turbulence - Theories, Observations and Models. Results of the CARTUM Project.* Cambridge Press, 2002.

Ooyama, K. On the Stability of the Baroclinic Circular Vortex: A Sufficient Condition for Instability. *J. Atmos. Sci.*, 23:43-53, 1966.

Paldor, N., and M. Ghil. Linear Instability of a Zonal Jet on an f Plane. *J. Phys. Ocean.*, 27:2361-2369, 1997.

Phillips, N. Geostrophic Motion. *Rev. Geophys.*, 1:123-176, 1963.

Ripa, P. General Stability Conditions for Zonal Flows in a One-layer Model on the Beta-plane or the Sphere. *J. Fluid Mech.*, 126:463-487, 1983.

Ripa, P. On the Stability of Ocean Vortices. In J. Nihoul and B. Jamart, editors, *Mesoscale/Synoptic Coherent Structures in Geophysical Turbulence*, pp. 167–179. Elsevier Oceanographic Series, Amsterdam, 1989.

Ripa, P. General Stability Conditions for a Multi-layer Model. *J. Fluid Mech.*, 222:119-137, 1991.

Ripa, P. Conservation Laws for Primitive Equations Models with Inhomogeneous Layers. *Geophys. Astrophys. Fluid Dyn.*, 70:85-111, 1993.

Ripa, P. On Improving a One-layer Ocean Model with Thermodynamics. *J. Fluid Mech.*, 303:169-201, 1995.

Ripa, P. Low Frequency Approximation of a Vertically Averaged Ocean Model with Thermodynamics. *Rev. Mex. de Físca*, 42:117-135, 1996.

Ripa, P. On the Validity of Layered Models of Ocean Dynamics and Thermodynamics with Reduced Vertical Resolution. *Dyn. Atmos. Oceans*, 29:1-40, 1999.

Salmon, R. Practical Use of Hamilton's Principle. *J. Fluid Mech.*, 132:431-444, 1983.

Salmon, R. Hamiltonian Fluid Mechanics. *Ann. Rev. Fluid Mech.*, 20:255-256, 1988.

Shapiro, L., and M. Montgomery. A Three-dimensional Balance Theory for Rapidly Rotating Vortices. *J. Atmos. Sci.*, 50:3322-3335, 1993.

Shepherd, T. Ripa's Theorem and its Relatives. This volume, 2003.

Stone, P. On Non-geostrophic Baroclinic Instability. *J. Atmos. Sci.*, 23:390-400, 1966.

Stone, P. On Non-geostrophic Baroclinic Instability: Part II. *J. Atmos. Sci.*, 27:721-726, 1970.

Vallis, G. Potential Vorticity Inversion and Balanced Equations of Motion for Rotating, Stratified Flow. *Q. J. Roy. Met. Soc.*, 291-322, 1996.

Vanneste, J., and I. Yavneh. Exponentially small inertia-gravity waves and the breakdown of quasi-geostrophic balance. Submitted for publication, 2002.

Vautard, R., and B. Legras. Invariant manifolds, quasi-geostrophy, and initialization. *J. Atmos. Sci.*, 43:565-584, 1986.

Yavneh, I., A. Shchepetkin, J. McWilliams, and P. Graves. Multigrid Solution of Rotating, Stably Stratified Flows: The Balance Equations and their Turbulent Dynamics. *J. Comp. Phys.*, 136:245-262, 1997.

Yavneh, I., J. McWilliams, and J. Molemaker. Non-axisymmetric Instability of Centrifugally Stable, Stratified Taylor-Couette Flow. *J. Fluid. Mech.*, 448:1-21, 2001.

A NOTE ON THE EFFECTS OF SOLID BOUNDARIES ON CONFINED DECAYING 2D TURBULENCE

G.J.F. VAN HEIJST, H.J.H. CLERCX AND S.R. MAASSEN
Fluid Dynamics Laboratory
Eindhoven University of Technology
PO Box 513, 5600 MB Eindhoven, The Netherlands

Abstract. This paper addresses the role of solid boundaries on the evolution of decaying 2D turbulence on a finite domain. It is demonstrated by laboratory experiments and numerical simulations of (quasi-) 2D flows on both square and circular domains that the lateral walls (i) act as sources of vorticity filaments which affect the flow evolution in the interior, and (ii) provide normal and shear stresses that may exert torques, thus causing changes in the net angular momentum of the contained fluid. It is also discussed how the net angular momentum L_0 affects the subsequent flow evolution and how it determines the structure of the 'final state'.

Key words: two-dimensional turbulence, inverse energy cascade, solid boundaries

1. Introduction

It is a well-established fact that two-dimensional (2D) turbulent flows are characterised by the inverse energy cascade, i.e. by a spectral kinetic energy flux to smaller wavenumbers. This is particularly valid for the situation in which the velocity field is being forced or "stirred" at a statistically steady rate, as contrasted with the decay situation, where the turbulence is set up at $t = 0$ and allowed to evolve. In the latter case also the selective decay mechanism, assuming that shorter wave lengths decay more rapidly than long wave length contributions (see Matthaeus and Montgomery, 1980), plays a prominent role during the decay process and competes with the inverse energy cascade. These remarkable properties were demonstrated for the case of decaying 2D turbulence by the numerical simulations performed by Matthaeus and Montgomery (1980), by McWilliams (1984) and by Santangelo, Benzi and Legras (1989): the action of the inverse cascade is manifest in the emergence of coherent vortex structures. These (and many subsequent) flow simulations were carried out on a double-periodic, square domain. It was shown by Li and Montgomery (1996) in a numerical study

O.U. Velasco Fuentes et al. (eds.), Nonlinear Processes in Geophysical Fluid Dynamics, 305-324.

of decaying 2D turbulence on a bounded, circular domain that the decay scenario is essentially different from that on a double-periodic domain, somewhat depending on the imposed boundary condition (stress-free or no-slip). A similar influence by the domain boundaries was observed by Clercx, Maassen and van Heijst (1999) in a numerical study of decaying 2D turbulence on a square domain with either stress-free or no-slip boundary conditions. In particular the no-slip calculations revealed the crucial role played by the boundaries, in the sense that they act as *sources of large-amplitude vorticity*: whenever a vortex structure approaches a wall, a boundary layer is formed with oppositely-signed vorticity, part of which is usually scraped off by the vortex flow, in the form of a thin filament of vorticity, which is hence advected into the interior of the flow domain. Obviously, this has a significant effect on the evolution and on the spectral characteristics of the flow, as was demonstrated in the numerical study by Clercx and van Heijst (2000).

Another remarkable effect the solid boundaries may have on the flow evolution was recently observed in the so-called 'spontaneous spin-up' of the fluid. Numerical simulations (Clercx *et al.* 1998) as well as laboratory experiments (Maassen, Clercx and van Heijst, 2002) on decaying 2D turbulent flow on a square domain with realistic (no-slip) walls revealed in a significant number of runs a substantial increase of the total angular momentum L with respect to the domain centre. Even in runs with negligibly small initial values of L (i.e. $L_0 \approx 0$) the flow acquired a substantial, non-zero angular momentum in the course of time. This change in L is established through the action of wall forces, which may exert a net torque on the fluid.

In this paper we will focus somewhat more on te role of the angular momentum in the flow evolution, both for square and for circular containers. We will discuss results obtained by laboratory experiments in a stratified fluid and by high-resolution numerical simulations based on a spectral method [for more background information on the numerical simulation technique the reader is referred to Clercx (1997)].

The paper is organized as follows. Some remarks about boundary conditions and their effects on global flow quantities are made in Section 2. After a brief description of the experimental facilities (Section 3), results of laboratory experiments and numerical simulations on decaying turbulence on a square domain are presented in Sections 4 and 5 for $L_0 \approx 0$ and $L_0 \neq 0$, respectively. Observations for the circular domain are described in Section 6. A discussion of the results and some general conclusions are given in Section 7.

2. Boundary conditions and some integral quantities

Consider the 2D motion of a viscous fluid on a bounded domain $\mathcal{D}$. Cartesian coordinates in a frame of reference are denoted by x and y. Let the (horizontal) flow field be given by $\mathbf{v} = (u, v, 0)$ with the vorticity $\boldsymbol{\omega} = \nabla \times \mathbf{v} = (0, 0, \omega)$, where $\omega = \frac{\partial v}{\partial x} - \frac{\partial u}{\partial y}$. This fluid motion is governed by the Navier-Stokes equation, which reads in non-dimensional form:

$$\frac{\partial \mathbf{v}}{\partial t} + (\mathbf{v} \cdot \nabla)\,\mathbf{v} = -\nabla p + \frac{1}{Re} \nabla^2 \mathbf{v} \, , \tag{1}$$

with t the time, p the pressure and $Re = VL/\nu$ the Reynolds number based on velocity and length scales V and L, respectively, and ν the kinematic fluid viscosity. By taking the curl, one obtains the vorticity equation

$$\frac{\partial \omega}{\partial t} + (\mathbf{v} \cdot \nabla)\,\omega = \frac{1}{Re} \nabla^2 \omega \, . \tag{2}$$

Either equation should be solved subject to conditions at the boundary $\partial\mathcal{D}$. Impermeability of this boundary implies that the stream function ψ satisfies

$$\psi = \text{constant} \quad \text{on } \partial\mathcal{D} \, . \tag{3}$$

In the case of a (physically realistic) *no-slip* boundary, the additional condition is

$$v_{||} = 0 \quad \text{on } \partial\mathcal{D} \, , \tag{4}$$

with $v_{||}$ the velocity component parallel to the boundary, or, taking (3) and (4) together:

$$\mathbf{v} = \mathbf{0} \quad \text{on } \partial\mathcal{D} \, . \tag{5}$$

The boundary value for the vorticity is not provided a priori for flows in bounded domains with no-slip walls. In order to solve the time-discretized version of (2) numerically, the boundary value of the vorticity is determined by means of an influence matrix method (for details see Clercx (1997)).

A *stress-free* boundary would imply

$$\left(\hat{\mathbf{n}} \cdot \underline{\underline{\tau}}\right)_{||} = 0 \quad \text{on } \partial\mathcal{D} \, , \tag{6}$$

with $\hat{\mathbf{n}}$ the unit vector normal to the boundary, and $\underline{\underline{\tau}}$ the viscous stress tensor.

For the specific cases of a square and a circular domain, the latter condition provides the following well-defined boundary conditions for the vorticity

$$\begin{aligned} \textit{square domain}: \ \omega &= 0 \quad \text{on } \partial\mathcal{D} \\ \textit{circular domain}: \ \omega &= \frac{2 v_\theta}{r} \quad \text{on } \partial\mathcal{D} \end{aligned} \tag{7}$$

with v_θ the azimuthal component of the velocity $\mathbf{v} = (v_r, v_\theta)$ in (r,θ)-coordinates.

2.1. CIRCULATION Γ

A useful integral quantity is the total *circulation* Γ of the flow,

$$\Gamma = \oint_{\partial\mathcal{D}} \mathbf{v}\cdot d\mathbf{s} = \int_{\mathcal{D}} \omega dA \ , \tag{8}$$

with $d\mathbf{s}$ a line element of the domain boundary. It is easy to verify that $\Gamma = 0$ for both the double-periodic and the no-slip boundaries, while Γ is not determined for the case of a stress-free boundary.

Term-by-term integration of the vorticity equation over the whole domain yields

$$\int_{\mathcal{D}} \frac{\partial\omega}{\partial t} dA + \int_{\mathcal{D}} (\mathbf{v}\cdot\nabla)\,\omega dA = \frac{1}{Re}\int_{\mathcal{D}} \nabla^2\omega dA \ , \tag{9}$$

in which the first term is equal to $\frac{d\Gamma}{dt}$. The second term can be written as

$$\int_{\mathcal{D}} (\mathbf{v}\cdot\nabla)\,\omega dA = \int_{\partial\mathcal{D}} \omega\,(\mathbf{v}\cdot\hat{\mathbf{n}})\,ds \ . \tag{10}$$

We see immediately that (10) is identically zero for the three different boundary conditions: for a double-periodic boundary because of periodicity, for a no-slip boundary because $\mathbf{v} = 0$ on $\partial\mathcal{D}$, and for a stress-free boundary because $\mathbf{v}\cdot\hat{\mathbf{n}} = 0$. The term on the r.h.s. of (9) can be written as a boundary integral, so that (9) becomes:

$$\frac{d\Gamma}{dt} = \frac{1}{Re}\int_{\partial\mathcal{D}} \mathbf{n}\cdot\nabla\omega ds \ . \tag{11}$$

Apparently, the total circulation Γ of the flow on a bounded domain $\mathcal{D}$ can only change through a net vorticity flux associated with diffusion through the boundary $\partial\mathcal{D}$. Because $\Gamma = 0$ in both the double-periodic and the no-slip boundary cases, however, we have $\frac{d\Gamma}{dt} = 0$ in those cases, implying zero net leakage of vorticity through $\partial\mathcal{D}$. In summary:

$$\begin{array}{lll} \textit{double-periodic}: & \Gamma = 0 \ , & \frac{d\Gamma}{dt} = 0 \\ \textit{no-slip}: & \Gamma = 0 \ , & \frac{d\Gamma}{dt} = 0 \\ \textit{stress-free}: & \Gamma = \ ? \ , & \frac{d\Gamma}{dt} = \ ? \end{array}$$

Only in the case of a stress-free boundary, diffusion of vorticity through $\partial\mathcal{D}$ may result in a net change in Γ.

2.2. ANGULAR MOMENTUM L

Another useful and relevant global quantity is the angular momentum L, defined with respect to the origin in the domain centre as

$$L = \int_{\mathcal{D}} \hat{\mathbf{k}} \cdot (\mathbf{r} \times \mathbf{v})\, dA \;, \tag{12}$$

with $\hat{\mathbf{k}}$ the unit vector perpendicular to the flow plane, and $\mathbf{r}$ the position vector. It is easy to verify that for the square domain (12) can be written as

$$L = \iint_{\mathcal{D}} (xv - yu)\, dxdy = 2 \iint_{\mathcal{D}} \psi dxdy \;, \tag{13}$$

with (u, v) the velocity components in x, y-direction, and ψ the stream function defined by $\mathbf{v} = -\hat{\mathbf{k}} \times \nabla\psi$.

The rate of change of the total angular momentum of the fluid on domain $\mathcal{D}$ can be written as

$$\frac{dL}{dt} = \frac{d}{dt} \int_{\mathcal{D}} \hat{\mathbf{k}} \cdot (\mathbf{r} \times \mathbf{v})\, dA = \int_{\mathcal{D}} \hat{\mathbf{k}} \cdot \left(\mathbf{r} \times \frac{\partial \mathbf{v}}{\partial t}\right) dA, \tag{14}$$

which becomes, after substitution of the Navier-Stokes equation (1):

$$\begin{aligned} \frac{dL}{dt} &= -\int_{\mathcal{D}} \hat{\mathbf{k}} \cdot [\mathbf{r} \times (\mathbf{v} \cdot \nabla)\, \mathbf{v}]\, dA \;-\; \int_{\mathcal{D}} \hat{\mathbf{k}} \cdot [\mathbf{r} \times \nabla p] + \\ &\qquad + \frac{1}{Re} \int_{\mathcal{D}} \hat{\mathbf{k}} \cdot \left(\mathbf{r} \times \nabla^2 \mathbf{v}\right) dA \\ &= -\int_{\mathcal{D}} \mathbf{v} \cdot \nabla \left[\hat{\mathbf{k}} \cdot (\mathbf{r} \times \mathbf{v})\right] dA \;+\; \int_{\mathcal{D}} \hat{\mathbf{k}} \cdot [\nabla \times p\mathbf{r}]\, dA \\ &\qquad + \frac{1}{Re} \int_{\mathcal{D}} \mathbf{r} \cdot \nabla\omega dA \;. \end{aligned} \tag{15}$$

The area integrals are now rewritten as contour integrals, resulting in

$$\begin{aligned} \frac{dL}{dt} &= -\int_{\partial\mathcal{D}} \hat{\mathbf{k}} \cdot (\mathbf{r} \times \mathbf{v})\, (\mathbf{v} \cdot \hat{\mathbf{n}})\, ds \;+ \oint_{\partial\mathcal{D}} p\mathbf{r} \cdot d\mathbf{s} \;+ \\ &\qquad + \frac{1}{Re} \oint_{\partial\mathcal{D}} \omega\, (\mathbf{r} \cdot \hat{\mathbf{n}})\, ds - \frac{2}{Re}\Gamma \;. \end{aligned} \tag{16}$$

The first term on the r.h.s. of this equation vanishes because $\mathbf{v} \cdot \hat{\mathbf{n}} = 0$ on the impermeable boundary of the domain. If the *no-slip* condition applies, $\Gamma = 0$, so that (16) becomes

$$\frac{dL}{dt} = \oint_{\partial\mathcal{D}} p\mathbf{r} \cdot d\mathbf{s} + \frac{1}{Re} \oint_{\partial\mathcal{D}} \omega\, (\mathbf{r} \cdot \hat{\mathbf{n}})\, ds \;. \tag{17}$$

This result expresses the (rather trivial) fact that the change in angular momentum is due to torques of wall forces, associated with the inviscid pressure (normal stress) and viscous stresses (normal and shear stresses). The circular domain is in this respect somewhat special, since the normal stresses do not produce a net torque relative to the centre of the domain. For this case (17) becomes:

$$\frac{dL}{dt} = \frac{1}{Re} \int_0^{2\pi} \omega\,(R, \theta)\, R d\theta \,, \tag{18}$$

with R the radius of the circular domain.

3. Experimental arrangement

Laboratory experiments were carried out in a square container of dimensions $100 \times 100 \times 30$ (length $\times$ width $\times$ depth) and in a circular container with diameter $2R = 92$ cm and depth 30 cm. In either case, the tank was filled with a two-layer stratification consisting of a fresh-water layer (density $\rho_1 \simeq 1.00$ g/cm^3) on top of a salty bottom layer (density ρ_2, varying between 1.08 en 1.12 g/cm^3), separated by an interfacial layer of typically a few centimeters depth.

Motion was introduced in the fluid by dragging a rake, consisting of a linear array of vertical bars (each with diameter $d = 3$ mm), horizontally at some prescribed speed V through the fluid. After the grid had moved from one side to the other side of the tank, it was removed by vertically lifting it out of the fluid. At large enough towing speeds ($V \simeq 15$ cm/sec) the motion in the wake of the grid was turbulent. In the non-stratified upper and lower layers, this motion was essentially 3D and thus decayed rapidly. In contrast, the motion in the stratified interfacial layer was quasi-2D, i.e. the flow field was planar, with substantial vertical gradients. This planar motion was visualised by adding small polystyrene particles (density 1.04 g/cm^2; a few mm in diameter) to the fluid. Illuminated by strip lights from the side, the motion of these tracer particles was monitored by a CCD camera mounted at some distance above the container, and the images were stored on a video tape recorder. After each experiment, the stored data were digitally processed using the software package *DigImage* (see Dalziel 1992), which provided quantitative information about the flow field, like e.g. the vorticity distribution.

Although not exactly 2D, the planar motion in the interfacial layer showed the phenomenological characteristics of 2D turbulence, namely the emergence of larger vortex structures. This motion was generally very persistent, the decay mainly governed by vertical diffusion. The dissipation of kinetic energy of the flow by vertical shearing (or vertical diffusion) in these experiments is discussed by Maassen *et al.* (2002) (see also the related studies on vertical shearing by Yap and Van Atta (1993) and Fincham *et al.* (1996)). Some results of the experiments will be discussed in the next section.

One of the questions addressed in the present study concerns the effect of the initial angular momentum L_0 on the subsequent evolution of the flow. The initial angular momentum could be controlled by changing the arrangement of the vertical rods in the linear array. The drag force F_D exerted on a cylindrical bar with diameter d and length l moving at constant speed V through a quiescent fluid of density ρ can be expressed as

$$F_D = \frac{1}{2} C_D \rho V^2 l d \ , \tag{19}$$

where C_D is the drag coefficient (see Blevins (1984) for a collection of empirical C_D-values for different Re-values). A rod moving in the x-direction along a straight line y = constant exerts a torque $T = yF_D$ with respect to the centre $(x, y) = (0, 0)$ of the domain, resulting in a contribution to the total angular momentum given by

$$L_{rod} = \frac{1}{\rho l} \int_0^\tau T dt = \frac{F_D y \tau}{\rho l} = \frac{1}{2} C_D V y a d \ , \tag{20}$$

with τ the forcing duration, and $a = V\tau$ the total displacement. By suitably arranging the rods at different positions y along the rake, one can control the net amount of angular momentum added initially to the flow. Using a rake that is symmetric with respect to the line $y = 0$ results in an initial flow with zero net angular momentum ($L_0 = 0$). In order to avoid such a symmetry, an initial flow with $L_0 \approx 0$ may also be created by using an asymmetric rake, in which the rods are arranged such that the sum of their y-coordinates vanishes. The flow generated by the grid forcing could be characterised by the Reynolds number based on the grid motion:

$$Re_M = \frac{VM}{\nu} \ , \tag{21}$$

with V the towing speed, and M the (average) grid spacing. In the experiments discussed here, these quantities had typical values $V = 15$ cm/sec and $M = 4$ cm, implying $Re_M \simeq 6000$, which is comparable to the grid-based Reynolds numbers in the experiments by Yap and Van Atta (1993) and Fincham *et al.* (1996). Alternatively, one could use the Reynolds number

$$Re^* = \frac{UW}{\nu} \tag{22}$$

based on the size of the tank (W is the half-width) and the r.m.s. velocity U of the initial flow field. In the numerical simulations, the initial field was characterised in this way. Experimentally, the r.m.s. velocity U would be measured (with an error of approximately 15%) typically 10 sec after the forcing was stopped, resulting in Re^*-values between 2000 and 5000. The decay rates of kinetic energy obtained from laboratory experiments with $Re^* \approx 5000$ are of the same order as those computed in numerical simulations with $Re \approx 1500 - 2000$ (Clercx *et al.*, 1999). Based on this observation we estimate that, for the present experiments, the characteristic decay time of 2D turbulence in an experiment with Reynolds number Re^* is comparable with the decay time in a numerical simulation with Reynolds number $Re \approx 0.4Re^*$.

4. Decaying 2D flow on a square domain with $\mathbf{L_0 \approx 0}$

Some experiments were conducted in which the induced angular momentum in the initial flow field was close to zero, i.e. $L_0 \approx 0$. Figure 1 shows a sequence of streak images taken during the course of such an experiment. One easily observes the small-scale motions in the earlier stages of the flow evolution, after which the flow becomes gradually dominated by bigger and bigger vortices. At some stage (t = 60 min) the flow consists of one large cell which is accompanied by a smaller cell with opposite circulation. In fact, this asymmetric dipolar structure is persistent in the late stages of the flow evolution: although the dipole slowly moves and rotates around in the domain, it retains its dipolar character while decaying. Only in the very late, 'final' stage the flow takes on the appearance of a single cell centered in the domain.

The evolution of the vorticity field derived from the experiment shown in Figure 1 is presented by the sequence of plots in Figure 2. One clearly observes the street-shaped vortex patterns in the early stage - soon after the grid forcing was stopped - and the gradual transition to the dipolar structure in the final stage. These vorticity plots provide clear evidence of the role of the no-slip boundaries: whenever a vortex approaches a boundary closely, a boundary layer is formed, containing oppositely-signed vorticity. Even in the later stages - when the flow has become very weak by vertical diffusion and dissipation - high vorticity values and strong gradients exist, in particular near the walls. Another feature worth noting is the 'shield' of the dipole in the later stages: each dipole half is shielded by a band of opposite vorticity. In the very late stages, the weaker part of this dipolar structure has disappeared, and the flow consists of a single central cell, surrounded by a ring of very weak, oppositely-signed vorticity. This final stage corresponds with the fundamental 'Stokes mode', for a square domain, in which the flow is governed by viscous diffusion (see also Van de Konijnenberg *et al.* (1998)). The significant role of the no-slip boundary throughout the flow evolution

(a) $t = 0.2$ (b) $t = 1$

(c) $t = 6$ (d) $t = 12$

(e) $t = 50$ (f) $t = 60$

Figure 1. Streak images of an experiment with $L_0 \approx 0$ and $Re^* \simeq 5000$ in a square container. The tails of the streaks are generated after digital processing of the images.

is even more clearly visible in the vorticity contour plots obtained from the numerical flow simulation as presented in Figure 3. Since this simulation has a higher resolution than the digital data analysis technique applied to the laboratory flows, Figure 3 shows far more detailed structures in the vorticity field than Figure 2. One clearly observes the abundance of vorticity

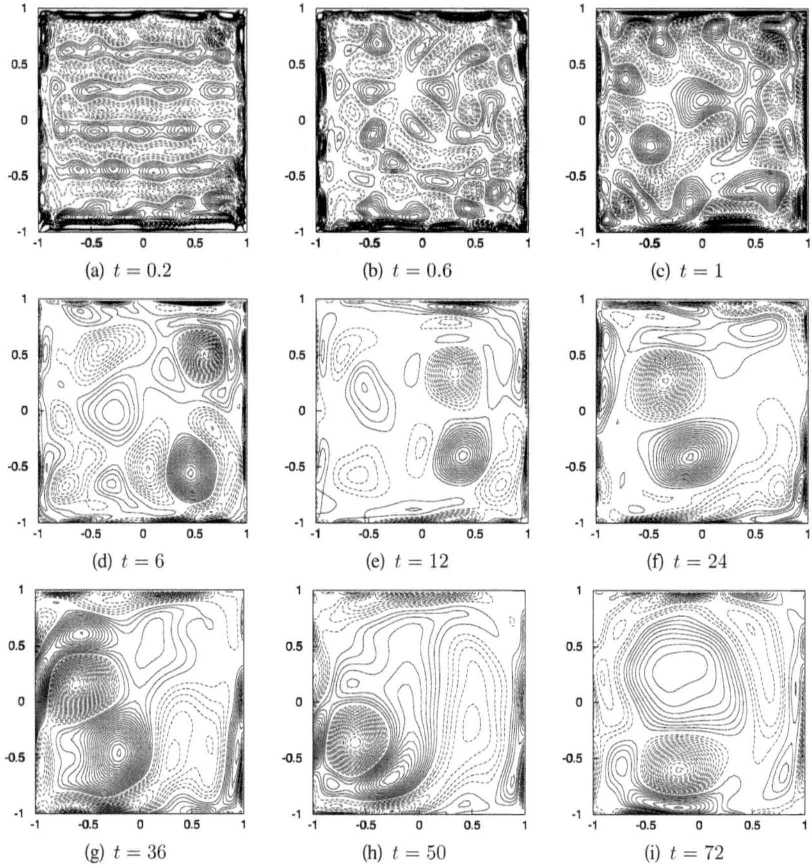

Figure 2. Vorticity contour plots of the same experiments as shown in Figure 1. Dashed contours represent negative ω-values, while solid contours represent positive values.

filaments in the flow field, even in the later stages of the flow evolution. These filaments originate from the walls, where viscously induced vorticity residing in boundary layers is 'peeled off' and advected by approaching vortex structures. It is thus clearly observed that the walls act as sources of vorticity filaments, in sharp contrast to the case of simulations with double-periodic boundary conditions (see e.g. McWilliams, 1984).

Another important feature of the flow evolution on a bounded domain is shown in Figure 4, which displays the angular momentum $L(t)$ of the contained fluid relative to the centre of the domain, calculated for two different numerical runs for $Re^* = 2000$, each starting from a more or less random initial state containing almost zero angular momentum, i.e. $L(t = 0) = L_0 \approx 0$. The graph clearly reveals that $L(t)$ changes significantly in both runs, either to a positive or to a negative value. The net angular

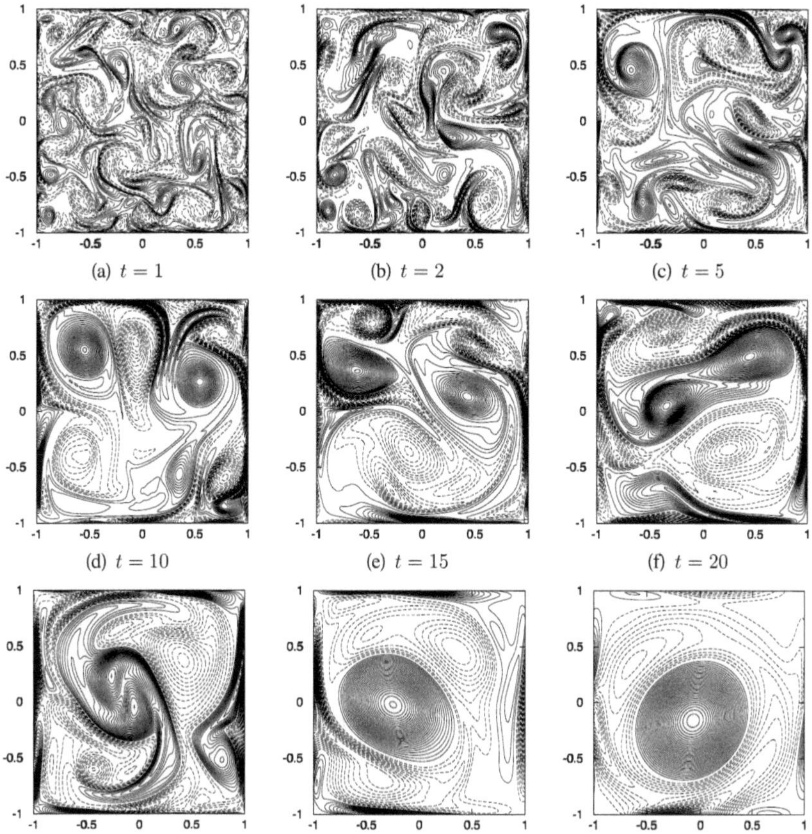

(a) $t = 1$ (b) $t = 2$ (c) $t = 5$

(d) $t = 10$ (e) $t = 15$ (f) $t = 20$

(g) $t = 30$ (h) $t = 50$ (i) $t = 100$

Figure 3. Vorticity contour plots for decaying 2D turbulence on a square domain as calculated numerically (taken from Clercx *et al.*, 1999). The calculations were carried out for $Re^* = 2000$.

momentum of the same amount of fluid with the same kinetic energy $E(t = 30) = 0.02$ is $L_{sb} = 0.33$. Nevertheless, the substantial increase of $|L(t)|$ is obvious: the fluid shows 'spontaneous spin-up'. This non-zero angular momentum is associated with the formation of a larger flow cell (see Figures 1-3) which tends to move towards and around the domain centre. As is evident from the analysis presented in Section 2, in particular equation (17), the net change in L can only be brought about by torques associated with wall stresses. This is once more evidencing the role played by the solid boundaries in the case of flow on a bounded domain.

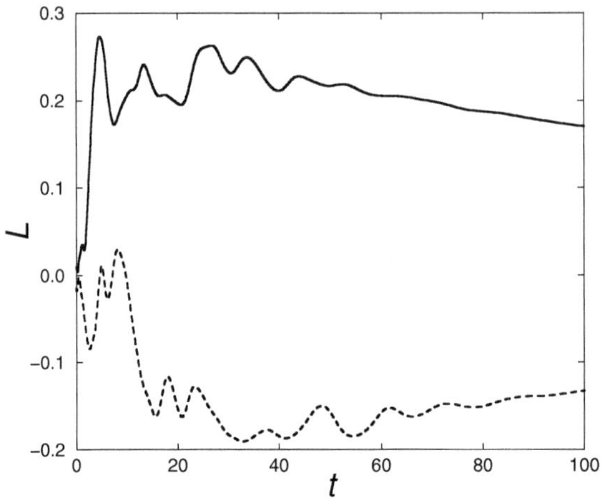

Figure 4. Dimensionless angular momentum $L(t)/L_{sb}L(t)$ calculated in two different numerical simulations with $Re^* = 2000$, starting from initial states with $L_0 \approx 0$.

5. Decaying 2D flows on a square domain with $\mathbf{L_0 \neq 0}$

A number of experiments were carried out in the square container with the initial flow generated by an asymmetric grid, with such an arrangement of the rods that a non-zero net angular momentum ($L_0 \neq 0$) is introduced to the fluid. A typical example of the flow evolution observed in such a case is shown by the sequence of vorticity contour plots presented in Figure 5, measured in an experiment with $|L_0| = 0.4$ and $Re^* = 4000$. It is easily seen how the flow soon becomes organised (from $t = 10$ onwards) with a larger cell in the centre, accompanied by some smaller cells around it. In the 'final' stage ($t \geq 50$) the flow has acquired a symmetric appearance: a central circular cell with a shielding ring of oppositely-signed vorticity around it. Apparently, the initial angular momentum $|L_0| \neq 0$ caused the flow to organize in a single-cell pattern quite rapidly, in contrast to the double-cell pattern usually observed in the later stage of flows generated with $L_0 \approx 0$ (see Section 4). It should be noted that strong spontaneous spin-up is observed in many numerical runs with $Re = 2000$ and $Re = 5000$ and initially $L_0 \approx 0$ (see e.g. Clercx *et al.*, 1998 and 2001). The late-time state in nearly all runs consisted of a rotating tripole with its core located in the centre of the container (see, for example, Figure 3). Experimental confirmation of this spin-up scenario is difficult due to the requirement that for suitable experiments we need at least $Re^* \approx 10000$. The crucial role of the initial angular momentum L_0 in the flow evolution is also clearly demonstrated by Figure 6, showing the total angular momentum, scaled by $L_{sb}(t)$, as a function of time for a number of experiments with $L_0 \approx 0$ (dashed lines) and $|L_0| > 0$ (solid lines). In all runs the fluid showed

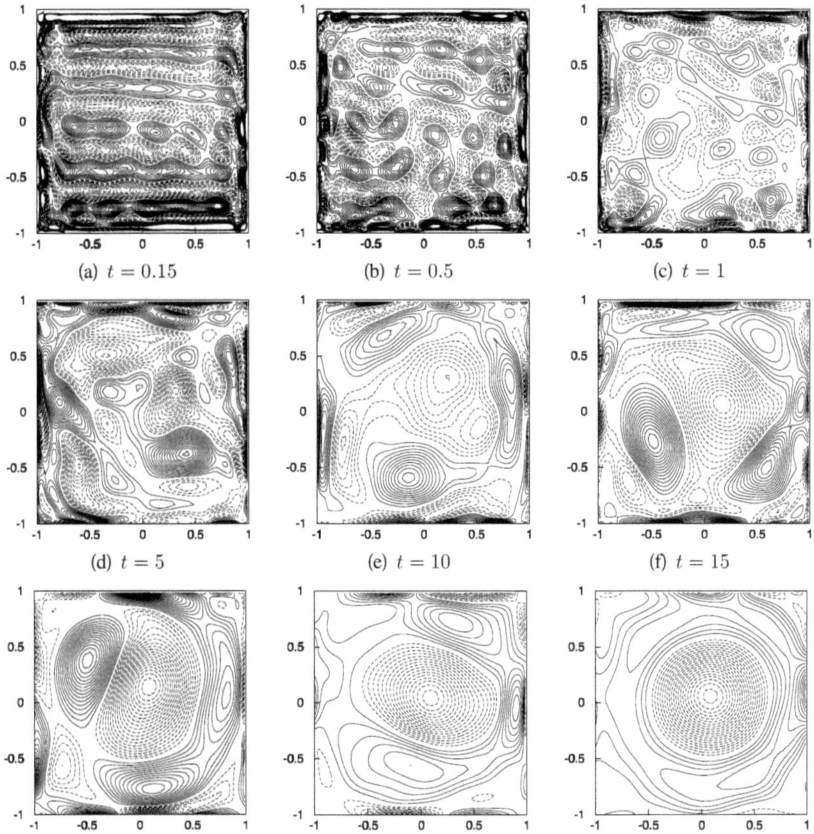

(a) $t = 0.15$ (b) $t = 0.5$ (c) $t = 1$

(d) $t = 5$ (e) $t = 10$ (f) $t = 15$

(g) $t = 20$ (h) $t = 30$ (i) $t = 55$

Figure 5. Vorticity contour plots measured in an experiment with $|L_0| = 0.4$ and $Re^* \simeq 4000$ in a square container. As before, dashed and solid contours represent negative and positive vorticity, respectively.

'spontaneous spin-up', but the rate at which this spin-up is accomplished as well as the final value of L is significantly higher in the cases with $|L_0| > 0$. These results demonstrate the importance of the initial angular momentum as well as the presence of the solid domain boundaries with respect to the organization of the 2D flow.

6. Decaying 2D flow on a circular domain

Experiments were also carried out in a circular container with the flow initialized by the rake moving along a straight diametrical path through the tank. Because of the circular geometry, the grid does not 'fill' the domain completely (as it did in the case of the square geometry), so that a residual dipolar cell pattern will result after passage of the grid. Additional

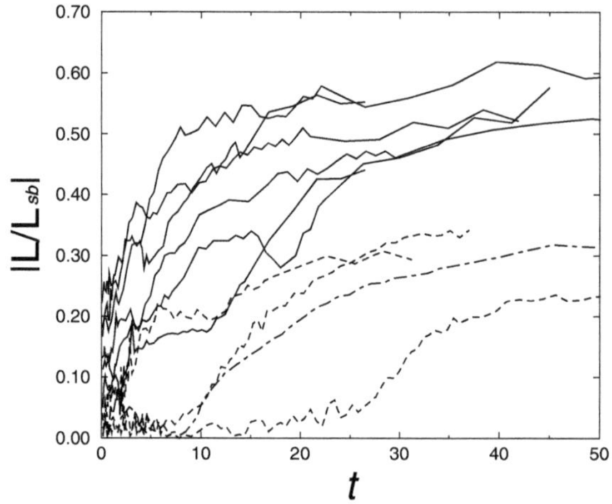

Figure 6. Normalised net angular momentum $L(t)/L_{sb}(t)$ measured in a number of experiments performed in a square container with $L_0 \approx 0$ (dashed lines) and $|L_0| > 0$ (solid lines).

angular momentum $L_0 \neq 0$ could by added by using an asymmetric grid configuration, as in the experiments described in Section 5.

Numerical analysis of the terms contributing to $\frac{dL}{dt}$ for the square-container case revealed that $|\oint_{\partial\mathcal{D}} p\mathbf{r} \cdot d\mathbf{s}| \gg 1/Re|\oint_{\partial\mathcal{D}} \omega(\mathbf{r} \cdot \hat{\mathbf{n}})ds|$. More specific, the viscous stress contribution appears to contribute negligibly to the spontaneous spin-up of the flow. Since normal (pressure and viscous) stresses at the circular wall do not produce any net torque relative to the tank centre, it can be expected from (17) that the wall effect in terms of the change in $L(t)$ will be rather modest compared to the previous case. In particular, spontaneous spin-up is virtually absent. Nevertheless, the influence of the no-slip boundary is crucial for the flow evolution, as in the square domain.

The experiments with $L_0 \approx 0$ showed the formation of a large dipolar cell structure, which was seen to move forward towards the wall. Owing to the no-slip condition, thin boundary layers are formed containing oppositely-signed vorticity. Subsequently, vorticity is advected away from the wall in the form of two filaments, which lead to the formation of two vorticity patches behind the dipole, giving the flow at this stage a quadrupolar appearance. Then, the 'primary' dipole becomes weaker, while the secondary structures become more pronounced. The newly formed dipole then starts to push against the wall as did the original one, and the process described above is repeated. In fact, no clear quasi-stationary 'final' state was reached in this set of experiments. The observed behaviour agrees with the findings of Li and Montgomery (1996) and Li, Montgomery and Jones (1996,

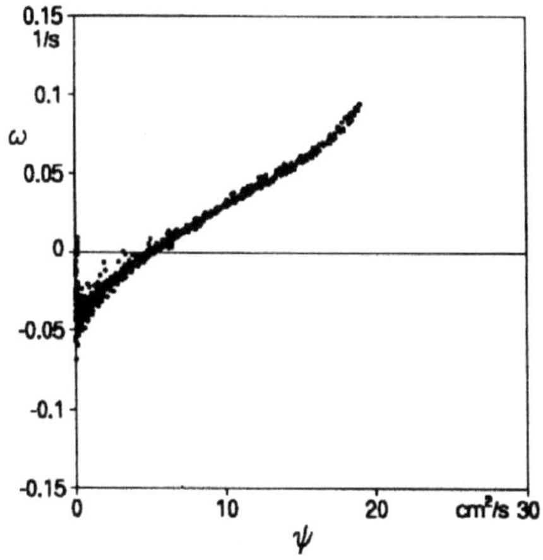

Figure 7. Scatter plot of ω versus ψ, measured in the 'final' state of the evolution of the flow in a circular tank, with $|L_0| > 0$. (taken from Flór, 1994)

1997) obtained from numerical simulations of decaying 2D turbulence on a circular domain. Laboratory experiments performed with $|L_0| > 0$ have revealed - as could be expected from the observed flow evolution in a square domain - that the flow rapidly evolved towards a monopolar structure (see Maassen, Clercx and van Heijst 1999, 2002). In terms of the vorticity: the 'final' state consists of a central cell accompanied by a ring of oppositely-signed vorticity around it. This behaviour is similar to that found for the case of a square domain with $|L_0| > 0$, in which the quasi-stationary 'final' state was also of the monopolar type. Useful information about such a quasi-steady state is obtained by plotting the so-called ω, ψ-scatter plot. After digitization of the flow field, the experimental velocity vector field is available on a grid overlying the domain, and by numerical manipulation the values of the vorticity ω and the streamfunction ψ can be determined in each grid point. A ω, ψ-scatterplot consists of all (or part of) the grid points plotted according to their ω- and ψ-values. An example of a scatterplot obtained for the case of the organised, final state in a circular domain with $|L_0| > 0$ is shown in Figure 7. Apart from the 'scatter' around $\psi = 0$, which represents the somewhat unsteady flow near the circular container wall, the points lie in a more or less straight band. The observed scatter in this band is partially due to experimental inaccuracy, partially due to the flow being not exactly steady. In order to derive an analytical representation of the flow at this stage, we approximate the measured relationship by

$$\omega = k^2\psi - \gamma \,, \tag{23}$$

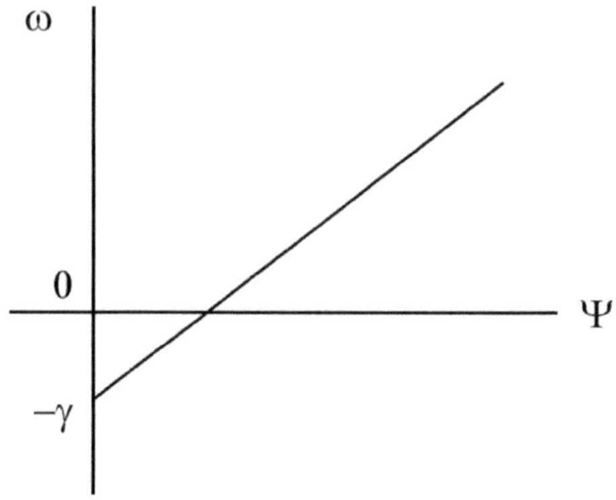

Figure 8. Idealized ω, ψ-scatter plot for the monopolar flow in a circular tank.

with γ the off-set at $\psi = 0$ and k^2 representing the slope of the straight line, see Figure 8.
This relationship (23) is now substituted into the Poisson equation

$$\nabla^2 \psi = -\omega \tag{24}$$

that relates ω and ψ (as follows from their definitions), yielding for the axisymmetric case

$$\frac{d^2\psi}{dr^2} + \frac{1}{r}\frac{d\psi}{dr} + k^2\psi = \gamma \ . \tag{25}$$

A particular solution of this inhomogeneous equation is $\psi_p = \gamma/k^2$, while the homogeneous equation has the general solution

$$\psi_h(r) = AJ_0(kr) + BY_0(kr) \ , \tag{26}$$

with A en B constants, and J_0 and Y_0 representing zeroth order Bessel functions of the first and second kind, respectively. Since singular behaviour in $r = 0$ is rejected, the constant B is taken zero. The complete solution is thus

$$\psi(r) = AJ_0(kr) + \gamma/k^2 \ , \tag{27}$$

and the corresponding expressions for the vorticity ω and the azimuthal velocity are

$$\begin{aligned} \omega(r) &= Ak^2 J_0(kr) \ , \\ v_\theta(r) &= AkJ_1(kr) \ . \end{aligned} \tag{28}$$

The no-slip condition $v_\theta(r = R) = 0$ requires $J_1(kR) = 0$, yielding

$$kR = \lambda_1 = 3.83 \ , \tag{29}$$

being the first zero of J_1. It is easily verified that the total vorticity on the domain $\{0 \le r \le R; 0 \le \theta \le 2\pi\}$ is

$$\int_0^R \omega(r) 2\pi r dr = 2\pi A k R J_1(kR) = 0 \ , \tag{30}$$

while the diffusive vorticity flux through $\partial\mathcal{D}$ is

$$\frac{d\omega}{dr}|_R = Ak^2 \frac{d}{dr} J_0(kr)|_R = -Ak^3 J_1(kR) = 0 \ , \tag{31}$$

in agreement with the requirements $\Gamma = 0$ and $\frac{d\Gamma}{dt} = 0$, see Section 2. Note that for this particular case condition (31) is stronger than the rather relaxed condition $\frac{d\Gamma}{dt} = 0$, which actually states that the net vorticity flux through $\partial\mathcal{D}$ should be zero.

In the case of purely 2D flow, the decay of the motion would be described by cross-diffusion of vorticity (positive and negative vorticity cancelling each other), governed by

$$\frac{\partial\omega}{\partial t} = \nu\nabla^2\omega = \nu\left(\frac{\partial^2\omega}{\partial r^2} + \frac{1}{r}\frac{\partial\omega}{\partial r}\right) \ . \tag{32}$$

The solution of this equation can be obtained by separation of variables, and proceeding as above one finds

$$\omega(r,t) = Ak^2 J_0(kr) \exp(-\nu k^2 t) \ . \tag{33}$$

Apparently, the vorticity shows a decay on the horizontal diffusion timescale

$$\tau_H = (\nu k^2)^{-1} = \frac{R^2}{\lambda_1^2 \nu} \ . \tag{34}$$

In the laboratory experiments described in this study, the planar motion was confined in a relatively shallow, stratified interfacial layer between two homogeneous fluid layers, implying that vertical diffusion is active as well. The typical timescale τ_v associated with the flow decay due to diffusion in the vertical direction is

$$\tau_v = \frac{\delta^2}{\nu} \ , \tag{35}$$

with δ the thickness of the interfacial region. With $\lambda_1 = 3.83$ and typical values $R = 46$ cm and $\delta \simeq 4$ cm, one finds

$$\tau_v \ll \tau_H \ , \tag{36}$$

expressing that the flow decay is essentially determined by vertical diffusion.

7. Conclusions

It has been shown both by laboratory experiments in a stratified fluid and by high-resolution numerical simulations on 2D flows that the solid boundaries play a crucial role in the evolution of confined, decaying 2D turbulence. The no-slip condition imposed by realistic lateral boundaries implies the generation of relatively thin boundary layers containing high-amplitude vorticity. Mainly due to approaching vortices, these boundary layers are seen to separate, leading to the formation of filamentary vorticity structures, which are subsequently advected into the interior of the flow domain. Apparently, lateral walls act as *sources of vorticity filaments.* In this respect, the flow evolution and its characteristics are essentially different from those in the case of double-periodic boundary conditions. Such differences are directly visible in the vorticity contour plots (higher filamentary activity in the no-slip simulations), but are also found in the energy spectra of the decaying flow (see Clercx and van Heijst, 2000).

A second important role the boundaries may play is in *providing normal and shear stresses* - and hence torques - that may change the net angular momentum of the contained fluid. This is commonly observed in the so-called spontaneous spin-up of fluid initially containing zero angular momentum ($L_0 \approx 0$). Experiments and numerical simulations of flows contained in both square and circular tanks have demonstrated this behaviour.

Moreover, it was found that the amount of initial net angular momentum L_0 is an important parameter in the flow evolution, and also in the establishment of the quasi-stationary 'final' state of the flow. In the case $L_0 \approx 0$ the flow in the later stages of decaying 2D turbulence in containers with a square geometry had a dipolar character for the experiments and numerical simulations in the small Reynolds-number limit ($Re^* \lesssim 4000$ and $Re \lesssim 1500$, respectively) or a tripolar character for the numerical simulations with $Re \gtrsim 1500$. Similar initial conditions ($L_0 \approx 0$) for decaying 2D turbulence in a circular geometry yield a dipolar flow structure in the later stage of the flow evolution. No monopolar structure is found to emerge, which reflects the absence of spontaneous spin-up in flows in a circular container. In the square domain, the dipolar flow structure is usually asymmetric, and eventually takes on the appearance of a single cell. In the case $|L_0| > 0$, decaying 2D turbulence in square and in circular containers showed a relatively rapid self-organisation of the flow which is directly associated with the formation of a persistent monopolar structure.

These findings reveal the essential differences between decaying 2D turbulence on an infinite domain (or a finite square domain with double-periodic boundaries) and decaying 2D flow on a bounded domain with solid lateral boundaries. For this reason, one should be cautious when comparing results of laboratory experiments on 2D turbulent flows (which are without exception bounded by solid walls) with theoretical or numerical results obtained for unbounded flows. This concerns aspects like the detailed flow evolution, the spectral characteristics of decaying or forced flows, and the nature of the 'final state' of decaying 2D turbulence, but also the transport properties of such flows.

References

Blevins, R.D. *Applied Fluid Dynamics Handbook.* Van Nostrand Reinhold, New York, 1984.

Clercx, H.J.H. A Spectral Solver for the Navier-Stokes Equations in the Velocity-vorticity Formulation for Flows with Two Non-periodic Directions. *J. Comput. Phys.*, 137:186–211, 1997.

Clercx, H.J.H., S.R. Maassen and G.J.F. van Heijst. Spontaneous Spin-up during the Decay of 2D Turbulence in a Square Container with Rigid Boundaries. *Phys. Rev. Lett.*, 80:5129–5132, 1998.

Clercx, H.J.H., S.R. Maassen and G.J.F. van Heijst. Decaying Two-dimensional Turbulence in Square Containers with no-slip or Stress-free Boundaries. *Phys. Fluids.*, 11:611–626, 1999.

Clercx, H.J.H. and G.J.F. van Heijst. Energy Spectra for Decaying 2D Turbulence in a Bounded Domain. *Phys. Rev. Lett.*, 85:306–309, 2000.

Clercx, H.J.H., A.H. Nielsen, D.J. Torres, and E.A. Coutsias. Two-dimensional Turbulence in Square and Circular Domains with no-slip Walls. *Eur. J. Mech. B Fluids*, 20:557–576, 2001.

Dalziel, S.B. *DigImage. Image Processing for Fluid Dynamics.* Cambridge Environmental Research Consultants Ltd., Cambridge, UK., 1992.

Fincham, A.M., T. Maxworthy and G.R. Spedding. Energy Dissipation and Vortex Structure in Freely Decaying, Stratified Grid Turbulence. *Dyn. Atmos. Oceans*, 23:155–169, 1996.

Flór, J.B. *Coherent Vortex Structures in Stratified Fluids.* PhD thesis, Eindhoven University of Technology, 1994.

Konijnenberg, J.A. van de, J.B. Flór and G.J.F. van Heijst. Decaying Quasi-two-dimensional Flow on a Square Domain. *Phys. Fluids*, 10:595–606, 1998.

Li, S. and D. Montgomery. Decaying Two-dimensional turbulence with rigid walls. *Phys. Lett.* A21:281–291, 1996.

Li, S., D. Montgomery and W.B. Jones. Inverse Cascades of Angular Momentum. *J. Plasma Phys.*, 56:615–639, 1996.

Li, S., D. Montgomery and W.B. Jones. Two-dimensional Turbulence with Rigid Circular Walls. *Theor. Comput. Fluid Dyn.*, 9:167–181, 1997.

McWilliams, J.C. The Emergence of Isolated Coherent Vortices in Turbulent Flows. *J. Fluid Mech.*, 146:21–43, 1984.

Maassen, S.R., H.J.H. Clercx and G.J.F. van Heijst. Decaying Quasi-2D Turbulence in a Stratified Fluid with Circular Boundaries. *Europhys. Lett.*, 46:339–345, 1999.

Maassen, S.R., H.J.H. Clercx and G.J.F. van Heijst. Self-organization of Quasi-two-dimensional Turbulence in Stratified Fluids in Square and Circular Containers *Phys. Fluids*, 14:2150–2169, 2002.

Matthaeus, W.H. and D.C. Montgomery. Selective Decay Hypothesis at High Mechanical and Magnetic Reynolds numbers. *Ann. (N.Y.) Acad. Sci.*, 357:203–222, 1980.

Santangelo, P., R. Benzi and B. Legras. The Generation of Vortices in High-resolution, Two-dimensional, Decaying Turbulence and the Influence of initial conditions on the breaking of self-similarity. *Phys. Fluids A*, 1:1027–1034, 1989.

Yap, C.T. and C.W. van Atta. Experimental Studies of the Development of Quasi-two-dimensional Turbulence in Stably Stratified Fluid. *Dyn. Atmos. Oceans*, 19:289–323, 1993.

EFFECTS OF ROTATION ON CONVECTIVE INSTABILITY

G. F. CARNEVALE
Scripps Institution of Oceanography
University of California, San Diego
9500 Gilman Drive
La Jolla, CA 92093–0213, U.S.A.

R. C. KLOOSTERZIEL
University of Hawaii, U.S.A.

P. ORLANDI
University of Rome, 'La Sapienza', Italy

Y. ZHOU
LLNL, Livermore, California, U.S.A.

Abstract. The ability of rotation to inhibit the development of Rayleigh-Taylor instability is examined in terms of the effect of the Coriolis force on the propagation of vortex rings. It is demonstrated through numerical simulation that, in the presence of non-zero diffusivity and viscosity, sufficiently strong ambient rotation can completely suppress Rayleigh-Taylor instability. An investigation of this effect of rotation in the context of the simpler Rayleigh-Bénard problem results in new closed-form expressions for the stability boundary.

Key words: convection, rotation

1. Introduction

A system with heavy fluid lying above light fluid is subject to Rayleigh-Taylor instability (RTI). If the density differences are not great, the instability will proceed with the propagation of bubbles of heavy fluid downward into the light fluid and bubbles of light fluid upward into the heavy fluid. Each bubble is associated with a vortex ring that propels it forward. In Verzicco *et al.* (1996), we found that a vortex ring propagating in the direction of an ambient rotation vector propagates more slowly than in the case of no rotation. Further, we demonstrated that sufficiently strong rotation could prevent the formation of vortex rings. It seemed natural then to consider the possibility of suppressing RTI by the application of rotation.

O.U. Velasco Fuentes et al. (eds.), Nonlinear Processes in Geophysical Fluid Dynamics, 325-337.

In Carnevale *et al.* (2002), we presented the results of a series of numerical simulations which demonstrated that with the ambient rotation vector anti-parallel to gravity, the onset of RTI is delayed and the growth rate is diminished relative to the non-rotating case. Indeed, with sufficiently strong background rotation, we found that the formation of bubbles was completely suppressed. This meant that instead of the formation of an efficient mixing zone created by the propagation and interaction of bubbles the density isosurfaces would remain flat, and the density gradient separating heavy from light fluid would merely diffuse away due to the presence of finite molecular diffusivity. We analyzed this result in terms of an inviscid non-diffusive model of vortex ring formation. This model did suggest that vortex rings with very large diameter could be prevented from forming by the agency of rotation. However, this model could not preclude the initiation of the instability with less symmetrical modes. Thus, since our numerical simulations of necessity had non zero viscosity and diffusivity, and since our theoretical model only considered a very idealized form of perturbation, we were left with the question of whether strong rotation alone was sufficient for complete suppression of RTI, or whether diffusivity and viscosity were important ingredients?

To analyze this question more thoroughly, we decided to examine the simpler case of a constant density gradient, in other words the Rayleigh-Bénard problem. This led us to the discovery of previously unknown closed-form expressions for the stability boundary in Rayleigh-Bénard convection. Furthermore, we found that, in contrast to the conclusion of Chandrasekhar (1961), the inviscid system is always unstable independent of the strength of rotation. Finally, we concluded that for the complete suppression of Rayleigh-Bénard instability, non-vanishing viscosity and diffusion are both necessary ingredients; rotation alone is insufficient. By analogy we suggest that the complete suppression of RTI requires non-vanishing viscosity and diffusivity along with sufficiently strong rotation.

2. Numerical method and unperturbed background

Our experiments are of the incompressible Boussinesq equations using a finite-difference staggered-mesh code with third-order Runge-Kutta time stepping. The details of the numerical methods employed are explained at length in Orlandi (2000). The computational domain is a cube, with vertical z-axis aligned along the direction of gravity $(-g\hat{\mathbf{z}})$ and the angular rotation vector $(\Omega\hat{\mathbf{z}})$. The lateral (x, y) boundary conditions are periodic, while at bottom and top there are flat free-slip boundaries. The unperturbed density profile is explicitly

$$\bar{\rho}(z) = \rho_1 + (\rho_2 - \rho_1)(1 + \tanh(z/l))/2, \tag{1}$$

with l fixed as $L/80$ where L is half of the domain height and $\rho_1 < \rho_2$. All of the results presented here are in non-dimensional units with length scaled by L (the vertical and horizontal coordinates run from -1 to 1), velocity scaled by $V_g = \sqrt{2LgA}$ where A is the Atwood number given by $A = (\rho_2 - \rho_1)/(\rho_2 + \rho_1)$, and time by $T = L/V_g$. To ensure the validity of the Boussinesq approximation, the Atwood number is assumed to be small. We will report the density ρ in terms of a scaled variable θ that varies between 0 and 1:

$$\theta = (\rho - \rho_1)/(\rho_2 - \rho_1). \tag{2}$$

An important non-dimensional quantity for this study is the Rossby number $\mathrm{Ro} = V_g/fL$, where $f = 2\Omega$. For all of the simulations presented here, the Reynolds number, which we define as $\mathrm{Re} = LV_g/\nu$ is 5000. We assume a single species flow with a Prandtl number $\nu/\kappa = 1$, where ν is the kinematic viscosity and κ the diffusivity, both assumed constant.

3. Rotational suppression of the growth of the mixing zone

The first simulations that we will present are designed to demonstrate the degree to which rotation delays RTI. Four simulations are presented with $\mathrm{Ro} = \infty$, 1/4, 1/8 and 1/12 (where $\mathrm{Ro} = \infty$ implies no rotation). In each case, the initial condition was a state of no motion. The initial density field was as described above plus uniformly distributed random noise on each grid point in the range $-.05 < z < .05$ with an rms level lest than 10% of the mean density. Figure 1 illustrates the efficacy of rotation in damping the growth of the turbulent mixing zone. The figure is a vertical cross section of the three-dimensional density field at time $t = 5$. Panel (a) shows the extent and nature of the mixing zone with no rotation. It shows a well developed turbulent mixed zone with many well formed mushroom shaped bubbles and elements of high/low density fluid that have been carried into the regions of low/high density. Similarly for panel (b) except that with Ro=1/4, the development of the mixing zone has not proceeded as far. As illustrated in panel (c), with Ro=1/8, no mixing zone has been established by $t = 5$; there are no fully developed bubbles, and the vertical extent of the perturbations is much reduced compared to that in panels (a) and (b). Finally in panel (d) we see that the effect of rotation with Ro=1/12 has diminished the growth of the perturbation so significantly that there is little that can yet be identified as bubble formation.

4. Suppression of the growth of a single bubble

In an attempt to understand the role of rotation in suppressing the growth of the mixing zone, we have performed a series of simulations of the growth

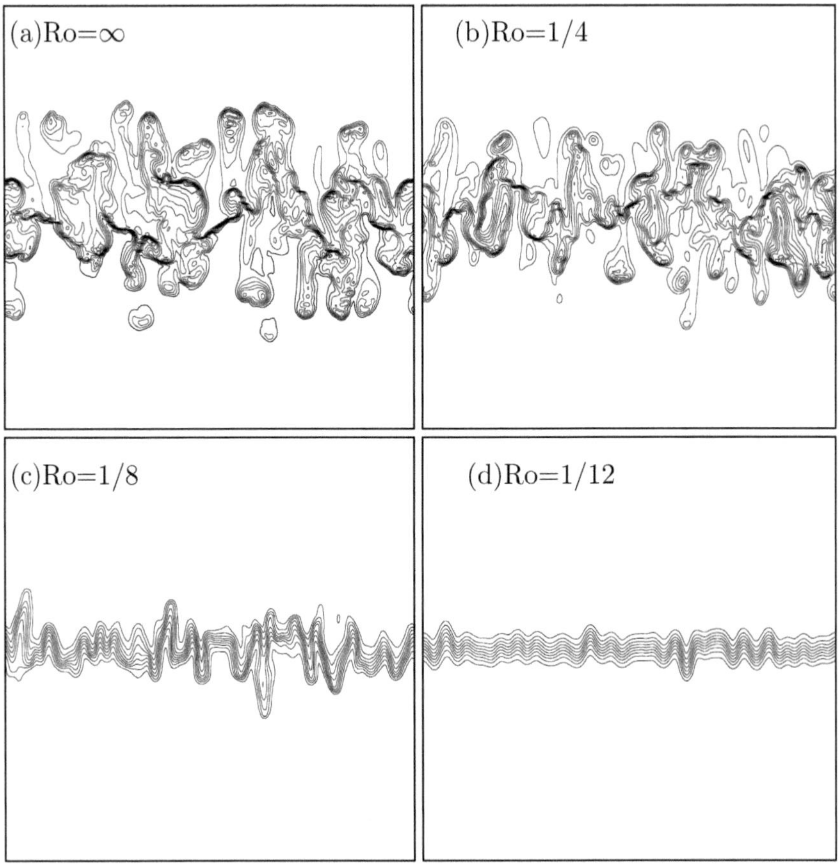

Figure 1. Contour plots of the scaled density θ in a vertical cross section ($y-z$ plane) at time t=5. The contour increment is $\Delta\theta = 0.1$. Both the vertical axis z and the horizontal axis y range from -1 to 1. (a) Ro=∞, (b) Ro=1/4, (c) Ro=1/8, (d) Ro=1/12.

of a single isolated bubble for different values of Ro. Each simulation was initialized with a circularly symmetric perturbation which bent the density iso-surfaces slightly upward in the middle of the domain. Specifically, we added the following small-amplitude perturbation to the unperturbed scaled density θ in (1):

$$\theta' = ae^{(-x^2-y^2)/b^2}e^{-z^2/c^2} \tag{3}$$

with $a = -0.025$, $b = 0.05$ and $c = 0.01$. These values were chosen through experimentation which showed that the proximity of the walls did not affect the growth of the bubble. In Figure 2, we show a comparison between the state of development for the resulting bubble at $t = 4$ for Ro=∞ and

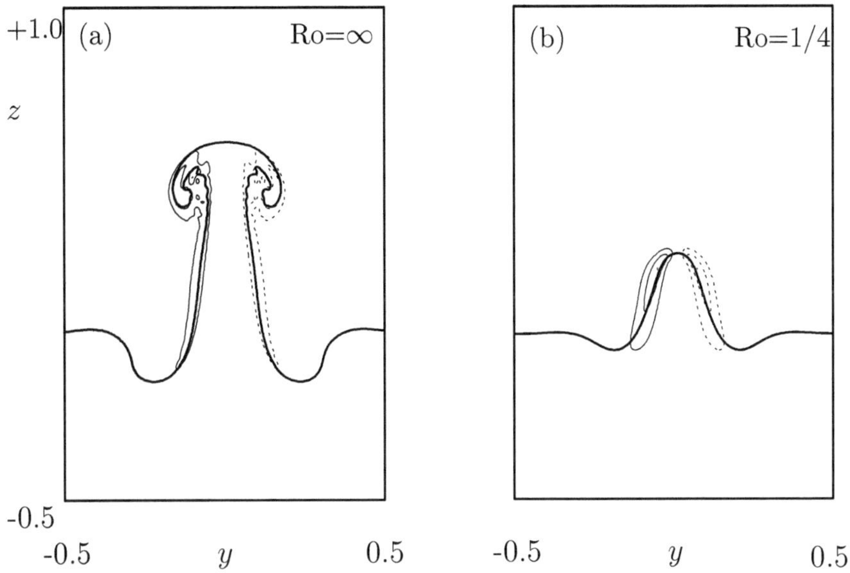

Figure 2. Isolated bubble and associated vorticity component ω_x in a $y-z$ cross section through the center ($x = 0$) of the bubble at t=4. The thick line is the $\theta = 0.5$ contour. The thin solid/dashed lines are contours of positive/negative ω_x. Only a portion of the computational domain is shown. (a) Ro=∞, vorticity contour interval 7.5 (b) Ro=1/4, contour interval 2.5.

Ro=1/4. The figures are vertical cross sections ($y - z$ plane) through the center of the bubbles. In addition to indicating the position of the middle of the density gradient (the θ=0.5 contour line), we also show contours of the x-component of the relative vorticity ω_x. In the case with no rotation (panel (a), Ro= ∞), by $t = 4$ a mushroom-like cap has formed on the advancing bubble and there is a strong vortex ring inside this cap. In contrast, in the case with rotation (panel (b), Ro=1/4), at $t = 4$ a cap structure has not yet formed, and the vorticity field is much weaker. Clearly rotation slows the growth of the bubble.

To gain some insight into the role that rotation plays in slowing the advance of a single bubble, it is convenient to introduce cylindrical coordinates as in Figure 3b. The bubble grows from the initial circularly symmetric (ϕ-independent) perturbation of the iso-density surface shown in Figure 3a. The arrows indicate the velocities induced by the action of gravity or, in other words, the flow induced by the baroclinic torque. This torque creates $\omega_\phi > 0$ in between the opposing arrows, which is the start of the formation of the vortex ring shown shown in panel (c). A vertical cross section is taken through the center of the ring to better reveal the flow structure. The small arrows on the leftmost circle, and the extended solid arrows on the rightmost circle show the direction of the flow induced by ω_ϕ. Once there is a horizontal flow component, the Coriolis force will act, and its effect will be to turn horizontal velocities to the right of their instantaneous direction. For the upper/lower arrow on the rightmost circle this means that the flow is turned out/in from the plane of the figure, as we have indicated with the dashed arrows. This secondary velocity field constitutes a new component of vorticity $\omega_r > 0$, which never occurs in the Ro$= \infty$ case. The Coriolis force continues to bend the flow 'to the right.' If we now imagine that the dashed arrows are bent to their right, we see that the Coriolis force induces a flow in a direction directly opposite to that associated with the original vortex. Thus the Coriolis force tends to create a vorticity component ω_ϕ of opposite sign to that of the ω_ϕ initially created by the density disturbance. Figure 3d displays the forces acting to create the flow shown in panels (c).

One can see this effect in the Boussinesq equations most easily when they are written in cylindrical coordinates. There is a simplification since the fields for the single bubble problem are independent of the angle ϕ for all time. A further simplification is the assumption that the the initial diameter of the bubble is so large that, in a vertical cross section, the flow near the center of the core of the initial vortex ring would be nearly circular. Furthermore, with this assumption we can ignore the self advection of the ring which diminishes with ring diameter. Finally, assuming the background density gradient to be uniform across the ring core, for flow near the center of the core we can write

$$\frac{\partial^2 \omega_\phi}{\partial t^2} = -\frac{1}{2}(N^2 + f^2)\omega_\phi. \tag{4}$$

Here we have used the traditional notation

$$N^2 = -\frac{g}{\rho_o}\frac{\partial \bar{\rho}}{\partial z}, \tag{5}$$

but must add the caution that in this case N^2 is negative (i.e. the Brunt-Väisällä frequency N is imaginary). Equation (4) suggests that for f^2 larger than $|N^2|$ growth of the perturbation will be suppressed. In our single bubble simulations, we found complete suppression for Ro$= 1/24$, and for the simulations with random initial density perturbations as in Figure 1, we found complete suppression with Ro$= 1/48$.

Although helpful in providing a mechanistic understanding of the initiation of the instability, (4) includes neither viscosity nor diffusion and is strictly valid only for very large idealized vortex rings in a uniform background density gradient. It remained unclear whether viscosity or diffusivity were necessary elements to achieve suppression. To understand this

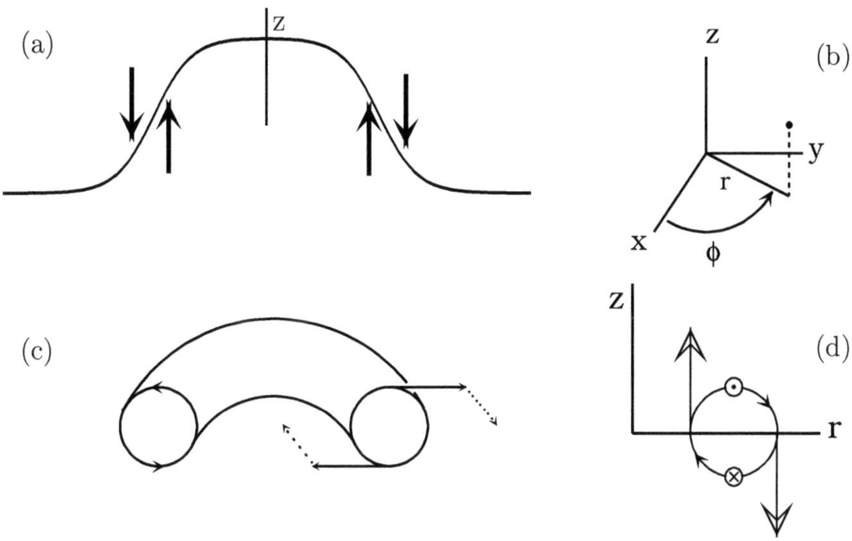

Figure 3. Schematic illustrating the effect of Coriolis force in the early evolution of a bubble. (a) is a vertical cross section illustrating the effect of the baroclinic torque exerted on the fluid due to the initial bending of an isopycnal. The solid arrows are the velocities induced by the action of gravity. This creates a vorticity component ω_ϕ around the interface in between each pair of arrows creating the vortex ring shown in (c), where a vertical cross section is taken to reveal the flow structure. In (c) the extended solid arrows on the rightmost circle, show the direction of flow induced by ω_ϕ. The dashed arrows, which represent flow directly into and out of the page, show the change in velocity induced by the Coriolis force. The application of this force to that secondary flow creates a flow directly counter to the original flow indicated by the solid arrows. (d) displays the forces acting to create the flow. The large circle is a vertical cross section of the vortex ring. The large arrows represent the buoyancy forces. $\odot$ and $\otimes$ indicate the Coriolis force out of and into the page respectively.

better we turned to the more tractable question of stability of a system with a constant temperature gradient β between free-slip top and bottom boundaries a vertical distance d apart. The lateral boundaries are taken to be infinitely far removed. This is the Rayleigh-Bénard problem. Chandrasekhar (1961) claimed that the inviscid Rayleigh-Bénard system could be stabilized with rotation, but Howard (1962) cast some doubt on this claim. Given this confusion, we decided to revisit the question of stability in the Rayleigh-Bénard problem.

5. Linear stability of the Rayleigh-Bénard system

The linear stability of this system is investigated analytically by introducing temperature and vertical velocity and perturbations proportional to

$$\exp\left[pt + \mathrm{i}(k_x x + k_y y)\right] \sin\left(n\pi z/d\right)$$

and horizontal velocity components that vary instead with $\cos(n\pi z/d)$. The vertical wavenumber takes the values $n = 1, 2, \cdots$. Chandrasekhar (1961) showed that linear stability is determined by a cubic polynomial in the exponential time-factor p with coefficients that are functions of $k_x, k_y, n, d, \Omega, g\alpha\beta$, where α is the coefficient of compressibility, κ the coefficient of thermal diffusivity and ν the kinematic viscosity. The cubic equation is (equations 239-240 in chapter III, §29)

$$\tilde{p}^3 + B\tilde{p}^2 + C\tilde{p} + D = 0, \tag{6}$$

where $\tilde{p} = (d^2/\nu)p$. The coefficients B, C and D are functions of

$$R = \frac{g\alpha\beta d^4}{\kappa\nu}, \quad T = \frac{(2\Omega)^2 d^4}{\nu^2} \quad \text{and} \quad P = \frac{\nu}{\kappa}, \tag{7}$$

the Rayleigh number, Taylor number and Prandtl number, respectively, n the vertical wavenumber and

$$a = |\mathbf{k}_\mathrm{h}|d = (k_x^2 + k_y^2)^{1/2} d$$

the non-dimensional horizontal wavenumber. Note that p has been non-dimensionalized with the viscous time scale d^2/ν, which was the choice Chandrasekhar made.

R, T and P characterize the system and a and n the perturbations. If for given $\{R, T, P\}$ for all perturbations the three roots of (6) have Re $p < 0$ there is stability. When for certain perturbations there is at least one root with Re $p > 0$ then there is instability. With each root of the cubic there is an associated combination of a flow field and temperature distribution, which we refer to as 'modes' although not explicitly considered here. The surface in the space spanned by R, T and P separating stable systems from unstable systems defines the marginally stable states. As fully explained by Chandrasekhar (1961), when crossing from the stable to the unstable side of this surface, instability can set in as stationary convection in which case one root of (6) is $p = 0$, or in an oscillatory fashion when there are two purely imaginary, complex conjugate roots. The latter case is in some places referred to by Chandrasekhar as a case of overstability, whereas in other places overstability means cases of complex conjugate roots with Im $p \neq 0$ and Re $p > 0$, i.e. instability in the form of oscillations of increasing amplitude. We will in what follows also say that there is overstability when Im $p \neq 0$ while Re $p = 0$ and call the associated modes 'overstable modes'. Modes associated with $p = 0$ we refer to as 'convective modes'.

5.1. THE MARGINAL STABILITY BOUNDARY

The marginal stability boundary in the parameter space spanned by $\{R, T, P\}$ can be determined by examination of the coefficients B, C and D without actually solving the cubic. In Kloostrerziel and Carnevale (2003), we showed how the stability of the system to a perturbation with given a and n is entirely determined by the signs of the coefficient D and the combination $BC - D$. Using this information when considering all possible values of n and a, we were able to divide up the parameter space into regions where the signs of D and $BC - D$ guaranteed stability or instability. The boundaries of these regions are determined by the surfaces on which D and $BC - D$ may vanish for some perturbation. Assuming a given value of P, the stability boundary can be represented as a curve in the $\{R, T\}$ plane. For P greater than a critical value P_c, the stability boundary is entirely given by the curve that is the locus of points for which it is possible for D to vanish for some value of the parameters. This curve is given by the simple expression

$$T = R\left[\sqrt{\frac{R}{R_c}} - 1\right] \tag{8}$$

where $R_c = (27/4)\pi^4 \approx 657.5$. On this curve, in addition to damped modes, one will find stationary convective modes $\tilde{p} = 0$, Hence, we will refer to this curve as the 'convection curve' and label is as $T_c^{(c)}$, the subscript c referring to the critical nature of this T for a given R and P, and the superscript refers to convection. For $\{R, T\}$ to the right of the curve (8) there are a and n for which $D < 0$ and there will be unstable modes $\tilde{p} > 0$. When $D > 0$ the stability of the flow is determined by the sign of $BC - D$. The curve that determines where this combination may vanish for some n and a is given by

$$T = R\left[\left(\frac{1+P}{2^3P^4}\right)^{1/2}\sqrt{\frac{R}{R_c}} - \left(\frac{1+P}{2P^2}\right)\right]. \tag{9}$$

On this curve one will find 'overstable modes' with zero growth rate, that is purely oscillatory modes, as well as damped modes. This will be referred to as the 'overstability curve' and denoted $T_c^{(o)}$.

How these two curves (8) and (9) determine the stability boundary is shown in Figures 4a–b. When $P \geq P_c$ the 'boundary' consists of the

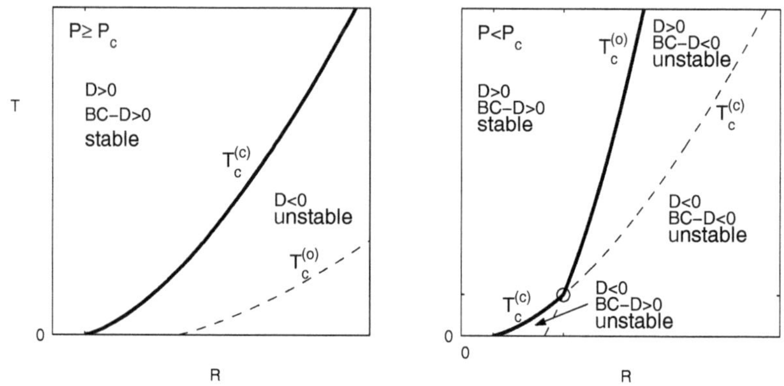

Figure 4. Graphs summarizing the stability properties of the system (a) for $P \geq P_c$ and (b) for $P < P_c$ where $\circ$ indicates the intersection point $\{R_i, T_i\}$. In each figure the solid curve is the marginal stability boundary.

convection curve alone. This is drawn as a thick line in Figure 4a. When $P < P_c$ the boundary (thick line in Figure 4b) is composed of the convection curve (8) for $R_c \leq R \leq R_i$ (defined below) and the overstability curve (9) for $R > R_i$. For all points $\{R, T\}$ to the left of this boundary $D > 0$ and $BC - D > 0$ for perturbations with any $\{a, n\}$, which implies stability.

Since

$$(1+P)^{1/2}/(2^3P^4)^{1/2} = 1 \quad \text{for} \quad P = P_c \approx 0.67659, \tag{10}$$

the coefficient multiplying $R^{3/2}$ in (9) is smaller than unity when $P > P_c$. In that case (9) stays to the right of the convection curve (8) as shown in Figure 4a. When $P < P_c$, (8) and (9) intersect at

$$R_i = (1+\gamma)^2 R_c, \quad T_i = \gamma(1+\gamma)^2 R_c, \quad \gamma = \frac{2^{1/2}(1+P) - (1+P)^{1/2}}{(1+P)^{1/2} - 2^{3/2}P^2}, \tag{11}$$

which is found by equating (8) to (9). Chandrasekhar (1961) believed that there was no simple formula like (11) for the intersection point. He calculated it numerically for several P-values. Comparison of his results with the exact values given by (11) revealed errors mostly on the order of 1% with one notable exception where it was almost 6%. More accurate results were provided recently by Cox and Matthews (2001).

Although Chandrasekhar (1961) was able to calculate these curves numerically, we have found it very useful to finally have analytic expressions for them. For example, in Kloosterziel and Carnevale (2003) we were able to derive the critical point on the boundary that determines whether an increase in viscosity with no other changes to the system will be stabilizing or destabilizing. Similarly we also found the critical point determining whether an increase in diffusivity would be stabilizing or destabilizing.

5.2. THE LIMIT $\nu \to 0$

In taking the limit $\nu \to 0$ with a non-zero, fixed κ, care must be exercised because R, T and the coefficients B, C and D become infinite. This singular behavior of the cubic occurs because we chose the time scale d^2/ν to non-dimensionalize p. This does not occur if the cubic is non-dimensionalized with the diffusive time scale d^2/κ. Then letting $\nu \to 0$ we get

$$\tilde{p}^3 + (a^2 + n^2\pi^2)\tilde{p}^2 + \left[\frac{T'n^2\pi^2 - Ga^2}{a^2 + n^2\pi^2}\right]\tilde{p} + T'n^2\pi^2 = 0 \quad \text{with} \quad \tilde{p} = (d^2/\kappa)p. \tag{12}$$

The new non-dimensional numbers are

$$T' = \frac{(2\Omega)^2 d^4}{\kappa^2} \quad \text{and} \quad G = \frac{g\alpha\beta d^4}{\kappa^2}. \tag{13}$$

By examining the coefficients, one can then show that the system is always unstable (there is always a real negative root $\tilde{p} = -d$ and two roots with Re $\tilde{p} > 0$). There are no convective modes ($\tilde{p} = 0$ is not a root because $D \neq 0$) and no purely oscillatory modes ($\tilde{p} = \pm i\omega$ are not roots because $BC - D \neq 0$). One can also show that in the limit of zero diffusivity ($\lim \kappa \to 0$) the system is always unstable.

6. Summary and discussion

We began by demonstrating numerically that background rotation could be used to completely suppress Rayleigh-Taylor instability. In other words, we showed that with sufficient rotation, no bubbles of heavy or light fluid would form although the density gradient would diffuse slowly due to the presence of Laplacian diffusivity in the simulations. We examined the effect of rotation on the early formation of bubbles, and showed that the Coriolis effect alone could prevent the growth of idealized large-diameter vortex rings in a simple inviscid model. This led to the question of whether rotation alone could suppress the Rayleigh-Taylor instability or was viscosity and or diffusivity also necessary?

To answer this question, we revisited the classical linear stability problem for Rayleigh-Bénard convection in a rotating system with flat stress-free boundaries. Chandrasekhar sought to describe the convection curve and overstability curve in the RT plane as curves $R_c^{(c)}(T)$ and $R_c^{(o)}(T, P)$, respectively. He took the Taylor number as the independent variable. No closed-form formulae for the curves defining the marginal stability boundary were noted by him for the stability boundary or the intersection point of the convection curve with the overstability curve. By switching to the Rayleigh number R as the independent variable, we have found rather simple expressions for the convection curve $T_c^{(c)}(R)$ (8) and the overstability curve $T_c^{(o)}(R, P)$ (9). Similarly, in Kloosterziel and Carnevale (2003), we also give closed form expressions for the critical horizontal wavenumber for the onset of stationary convection and oscillatory convection, as well as the frequency of the oscillations. None of these simple expressions appear to have been noted before. They enabled us to derive the exact expression for the point $\{R_i, T_i\}$ (11) beyond which the overstability curve determines the stability boundary and the critical points $\{R_c^\nu, T_c^\nu\}$ and $\{R_c^\kappa, T_c^\kappa\}$ on the stability boundary beyond which an increase in viscosity or diffusivity destabilizes marginally stable systems.

When the Taylor number is used as the independent variable, the convection curve and the overstability curve can also be expressed in closed-form, but the formulae are much more complicated. For example, the equation (8) can be written as a cubic in $\sqrt{R}$. This cubic can then be solved by the method of Tartaglia yielding

$$R = R_c \left(\frac{1}{3} + \frac{1}{6} F^{1/3} + \frac{2}{3} F^{-1/3} \right)^2 \tag{14}$$

where

$$F = 8 + 108(T/R_c) + 12\sqrt{81(T/R_c)^2 + 12(T/R_c)}.$$

Alternatively, another form of the solution to the cubic can be obtained by using the properties of the hyperbolic functions (cf. Kloosterziel and Carnevale, 2003):

$$R = R_c \left\{ \frac{1}{3} + \frac{2}{3} \left(1 + 6(T/R_c)\right)^{1/2} \cosh \left[\frac{1}{3} \operatorname{arccosh} \left(\frac{2 + 18(T/R_c) + 27(T/R_c)^2}{2(1 + 6(T/R_c))^{3/2}} \right) \right] \right\}. \tag{15}$$

These expressions are both equivalent to our simple result (8), although clearly more complicated in form. The expression for the overstability curve with R given as a function of (T, P) follows from (14) or (15) by replacing R and T with $R/(2+2P)$ and $P^2T/(1+P)^2$, respectively. It would be difficult to find the intersection point (11) by equating these new expressions for the convection and overstability curves, whereas it was rather straightforward when they are expressed with R and P as the independent variables.

Chandrasekhar (1961) in §24 stated that "in contrast to non-rotating fluids, an inviscid fluid in rotation should be expected to be thermally stable for *all* adverse temperature gradients. Indeed, only in the presence of viscosity can thermal instability arise", but this is wrong because as we have shown in §5.2 the inviscid system is *always* unstable. Chandrasekhar (cf.

§32) treated R and T in this limit as numbers that remain finite, while the Prandtl number $P = \nu/\kappa$ tends to zero. The problem is that the time scale d^2/ν and the Rayleigh and Taylor number become indeterminate in the limit $\nu \to 0$. In his review of Chandrasekhar's monograph (Howard, 1962) already voiced some doubts about Chandrasekhar's assertion. We have found that if either viscosity or diffusivity vanishes, the Rayleigh-Bénard system is always unstable, no matter how large the rotation. However, for nonzero viscosity and diffusivity, there is always a critical value above which rotation will suppress convective instability. This also suggests that diffusivity and viscosity are both necessary for the complete suppression of Rayleigh-Taylor instability with rotation in the manner discussed above.

Acknowledgements

We thank a very helpful anonymous reviewer for suggesting the inclusion of (14) and for other useful comments. This work has been supported by Office of Naval Research grants N00014-97-1-0095 and N00014-96-0762, National Science Foundation grants OCE 97-30843, OCE 97-30843, OCE 01-28991 and OCE 01-29301, the U.S. Department of Energy through the University of California Lawrence Livermore National Laboratory under contract No. W-7405-Eng-48, the Università di Roma "La Sapienza" under the grant MURST 60% and the San Diego Supercomputer Center.

References

Carnevale, G.F., P. Orlandi, Y. Zhou and R. C. Kloosterziel. Rotational Suppression of Rayleigh-Taylor Instability. *J. Fluid Mech.*, 457:181–190, 2002.

Cox, S.M. and P. C. Matthews. New Instabilities in Two-dimensional Rotating Convection and Magnetoconvection. *Physica D.*, 149:210–229, 2001.

Chandrasekhar, S. *Hydrodynamic and Hydromagnetic Stability.* Oxford University Press, 1961.

Howard, L.N. Review of: *Hydrodynamic and Hydromagnetic Stability* by S. Chandrasekhar. *J. Fluid Mech.*, 12:158–160, 1962.

Kloosterziel, R.C. and G. F. Carnevale. Closed-form Linear Stability Conditions for Rotating Rayleigh-Bénard Convection with Rigid Stress-free Upper and Lower Boundaries. *J. Fluid Mech.*, 480:25–42, 2003.

Orlandi, P. *Fluid Flow Phenomena: A Numerical Toolkit.* Kluwer Academic Publishers, Boston, 356 pp., 2000.

Verzicco, R., P. Orlandi, A. H. M. Eisenga, G. J. F. van Heijst, and G. F. Carnevale. Dynamics of a Vortex Ring in a Rotating Fluid. *J. Fluid Mech.*, 317:215–39, 1996.

ADVECTION BY INTERACTING VORTICES ON A β PLANE

O. U. VELASCO FUENTES
Departamento de Oceanografía Física, CICESE
Ensenada, Baja California, México

Abstract. Particle advection by two equal vortices under the influence of a background-vorticity gradient (β) is studied using a dynamical systems approach. The velocity field is a data set obtained by numerically solving the Euler equation with a vortex-in-cell model. Two methods are used to identify finite-time invariant manifolds: the first one relies on the use of Eulerian information and applies to flows with slowly moving stagnation points; the second one combines Eulerian and Lagrangian information and does not depend on the existence of stagnation points. The invariant manifolds are used to quantify the efficiency of merger, which is defined as the ratio of the area of the resultant vortex to the total area of the original vortices. It is found that the original vortices always contribute unequally to the merger or exchange processes. When the vortices are cyclonic the one located to the pole or west dominates the interaction; and when the vortices are anticyclonic, the one located to the equator or west is dominant.

Key words: Vortices, merger, β plane, chaotic advection.

1. Introduction

The interaction of two vortices in a two-dimensional fluid has been the subject of intense research for the last three decades. Early numerical studies considered equal vortices with step-like vorticity distribution (i.e. uniform vorticity inside the vortex and null outside) and were mainly concerned with determining the critical distance for merger (Zabusky *et al.*, 1979). Later studies have considered less idealized conditions; for instance the asymmetry of the interacting vortices, either in the vortex size or the amplitude of the vorticity (Melander *et al.*, 1987; Dritschel and Waugh, 1992; Yasuda and Flierl, 1997).

In a previous paper (Velasco Fuentes and Velázquez Muñoz, 2003, hereafter referred to as VFVM) we studied the effect of a gradient of background vorticity (β) on the interaction of two equal vortices. We identified and characterized regimes of behavior, and explained some of the mechanisms that lead to their existence. In the present paper I will study the advection of particles in the velocity field of these vortices, using ideas and methods

O.U. Velasco Fuentes et al. (eds.), Nonlinear Processes in Geophysical Fluid Dynamics, 339-353.

from the theory of transport in dynamical systems. The main objective is to understand the process of mass exchange between the vortices and to quantify it. In this respect, this work is related to previous studies of efficiency of vortex merger (Waugh, 1992; Dritschel and Waugh, 1992) and, in particular, with my previous work on the chaotic advection by equal vortices (Velasco Fuentes, 2001, hereafter referred to as VF).

Although here ($\beta \neq 0$) as well as in VF ($\beta = 0$) the flow is aperiodic, there are essential differences in the flow evolution which call for the use of different methods of analysis. When $\beta = 0$ the geometry of the instantaneous velocity field is known, in a qualitative way, for the whole evolution. In the initial stages the flow has three stagnation points of hyperbolic type when viewed in a frame of reference moving with the vortices. One of these stagnation points holds a fixed position between the vortices while the other two rotate quasi-steadily around it at an approximately constant distance. The flow will behave in this way permanently if the vortices are located beyond the critical distance for merger. Otherwise, the flow behaves as described above only in the initial stage, which is followed by a rapid transition leading to the destruction of the central hyperbolic stagnation point (the merger event). Finally there is a stage when the flow evolves slowly again. In this stage two stagnation points rotate quasi-steadily around the new vortex at an approximately constant distance.

The persistence of the stagnation points facilitates the analysis of transport. For in most cases the flow can be considered as the sum of a steady part and a small perturbation (with the time dependency being either quasi-periodic or arbitrary). In such cases some analytic results guarantee that, under certain conditions, hyperbolic trajectories exist in the neighborhood of the stagnation points (Haller and Poje, 1998; and Malhotra and Wiggins, 1998). This approach has been very succesful for computing transport templates for vortex-jet and vortex-vortex interaction problems (see e.g. Poje and Haller, 1999; and VF).

In contrast, when $\beta \neq 0$ the instantaneous flow geometry undergoes endless qualitative changes, with stagnation points ceaselessly appearing and disappearing at different locations. Most of these stagnation points will not leave their mark on the dynamics of particles, whereas some hyperbolic trajectories will exist where no stagnation points of the instantaneous velocity field are observed. Several methods have been proposed to deal with these types of flow (Mezić *et al.*, 1999; Haller, 2000; Haller, 2001). Here, I will use the theoretical results of Haller (2000) in order to detect the finite-time invariant manifolds.

The rest of the paper is organized as follows: section 2 summarizes the model equations, the numerical method and the regime definitions of VFVM, section 3 describes the techniques (and their numerical implement-

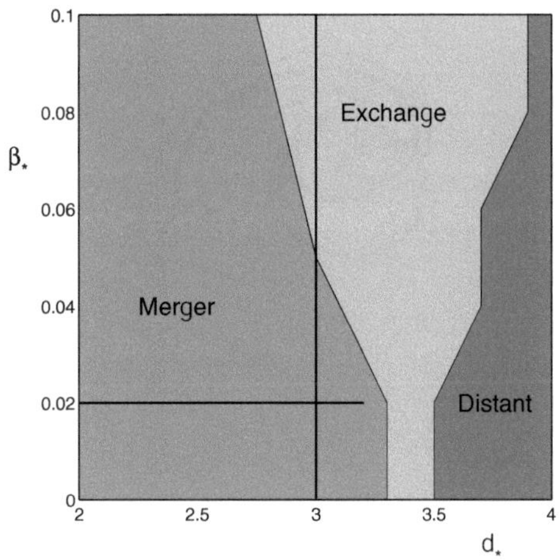

Figure 1. The regimes of interaction of a pair of equal vortices on the β plane according to VFVM (representative cases are shown in Fig. 2). The straight lines represent the initial conditions chosen here for the analysis of mass transport.

ation) used to construct the template for transport. This template is used in section 4 to compute the efficiency of vortex merger. Finally, I summarize the results and give some conclusions in section 5.

2. Regimes of interaction on the β plane

The β-plane approximation incorporates the combined effects of the Earth's rotation and curvature, but using a Cartesian geometry instead of the more complicated spherical one. It is obtained by expanding the equations of motion, in inverse powers of the Earth's mean radius, about a fixed point on the Earth's surface. The expansion is truncated and the first order variation of the Coriolis parameter is retained (the β term) and all metric terms are neglected (see e.g. Pedlosky, 1987; and Ripa, 1997).

The conservation of potential vorticity of an unforced, ideal fluid of uniform depth on the β plane is expressed by the following equation:

$$\frac{D}{Dt}(\omega + \beta y) = 0 \tag{1}$$

where $D/Dt = \partial/\partial t + u\partial/\partial x + v\partial/\partial y$ is the material derivative [with u and v the velocities in east (x) and north (y) directions, respectively], $\omega = \partial v/\partial x - \partial u/\partial y$ is the relative vorticity, and $\beta = 2\Omega_E \cos\phi_0/R_E$ (with Ω_E and R_E the Earth's angular speed and mean radius, respectively, and ϕ_0 the geographic latitude of reference).

Equation (1) is solved using the vortex-in-cell method (VIC). See Hockney and Eastwood (1981), for a general account of particle methods and

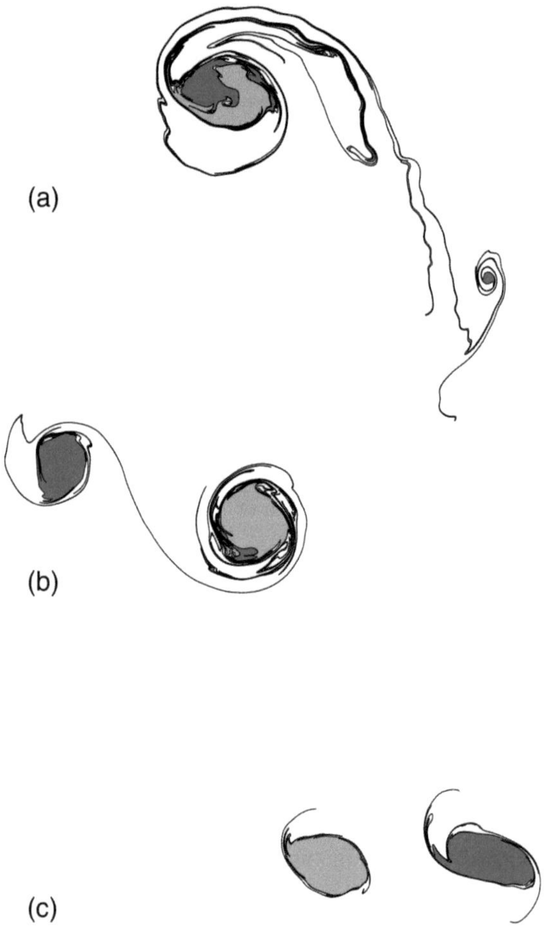

Figure 2. Typical outcome for initial conditions in each regime of behaviour. (a) Merger $[(d_*, \beta_*) = (3, 0.04)]$, (b) exchange $[(d_*, \beta_*) = (3.4, 0.02)]$ and (c) distant interaction $[(d_*, \beta_*) = (4, 0.04)]$. The west and east vortex are represented by light and dark gray, respectively. The elapsed time is $t = 5.6\tau$, where τ is the eddy turnover time.

VFVM for details of the implementation used here. In the initial condition the vortices are aligned in west-east direction, they are circular and have equal radius and relative vorticity (which is uniform inside the vortices and null outside). Under these conditions two parameters determine the evolution: the initial intercentroid distance d normalized by the vortex radius R ($d_* = d/R$), and the strength of β with respect to the intensity of the vortices ($\beta_* = \beta R/\omega$, where ω is the relative vorticity of the initial condition). Furthermore, in all numerical experiments the vortices have cyclonic relative vorticity and $\beta > 0$ (northern hemisphere convention). The effect of the vortices having different orientations and being initialized with equal absolute vorticity will be briefly discussed in section 4. The results are translated to anticyclonic vortices and Southern hemisphere convention in section 5.

VFVM identified three regimes in the parameter plane (d_*, β_*), as shown in Figure 1. The regimes are characterized as follows:

- Merger. The vortices rotate rapidly about each other while getting closer until a single vortex is formed. In the process, the two original vortices eject vorticity filaments (see Fig. 2a).
- Exchange. The vortices rotate about one other, they get closer and form filaments which are rolled-up around the partner. Later the vortices separate and their direction of rotation is usually reversed (see Fig. 2b).
- Distant interaction. The two vortices wobble about each other, they move farther apart and undergo changes in shape. Sometimes they form filaments but these remain in the vicinity of the vortex from which they originate (see Fig. 2c).

3. The template for transport

3.1. COMPUTATION OF PARTICLE TRAJECTORIES

The trajectories of fluid particles are the solutions of the following pair of ordinary differential equations:

$$\frac{dx}{dt} = \frac{\partial\psi(x,y,t)}{\partial y}, \qquad \frac{dy}{dt} = -\frac{\partial\psi(x,y,t)}{\partial x}. \tag{2}$$

where ψ is the stream function obtained by numerically solving equation (1) with the VIC method. Therefore, equation 2 must be integrated when the right hand side is only known at discrete spatial and temporal grid points. The database consists of $N+1$ *slices* of data, where the nth slice is the computed solution at time $t_n = n \cdot \Delta t$, for $n = 0, 1, \ldots, N$. Each slice is a 2D array of data defining the streamfunction on $M \times M$ grid points. In this study $N = 2000$, $\Delta t \approx \tau/125$ (where τ is the eddy turnover time), and $M = 256$ (with 10 grid points per vortex radius).

The integration of equation (2) thus requires interpolation in three dimensions to find the velocity at arbitrary points (x, y, t). For consistency, I made the interpolation in space with the same biquadratic scheme used in the VIC model, and the time integration with the same second-order Runge-Kutta scheme. In the latter case, however, the time integration requires either the use of a time-step which is twice as large as that used in the VIC model or the interpolation of the data slices, even if the solution of

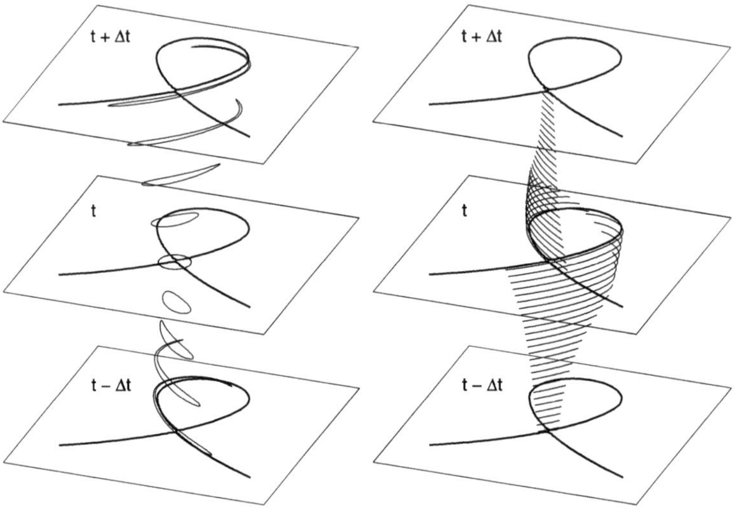

Figure 3. (a) Testing the existence of a hyperbolic trajectory. The streamfunction in a co-moving frame is computed at time $t = t_0$ and a small circle of particles is placed around the saddle stagnation point. If the contour is stretched when advected by the flow toward time $t = t_0 - \Delta t$ and $t = t_0 + \Delta t$ then there is a hyperbolic trajectory *close* to the saddle point. (b) Numerical computation of the manifolds at time $t = t_0$. Small line segments are placed along the unstable and stable directions of the saddle point of the Eulerian flow at times $t = t_0 - \Delta t$ and $t = t_0 + \Delta t$, respectively. The evolution of this segments toward time $t = t_0$ give the unstable and stable manifolds.

equation (1) is output at every time step available in the model. At the space and time resolution used here, the results of integrating equation (2) with a time step Δt and a linear interpolation of consecutive data slices are practically equal to results obtained with a time step $2\Delta t$ and no interpolation of data slices.

3.2. FINDING HYPERBOLIC TRAJECTORIES

3.2.1. *An Eulerian approach*

The starting point is the streamfunction $\psi(x, y, t_0)$ computed by the VIC model at some time t_0. This ψ is the one observed in a frame fixed to Earth. The stream function Ψ observed in a system that rotates with the vortices is given by the simple transformation $\Psi(x, y, t_0) = \psi(x, y, t_0) + \frac{1}{2}\Omega\left((x - x_c)^2 + (y - y_c)^2\right)$, where Ω is the angular velocity and (x_c, y_c) is the center of rotation of the vortices. A good estimation for the center of rotation is the middle point between the vortices (before merger takes place) or the center of the new vortex (after merger). A good estimation of Ω is obtained making any of the following assumptions: (a) Before merger the two vortices are taken to be point vortices located at the vorticity centroids, or (b) After merger the new vortex is taken to be an elliptic patch of uniform vorticity. In each case the angular velocity Ω is evaluated from the corresponding analytic expression: (a) $\Omega = \Gamma/(\pi d^2)$, where Γ is the circulation of each vortex and d is the intercentroid distance; and (b) $\Omega = ab\omega/(a+b)^2$, where ω is the relative vorticity and a and b the semiaxes of the approximated ellipse. More elaborated methods have been proposed to find the co-moving frame (see, e.g. Dritschel, 1995), but it must be stresses that, since the flow is time-dependent, there is no unique way to define a co-moving frame. Furthermore, all methods proposed in the litarature give similar results; in particular, the locations of the stagnation points do not change significantly with the method used. This is all we need to know at this stage: The neighborhood where a Lagrangian hyperbolic trajectory might exist.

The next step is to determine the geometry of the co-rotating stream function Ψ; that is to say, locate the stagnation points of center and saddle type and their associated manifolds. A wide variety of methods and tools for obtaining these geometric elements from a numerically generated vector field are available (see e.g. Helman and Hesselink, 1991, and VF for the implementation used here). Since Ψ is time dependent, the relation between the trajectories of fluid particles (Lagrangian dynamics) and the geometry of the instantaneous velocity field (Eulerian flow) is not obvious. We expect, however, that under certain conditions the character of the stagnation points of $\Psi(x, y, t_0)$ will manifest itself in the particle motion. Loosely speaking, this implies the existence of particle trajectories which behave like saddles (i.e. they exponentially attract a set of particles and exponentially repel another set) and of particle trajectories that behave like centers (i.e. they induce a swirling motion on particles around them).

If the stagnation point exists long enough and the velocity field around it changes slowly there exists a hyperbolic trajectory in its neighborhood. This can be proved rigorously (see Haller and Poje, 1998, for the theory, and Poje and Haller, 1999 and VF for examples of the use of the criteria). Alternately, it can be directly tested if the expected behavior takes place. Let us assume that we have found that $\Psi(x, y, t_0)$ possesses a saddle stagnation point $\vec{x}_0$ in the interval $t - \Delta t < t < t + \Delta t$ (where $\Delta t > 0$ is a small time interval). The existence of the hyperbolic trajectory can be tested by computing the evolution under $\vec{u}(x, y, t)$ of a set of initial conditions on a circle centered at $\vec{x}_0$ (see Figure 3a). Its evolution is computed from time t_0 to time $t_0 + \Delta t$ and from time t_0 to time $t_0 - \Delta t$. If the contour is subjected to stretching in transverse directions then there is a hyperbolic trajectory in the neighborhood of $\vec{x}_0$.

The actual location of the hyperbolic trajectory and its associated finite-time manifolds in the time slice $t = t_0$ can be determined in an equivalent way. The stable manifold is obtained computing the evolution, from time $t_0 + \Delta t$ to time t_0, of a short line which crosses the stagnation point of $\Psi(x, y, t_0 + \Delta t)$ in the attracting direction; and the unstable manifold is obtained computing the evolution, from time $t_0 - \Delta t$ to time t_0, of a short line which crosses the stagnation point of $\Psi(x, y, t_0 - \Delta t)$ in the repelling direction (see Figure 3b).

3.2.2. *A combined Lagrangian-Eulerian approach*

The main shortcoming of the method described above is that it relies completely on the existence of stagnation points. It works best for flows which can be divided into a steady state and an arbitrary perturbation which evolves slowly with respect to the particle motion. These conditions are valid (if not throughout the whole evolution, at least during significant time spans) for some realistic flows like the jet-vortex interaction studied by Poje and Haller (1999) or the vortex-vortex interaction on the f plane ($\beta = 0$) discussed in VF. More complicated flows, however, do not satisfy these conditions and call for the use of different approaches. Several methods have been proposed (Mezić *et al.*, 1999; Haller, 2000, Haller, 2001); here I will use the theoretical results of Haller (2000), which are based on the analysis of the linearized velocity field along particle trajectories. Before presenting Haller's algorithm, it is useful to define the Jacobian determinant along a particle trajectory:

$$J(t; \vec{x}_0) = \left| \frac{\partial \vec{u}[\vec{x}(t; \vec{x}_0), t]}{\partial \vec{x}} \right|$$

where $|\,\boldsymbol{.}\,|$ denotes the determinant and $\partial\vec{u}(\vec{x})/\partial\vec{x} \equiv \partial u_i/\partial x_j$ is the time-dependent Jacobian matrix of the velocity field $\vec{u}$, computed along the particle trajectory $\vec{x}(t)$ [by definition $\vec{x}(t_0) = \vec{x}_0$]. Haller's algorithm can be summarized as follows: (1) Consider a grid of initial conditions in the region of interest. (2) Integrate each initial condition $\vec{x}_0$ forward in time as long as $J(t; \vec{x}_0) < 0$. (3) Let $T_p(\vec{x}_0)$ denote the time for which $J(t; \vec{x}_0)$ stays negative [we define $T_p(\vec{x}_0) = 0$ for initial conditions having $J(t; \vec{x}_0) > 0$]. (4) Local extrema of the scalar field $T_p(\vec{x}_0)$ are candidates for the $t = t_0$ slices of local stable manifolds. If the integration is made backward in time from $t = t_0$, the local extrema of $T_n(\vec{x}_0)$ are the $t = t_0$ slices of local unstable manifolds.

4. Results

4.1. COMPARISON OF THE TWO METHODS

Figure 4 shows the stable manifolds at time $t = t_0$ for cases representative of each regime of interaction. The gray level is proportional to the time T_p

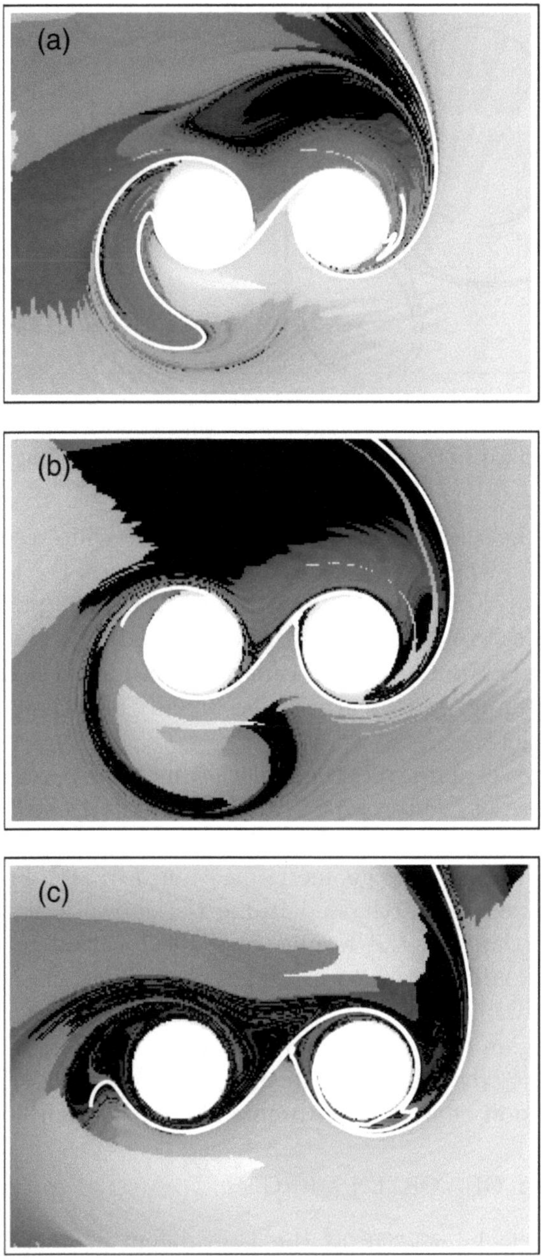

Figure 4. Map of T_p at $t = 0$ for various initial conditions (d_*, β_*), (a) (3, 0.04), (b) (3.4, 0.02), and (c) (4, 0.04). The darkest region indicates the longest time; the white lines indicate the stable manifolds computed with the Eulerian method (see text).

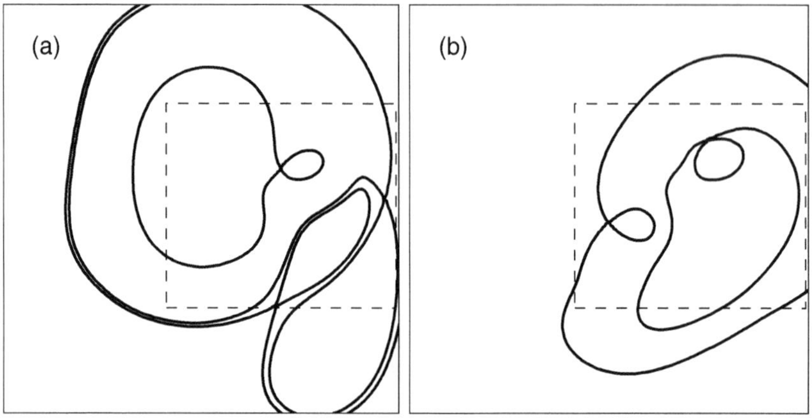

Figure 5. Instantaneous Eulerian flow geometry for a vortex pair in the distant interaction regime $[(d_*, \beta_*) = (4, 0.04)]$ at time $t = 4\tau$, where τ is the eddy turnover time. (a) As seen in a frame fixed to Earth, (b) as seen in a system co-rotating with the vortices.

computed with the Eulerian-Lagrangian method (white represents $T_p = 0$ and black $T_p = 8\tau$, where τ is the eddy turnover time). For the computation of each graph, a grid of initial conditions with 280×200 points was defined over a region $11R \times 8R$ (where R is the vortex radius). The white lines are the manifolds as computed by the Eulerian method with $\Delta t = 2.4\tau$. It is clear that at this stage both methods are equally effective at detecting the Lagrangian geometry.

As time increases, however, the Eulerian method becomes increasingly ineffective because of the secondary vorticity field (Rossby waves). This is specially true for larger values of β_*. As an example, Figure 5 shows the instantaneous Eulerian flow geometry at time $t = 4\tau$. Frame (a) shows it in a system fixed to Earth, and frame (b) shows it in a system moving with the vortices. In the former case the location of the vortices is not obvious; in contrast, the co-moving streamfunction shows them very clearly, together with two hyperbolic stagnation points. The results of the Eulerian-Lagrangian method, however, shows four points where hyperbolic trajectories might cross this time slice (Figure 5). None of these points is in the neighborhood of stagnation points of the Eulerian flow.

4.2. EFFICIENCY OF VORTEX MERGER

In this section the knowledge of the Lagrangian geometry is applied to the computation of efficiency of vortex merger and exchange. The relative position of a particle with respect to the manifolds at any given time determines the qualitative character of its whole evolution, past and future. In particular, a detailed analysis of the intersections of the stable manifold

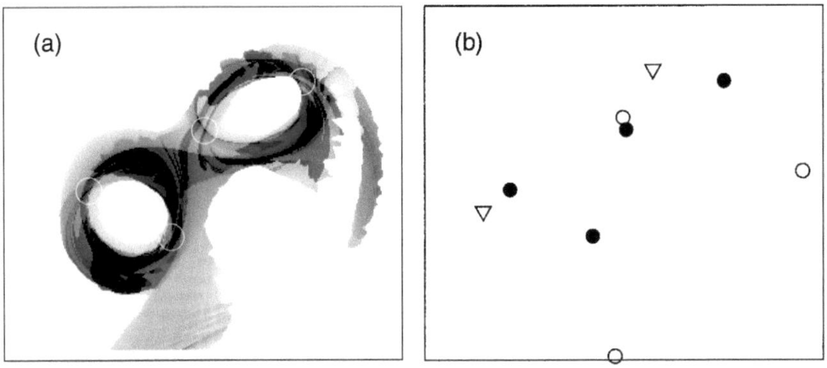

Figure 6. Location of hyperbolic trajectories for a vortex pair in the distant interaction regime $[(d_*, \beta_*) = (4, 0.04)]$. (a) Map of $T_p - T_n$; the white circles indicate the likely position of the hyperbolic points. (b) Hyperbolic points of the Lagrangian flow (black dots) and stagnation points of the Eulerian flow (circles: flow seen in a frame fixed to Earth, Figure 5a; triangles: flow seen in the co-rotating frame, Figure 5b). Time is $t = 4\tau$, where τ is the eddy turnover time.

with the vortices at time $t = 0$ enables us to compute the amount of fluid that is detrained as filaments. In this way it is possible to compute the efficiency of vortex merger or of mass exchange (depending on the regime in which a particular initial condition is located). Figure 7a shows the stable and unstable manifolds of the eastward hyperbolic point at time $t = 0$. Since the flow preserves topology any particle located to the right-hand side (when looking towards the hyperbolic point in the particle-flight direction) will remain on that side and, therefore, will be ejected from the vortex along the unstable manifold. Figure 7b shows, in darker gray, the patch that will be expelled from each vortex to form the filaments.

Figure 8a shows the area A expelled from the vortices as a function of the gradient of background vorticity β_* (for a fixed initial intercentroid distance $d_* = 3$). As β_* grows the area lost by the west vortex decreases while that lost by the east vortex increases. In this graph two regimes are represented: merger ($\beta_* < 0.05$) and exchange ($\beta_* > 0.05$). Note that the area detrained gives no indication of a regime transition. The efficiency ϵ (Figure 8b) is defined as follows: (a) in the merger regime it is the ratio of the area of the new vortex to the sum of the areas of the initial vortices, and (b) in the exchange regime it is the ratio of the area of the larger vortex to the sum of the areas of the initial vortices. The strong asymmetry in the mass expelled by each vortex as β_* increases cancels out and the efficiency ϵ remains close to its known value for $\beta_* = 0$ (0.8).

Figure 9a shows the area expelled from the vortices as a function of d_* (for a fixed $\beta_* = 0.02$). As d_* grows the area lost by the East vortex

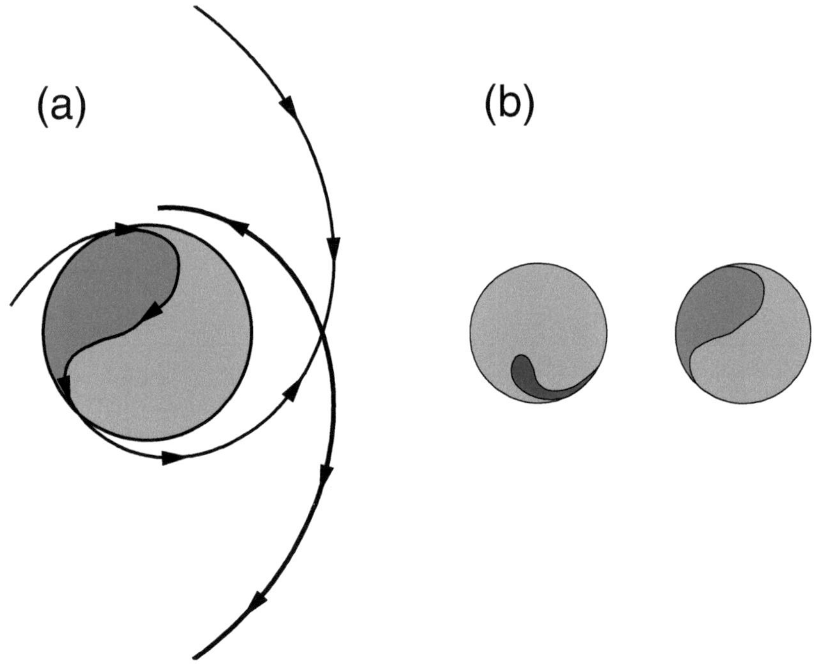

Figure 7. (a) Template for transport. Unstable and stable manifolds are represented by thick and thin lines, respectively; the arrows indicate flow direction. The dark gray area is the regions due to be detrained. (b) Detrainment regions of west and east vortex.

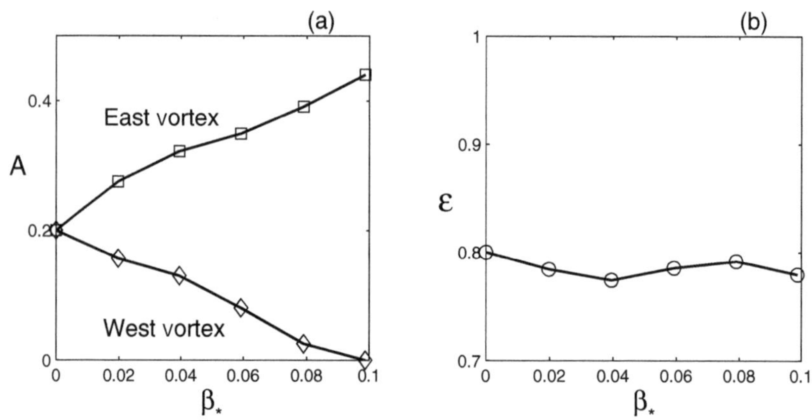

Figure 8. (a) Area A expelled from the vortices and (b) efficiency ϵ, as a function of β_* (and $d_* = 3$). Squares and diamonds indicate the vortex that is initially located to the east and west, respectively. A is given as a fraction of the initial area of each vortex, the efficiency ϵ is the fraction of the combined area that ends within the merged vortex or within the largest vortex (in the exchange regime).

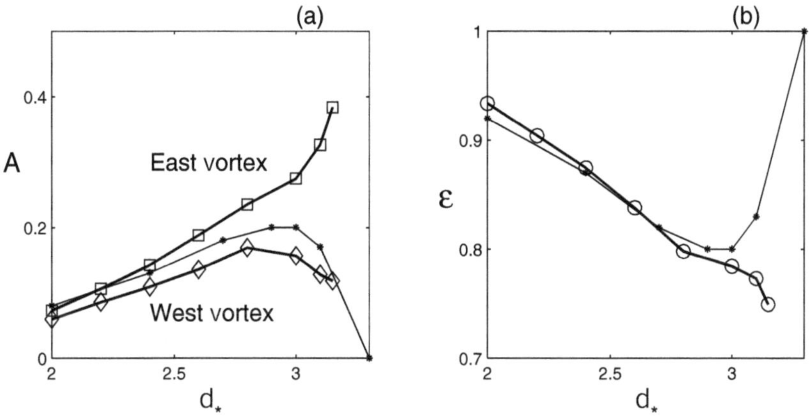

Figure 9. Same as Figure 8 but now as a function of the initial intercentroid distance d_*. The thick lines are for $\beta_* = 0.02$ and the thin lines for $\beta_* = 0$.

increases continuously, whereas the area lost by the west vortex increases to its maximum value at about $d_* = 2.8$ and the decreases. The efficiency of merger ϵ, also decreases continuously (Figure 9b).

The effect of the orientation of the vortices on the exchange of mass was investigated with two sets of experiments with vortices oriented in north-south direction. In the first one, the vortices were initialized with equal relative vorticity (thin lines in Figure 10) and in the second one the vortices were initialized with equal absolute vorticity (thick lines in Figure 10). The behavior is similar to the one described for WE oriented pairs, with the north vortex taking the dominant role. This happens for both types of initialization, although the asymmetry is stronger when the vortices have equal relative vorticity.

5. Conclusions

The exchange of mass between interacting vortices on the β plane has been studied using the theory of transport in dynamical systems. Two methods were used to construct the template for transport (i.e. the finite-time invariant manifolds). The first method uses purely Eulerian information and applies to flows with slowly moving stagnation points; the second one combines Eulerian and Lagrangian information and does not depend on the existence of stagnation points. The characteristics of the flow evolution make the Eulerian method best suited for the analysis of early stages, when the flow field is determined primarily by the vortices. But as time goes on the secondary vorticity field (Rossby waves) makes it increasingly difficult

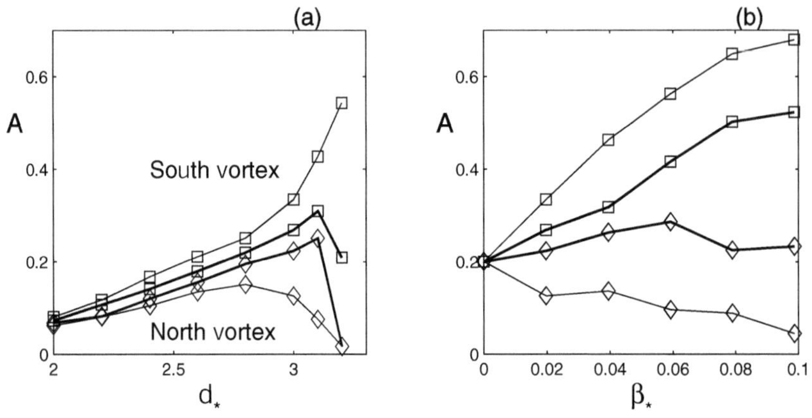

Figure 10. (a) Area A expelled from the vortices as a function of the initial intercentroid distance d_* (and $\beta_* = 0.02$). (b) A as a function of β_* (and $d_* = 3$). Squares and diamonds indicate the vortex that is initially located to the south and north, respectively; the thick and thin lines indicate experiments where vortices were initialized with equal absolute vorticity and equal relative vorticity, respectively. A is given as a fraction of the initial area of each vortex.

to apply it. In those stages one must resort to the Eulerian-Lagrangian method.

Once the invariant manifolds are identified, their geometry is analyzed and used to quantify the efficiency of merger. It is found that the interaction of equal vortices on the β plane is always asymmetric. During the interaction the vortices always are displaced from their original latitude, therefore one becomes stronger than the other owing to conservation of potential vorticity $(\omega + \beta y)$. The strongest vortex then dominates the interaction: In the merger regime the strongest vortex contributes with more mass and occupies the central region of the merged vortex; whereas in the exchange regime the dominant vortex takes more mass from the partner than the mass that is taken from it. A similar behaviour is observed in the f plane $(\beta = 0)$ when the vortices are initially unequal (Dritschel and Waugh, 1992).

When the vortices are cyclonic, the one located towards the west and pole dominates the interaction. When the vortices are anticyclonic, the one located towards the west and equator is the dominant one. Vortices which are initialized with equal absolute vorticity also behave as described above, although their asymmetry is smaller.

Acknowledgements

I thank José Luis Ochoa for useful comments on an earlier version of this paper. This work was partially supported by CONACyT (México) through Grant No. 28137-T.

References

Dritschel, D. A general theory for two-dimensional vortex interactions. *J. Fluid Mech.*, 293:269–303, 1995.

Dritschel, D. and D. Waugh. Quantification of the inelastic interaction of unequal vortices in two-dimensional vortex dynamics. *Phys. Fluids A*, 4:1737–1744, 1992.

Haller, G. Finding finite-time invariant manifolds in two-dimensional velocity fields. *Chaos*, 10(1):99–108, 2000.

Haller, G. Lagrangian structures and the rate of strain in a partition of two-dimensional turbulence. *Phys. Fluids*, 13(1):3376–3385, 2001.

Haller, G. and A. Poje. Finite time transport in aperiodic flows. *Physica D*, pp. 352–380, 1998.

Helman, J. and L. Hesselink. Visualizing vector field topology in fluid flows. *IEEE Comp. Graph. Appl.*, pp. 36–46, 1991.

Hockney, R. and J. Eastwood. *Computer Simulation Using Particles.* McGraw-Hill. 1981.

Malhotra, N. and S. Wiggins. Geometric structures, lobe dynamics, and Lagrangian transport in flows with aperiodic time-dependence with applications to Rossby wave flow. *J. Nonlin. Sci.*, 8:401–456, 1998.

Melander, M., N. Zabusky, and J. McWilliams. Asymmetric vortex merger in two dimensions: which vortex is "victorious?" *Phys. Fluids*, 30:2610–2612, 1987.

Mezić, I., and S. Wiggins. A method for visualization of invariant sets of dynamical systems based on the ergodic partition. *Chaos*, 9:213–218, 1999.

Pedlosky, J. *Geophysical Fluid Dynamics.* Springer-Verlag, 1987.

Poje, A. and G. Haller. Geometry of cross-stream mixing in a double-gyre ocean model. *J. Phys. Oceanogr.*, 29:1649–1665, 1999.

Ripa, P. Inertial oscillations and the β-plane approximation(s). *J. Phys. Oceanogr.*, 27:633–647, 1997

Velasco Fuentes, O.U. Chaotic advection by two interacting finite-area vortices. *Phys. Fluids*, 13:901–912, 2001.

Velasco Fuentes, O.U. and F.A. Velázquez Muñoz. Interaction of two equal vortices on a β plane. *Phys. Fluids*, 15:1021–1032, 2003.

Waugh, D. The efficiency of symmetric vortex merger. *Phys. Fluids A*, 4:1745–1758, 1992.

Yasuda, I. and G. Flierl. Two-dimensional asymmetric vortex merger: merger dynamics and critical merger distance. *Dyn. Atmos. Oc.*, 26:159–181, 1997.

Zabusky, N., M. Hughes, and K. Roberts. Contour dynamics for the Euler equations in two dimensions. *J. Comp. Phys.*, 30:96–106, 1979.

A LOW-DIMENSIONAL DYNAMICAL SYSTEM FOR TRIPOLE FORMATION

R. C. KLOOSTERZIEL
School of Ocean and Earth Science and Technology
University of Hawaii at Manoa
1000 Pope Road
Honolulu, HI 96822, U.S.A.

G. F. CARNEVALE
Scripps Institution of Oceanography, U.S.A.

Abstract. Laboratory observations and numerical experiments have shown that a variety of compound vortices can emerge in two-dimensional flows due to the instability of isolated circular vortices. Their simple geometrical features suggest that their description may take a simple form if an appropriate set of functions is used. We employ a set which is complete on the infinite plane for vorticity distributions with finite total enstrophy. Through projection of the vorticity equation (Galerkin method) and subsequent truncation we derive a dynamical system which is used to model tripole formation. It is found that at low-order truncations the observed behavior is qualitatively captured by the dynamical system. We determine the necessary ingredients for saturation of the instability at finite amplitude in terms of nonlinear interactions between various azimuthal components of the vorticity field.

Key words: dynamical system, tripole

1. Introduction

In two-dimensional or quasi-geostrophic fluid dynamics, several types of coherent flow structures are known. The most common is the simple monopolar vortex, often circularly symmetric in the absence of external strain. Chance encounters of such vortices with oppositely-signed circulations can lead to the formation of a 'dipole' which is a self-propelling compound vortex. In forced two-dimensional turbulence, Legras *et al.* (1988) observed a more-complicated coherent compound vortex, later called a 'tripole'. Laboratory experiments (van Heijst & Kloosterziel, 1989) and numerical simulations (Carton *et al.*, 1989) showed tripole formation due to the growth of an azimuthal wavenumber 2 instability of an unstable isolated circular

O.U. Velasco Fuentes et al. (eds.), Nonlinear Processes in Geophysical Fluid Dynamics, 355-374.

vortex. Laboratory experiments revealed that wavenumber 3 instabilities can lead to another compound vortex, called the 'triangular vortex' (see Kloosterziel & van Heijst, 1991). Carnevale & Kloosterziel (1994) and Morel & Carton (1994) investigated whether compound vortices could result from instabilities associated with even higher azimuthal wavenumbers. This was found unlikely because these vortices are unstable to infinitesimally small perturbations. The tripole and triangular vortex have simple symmetric vorticity distributions. This suggests that these vorticity patterns can be closely approximated by sums of a small number of appropriately chosen functions and that the dynamics describing the evolution might be modeled in a simple fashion. These issues are addressed here, but only regarding tripole formation. For a more detailed discussion and technical details see Kloosterziel & Carnevale (1999).

In §2 we present graphs from numerical simulations showing the instability of a circular vortex leading to tripole formation. The vorticity field ω is decomposed into an axisymmetric component ω_0 and azimuthal deviations $\omega_{k>0}$, i.e. $\omega = \sum_k \omega_k$ with $k = 0, 1, 2, \cdots$. We determine the temporal evolution of the components and show that there is a strict hierarchy in their amplitudes. A small number of azimuthal components ω_k are found to dominate during the evolution towards the steadily rotating tripole state. In §3 we derive through projection of the vorticity equation (Galerkin method) and truncation a finite-dimensional dynamical system. The best known example of a dynamical system thus derived is the celebrated Lorenz (1963) model. It has not been used before to study the evolution of unstable 2D-vortices, in particular the seemingly complex metamorphosis from circular to compound. It differs also from previous studies (in all areas of physics) in that we apply it to the evolution on an infinite domain. Usually box-like or compact domains like a sphere are considered and expansions in eigenfunctions of the Laplace operator are used. On such domains the eigenvalues (e.g. wavenumbers) of the Laplacian are discrete, and a spectral truncation usually involves eliminating all modes of length scale smaller than some prescribed value. On the infinite plane, however, the spectrum is continuous so that no finite-dimensional system can be derived by truncation. We use functions which are not eigenfunctions of the Laplace operator. A low-order model can be obtained by truncation because they form a discrete set. We discuss the non-linear dynamics in §4. We determine what the necessary ingredients are for saturation at finite amplitude in terms of feedback between various azimuthal components ω_k and the generation of higher harmonics. It is found that the first harmonic ω_{2m} ($m = 2$ for the tripole) is of fundamental importance in the formation process, but only for a limited time. Laplacian diffusion is further used to mimic the flow of enstrophy to smaller scales which are not resolved at low-order truncations.

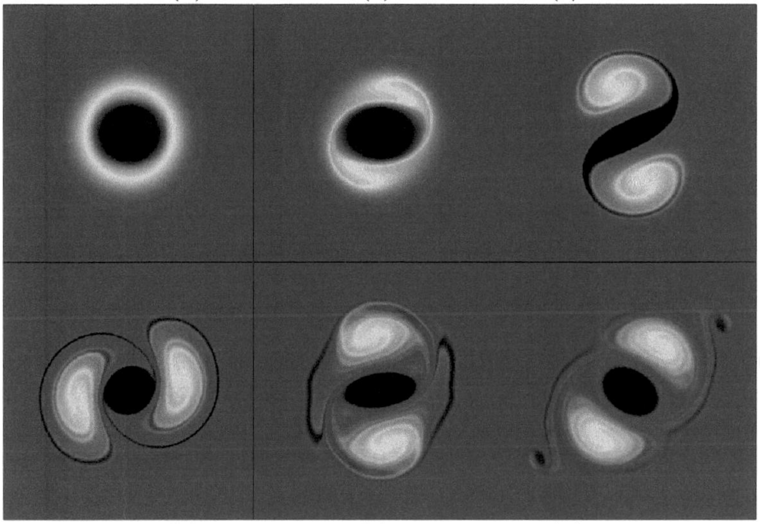

Figure 1. Shaded contour plots of vorticity in a numerical simulation of the evolution of $\bar{\omega}(r)$ given by (1) plus a small wavenumber $k = 2$ perturbation. Non-dimensional times are (*a*) $t = 0$, (*b*) $t = 25$, (*c*) $t = 50$, (*d*) $t = 75$, (*e*) $t = 100$ and (*f*) $t = 200$. Uniform gray away from the central region indicates zero vorticity, black and darker gray indicate negative vorticity, white and lighter gray indicate positive vorticity.

Through an example we show that this diminishes shape vacillations of the pattern not unlike observed in the laboratory and numerical simulations. In §5 the main results and conclusions are summarized.

2. Tripole formation

In Figure 1, snapshots are shown of a numerical simulation of the 2-D vorticity equation. The method of simulation we used is that of Patterson & Orszag (1971) on a doubly periodic domain of $N \times N$ grid points. We employed an isotropic spectral truncation at wavenumber $k_{\text{trunc}} = (8/9)^{1/2}N/2$, with a resolution $N = 256$. In the simulation, hyper-viscosity was used to prevent build-up of small-scale enstrophy due to finite resolution (256^2). The initial condition ($t = 0$) was a circularly symmetric vortex with a vorticity distribution

$$\bar{\omega}(r) = (\tfrac{3}{2}r^3 - 1)\mathrm{e}^{-r^3} \tag{1}$$

plus a small perturbation of the form $\omega'(r, \theta) = f(r)\cos(k\theta)$ where $k = 2$ and $\max_r|f(r)| = O(10^{-1})$. Polar coordinates (r, θ) are used with the origin ($r = 0$) at the center of the vortex (Figure 2*a*). The azimuthal (swirl)

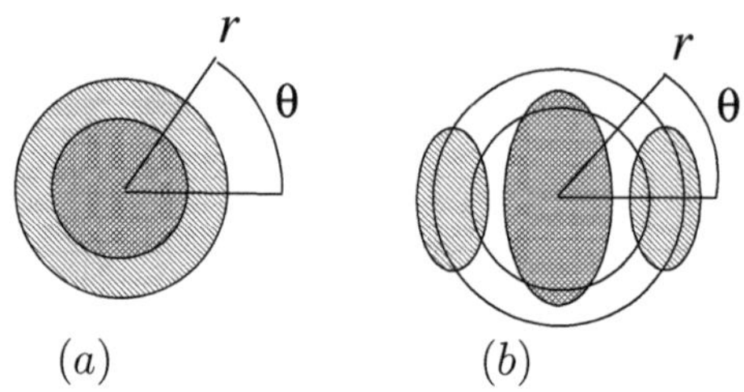

Figure 2. (*a*) The polar coordinate system (r, θ) with the origin at the center of a circular vortex and (*b*) with the origin at the center of a tripole.

velocity field associated with (1) is $\bar{v}(r) = -\frac{1}{2} r \exp(-r^3)$. The flow of the unperturbed vortex is everywhere clockwise, the vorticity is negative in the core and positive further out. Time is scaled by $|\bar{\omega}(0)^{-1}|$. The elongated shape of the negative core vorticity at $t = 25$ seen in Figure 1(*b*) indicates that a $k = 2$ mode has attained an appreciable amplitude. Two semi-circular regions of positive vorticity have formed ('satellites') at $t = 50$ around which tendrils of core vorticity have been wrapped. Thin core vorticity filaments are observed at $t = 75$ as the wrapping and stretching continues. The core is now almost circular but returns to an ellipsoidal shape as time increases to $t = 100$. At this time, the tripole has clearly formed. The last panel in the Figure ($t = 200$) is representative for all subsequent times when the simulation is continued. The tripole rotates clockwise about its center without any further substantial shape variations. The simulation was continued until $t = 600$ and in that time span total energy was conserved to the fourth significant digit whereas enstrophy decayed roughly 10%.

The vorticity fields shown in Figure 1 can be expressed as $\omega = \sum_{k=0}^{\infty} \omega_k$ where $\omega_k = f_k(r;t)\mathrm{Re}(\mathrm{e}^{\mathrm{i}k\theta + \mathrm{i}\phi_k(r;t)})$ ($\mathrm{Re}(\cdot)$ denotes real part). The $f_k(r;t)$ and the phase-factors are found using $C_k = f_k \cos\phi_k$, $S_k = -f_k \sin\phi_k$ where $\{C_k(r;t), S_k(r;t)\} = \frac{1}{\pi}\int_0^{2\pi} \omega(r,\theta;t)\{\cos k\theta, \sin k\theta\}\,\mathrm{d}\theta$ ($\phi_k = 0$ for $k = 0$). This extracts from a vorticity distribution its azimuthal components ω_k. A measure for the amplitude of ω_k at a given time is

$$A_k(t) = {Q_k}^{1/2} = \left(\int_0^{2\pi} \int_0^{\infty} \omega_k^2(r,\theta;t) r \mathrm{d}r \mathrm{d}\theta \right)^{1/2}. \tag{2}$$

Q_k is the enstrophy associated with wavenumber k. The integral converges rapidly because the vorticity amplitudes drop quickly off to zero for large r, and the actual integral is taken over the finite-sized computational domain. For a circularly symmetric vortex all ω_k except ω_0 are zero but Figure 2(*b*) indicates that for a tripole there will be non-zero ω_k for various k.

In Figure 3(*a*) we show the evolution of $A_k(t)$ for several even wavenumbers k in the experiment of Figure 1. Black dots indicate the times for which

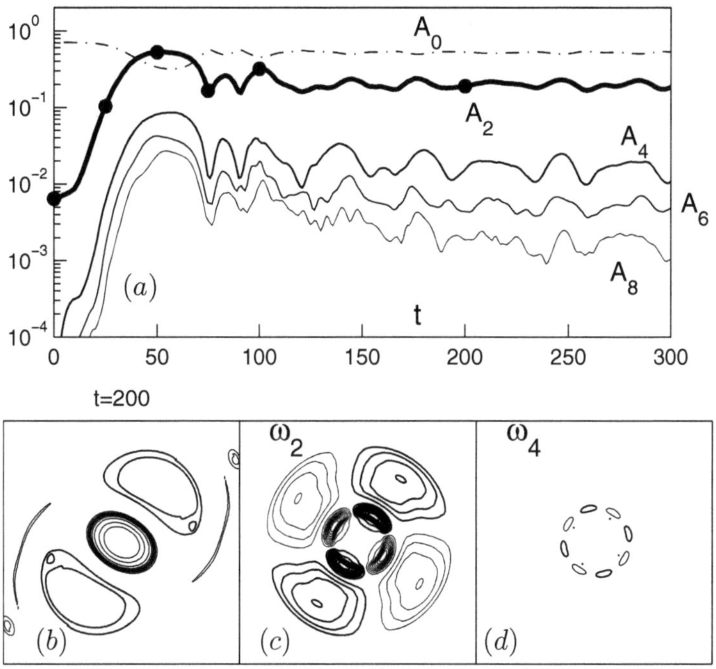

Figure 3. (a) Evolution of the amplitudes of various azimuthal components ω_k during the tripole formation shown in Figure 1. For a definition of A_k see text. (b) Contour plot of the vorticity of the tripole at $t = 200$ and (c) of the ω_2 component and (d) of the ω_4 component. Contour levels are the same in (c) and (d). Thin lines indicate negative values, thick lines positive values.

the vorticity fields were shown in Figure 1. Amplitudes with odd k remained negligibly small which implies that during the transformation the center of the vortex did not drift. Amplitudes for $k > 8$ are at all times smaller than $A_8(t)$. On the logarithmic scale of Figure 3(a) after a brief transient period $A_2(t)$ is seen to grow linearly with time. This is the period of exponential growth of unstable $k = 2$ normal modes as in linearized dynamics. A_4, A_6 and A_8 are initially zero but eventually grow to finite amplitudes, which is a nonlinear effect. Around $t = 50$ all amplitudes attain a maximum except for A_0, which is then at a minimum. Small oscillations appear in all A_k which persist up to $t = 300$ and beyond. The relative maximum in A_2 at $t = 50$ is when in Figure 1(c) the core has become highly elongated. The following relative minimum at $t = 75$ is when in Figure 1(d) the core is momentarily far less elongated. The levels about which the amplitudes oscillate afterwards correspond roughly to the vorticity distribution shown in Figure 1(f). Contours of constant vorticity of the tripole at $t = 200$ are shown in Figure 3(b) while its components ω_2 and ω_4 are shown in Figure

3(c) and (d), respectively. Clearly the ω_4 component has a much smaller amplitude than ω_2. By adding $\omega_0 + \omega_2 + \omega_4$ we found that the shape of the core and the satellites compared well with that of the original. Also the positions of maximum amplitude in the satellites coincided nicely. The amplitudes differed by roughly 10%. With $\omega_0 + \omega_2$ the main features of the tripole were also recovered. The higher azimuthal harmonics ($k = 4, 6, \cdots$) only contribute to the finer details at this point. However, to get the main features seen in Figure 1(c) at $t = 50$ for example, i.e. the highly elongated core and the thin tendrils of core vorticity wrapping around the satellites, also ω_6 and ω_8 had to be added.

3. A finite-dimensional dynamical system

Since only a small number of components ω_k dominate during the unfolding of the instability, we investigate whether the same is true *dynamically.* That is, we question whether the instability and saturation can be described using a small number of azimuthal wavenumbers. The starting point is the 2-D inviscid vorticity equation in polar coordinates

$$\frac{\partial \omega}{\partial t} + \frac{1}{r}\frac{\partial \psi}{\partial \theta}\frac{\partial \omega}{\partial r} - \frac{1}{r}\frac{\partial \psi}{\partial r}\frac{\partial \omega}{\partial \theta} = 0, \tag{3}$$

where ψ is the streamfunction so that in polar coordinates (r, θ) the radial and azimuthal velocity components are $u = r^{-1}\partial_\theta \psi$ and $v = -\partial_r \psi$, respectively, while $\omega = -\nabla^2 \psi$, with ∇^2 the Laplace operator. In (3) we substitute expansions of the form

$$\psi(r, \theta; t) = \sum_{n=0}^{\infty} \sum_{k=-\infty}^{+\infty} a_n^k(t)\varphi_n^k(r, \theta), \qquad \omega(r, \theta; t) = \sum_{n=0}^{\infty} \sum_{k=-\infty}^{+\infty} b_n^k(t)\varphi_n^k(r, \theta), \tag{4}$$

where the φ_n^k are orthonormal: $\int_0^\infty \int_0^{2\pi} \varphi_n^k \varphi_{n'}^{k'\star} \mathrm{d}\theta r \mathrm{d}r = \delta_{nn'}\delta_{kk'}$ (a '$\star$' denotes complex conjugate). Then formally

$$\{a_n^k, b_n^k\} = \langle \{\psi, \omega\}, \varphi_n^k \rangle, \qquad \langle f, g \rangle \equiv \int_0^{+\infty} \int_0^{2\pi} f g^\star \mathrm{d}\theta r \mathrm{d}r. \tag{5}$$

Through projection (Galerkin method) of (3) on the φ_r^m and elimination of the a_n^k through use of the relation $\omega = -\nabla^2 \psi$, a system of coupled ODE's can be derived for the evolution of the vorticity expansion coefficients

$$\frac{\mathrm{d}b_n^m}{\mathrm{d}t} + \mathrm{i} \sum_{p=0} \sum_{q=0} \sum_{k+l=m} I\begin{pmatrix} m & k & l \\ n & p & q \end{pmatrix} b_p^k b_q^l = 0. \tag{6}$$

We used the functions

$$\varphi_n^k(r, \theta) = c_n^k \times (\tfrac{1}{2}r^2)^{\frac{1}{2}|k|} \mathrm{e}^{-r^2/4} L_{|k|+n}^{|k|}(\tfrac{1}{2}r^2) \mathrm{e}^{\mathrm{i}k\theta} \tag{7}$$

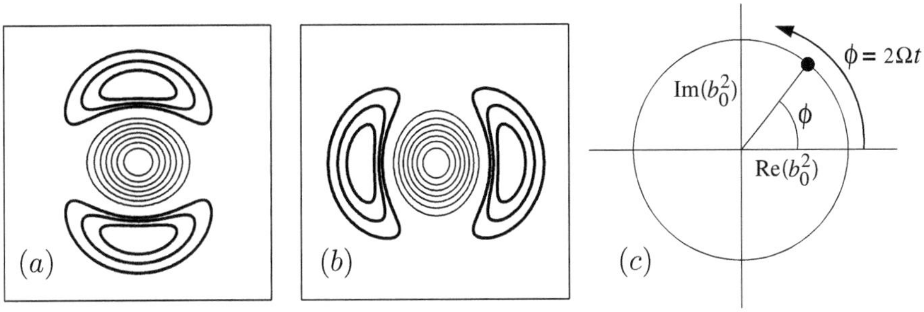

Figure 4. Contours of a tripole-like pattern formed with a few φ_p^k. The amplitudes are $b_0^1 = -1$, $b_1^0 = -1$, $b_0^2 = 0.4\exp(\mathrm{i}\phi)$ plus $b_0^{2\star}$ with (*a*) $\phi = 0$ and (*b*) $\phi = \pi/2$. Thin lines indicate negative values, thick lines positive values. (*c*) When $\phi = 2\Omega t$ and $\Omega > 0$, the tripole patterns rotate in clockwise direction with increasing time t.

where $c_n^k = (n!/2\pi(|k|+n)!^3)^{1/2}$. The L_n^m (with $n \geq m$) are associated Laguerre polynomials of order $n-m$, defined as the mth derivative of the ordinary Laguerre polynomial: $L_n^m(x) \equiv \mathrm{d}_x^m L_n(x)$, $L_n(x) \equiv \mathrm{e}^x \mathrm{d}_x^n x^n \mathrm{e}^{-x}$. We call n the radial wavenumber since for increasing n the functions become more oscillatory. When $\int\int_{\mathbf{R}^2} \psi^2 \mathrm{d}\theta r \mathrm{d}r < \infty$ and $\int\int_{\mathbf{R}^2} \omega^2 \mathrm{d}\theta r \mathrm{d}r < \infty$, the expansions (4) are possible, i.e. the set $\{\varphi_n^k\}$ is complete on $\mathbf{R}^2$ in a square-integrable sense (see Higgins, 1977). The φ_n^k are *not* eigenfunctions of the Laplace operator since $\nabla^2\varphi_n^k = \frac{1}{4}r^2\varphi_n^k - (2n+1+|k|)\varphi_n^k$. This makes the calculation of the interaction coefficients $I(\because\because)$ computationally extremely costly and complicated. Note that in expansions like (4) we get for $k \neq 0$ terms $b_n^k\varphi_n^k + b_n^{-k}\varphi_n^{k\star}$ where $\varphi_n^{k\star} = \varphi_n^{-k}$. Since ω is real, therefore $b_n^{-k} = b_n^{k\star}$, while the b_n^0 are always real. The interaction coefficients are also real.

Since the φ_n^k decay rapidly with increasing r they are well-suited to represent compact vorticity distributions. This is illustrated in Figure 4 where with a few φ_n^k's a pattern has been created that resembles that of the tripole. We used only φ_0^0, φ_1^0 plus φ_0^2. If $b_0^2(t) = b_0^2(0)\exp(\mathrm{i}2\Omega t)$ was a solution of (6) with $b_0^0 = b_1^0 = -1$, $b_0^2(0) = 0.4$ and all other $b_n^k(t) = 0$, the tripole-like pattern would rotate clockwise at a constant angular velocity $|\Omega|$ when $\Omega > 0$, as observed in the numerical simulations. This is impossible, however, since higher harmonics (b_p^k with $k = 4, 6, \cdots$) will be generated unless the system is truncated at wavenumber $k = 2$.

In the simulations discussed below we integrated (6) forward in time after assigning initial values to the expansion coefficients b_p^k. We limit the azimuthal wavenumbers k to a finite range and for each k in this range we also truncate at finite radial wavenumbers N_k. The truncations are symmetric, i.e. for each k used also $-k$ is used and $N_{-k} = N_k$. Further, the numerical experiment suggests that the basic instability and tripole formation involves only the component ω_m ($m = 2$ for the tripole), its harmonics

$(\omega_{2m}, \omega_{3m}, \cdots)$ and the circular component ω_0. For this reason the dynamical system we discuss below only involves wavenumbers $k = 0, m, 2m, \cdots$. Azimuthal truncations were thus taken at $k = m$, $k = 2m$, etc. None of the intermediate wavenumbers were used, this allows us to focus entirely on the formation process and disregard possible instabilities due to perturbations with azimuthal wavenumbers other than $k = m, 2m, \cdots$. In such truncated models the following set of equations governs the dynamics:

$$\begin{aligned}\frac{\mathrm{d}b_n^0}{\mathrm{d}t} &= 2\sum_{p=0}^{N_m}\sum_{q=0}^{N_m}\overbrace{I\begin{pmatrix}0 & m & -m\\ n & p & q\end{pmatrix}\mathrm{Im}\left(b_p^m b_q^{m\star}\right)}^{\text{feedback}} \qquad (n = 0, \cdots, N_0)\\ &+ 2\sum_{p=0}^{N_{2m}}\sum_{q=0}^{N_{2m}}\underbrace{I\begin{pmatrix}0 & 2m & -2m\\ n & p & q\end{pmatrix}\mathrm{Im}\left(b_p^{2m} b_q^{2m\star}\right)}_{\text{feedback}} + 2\sum_{p=0}^{N_{3m}}\sum_{q=0}^{N_{3m}}\underbrace{\{\cdots\}}_{\text{fdbck}} - \cdots\end{aligned} \tag{8}$$

$$\begin{aligned}\frac{\mathrm{d}b_n^m}{\mathrm{d}t} &= -2\mathrm{i}\sum_{p=0}^{N_0}\sum_{q=0}^{N_m}\overbrace{I\begin{pmatrix}m & 0 & m\\ n & p & q\end{pmatrix}b_p^0 b_q^m}^{\text{advection}} \qquad (n = 0, \cdots, N_m)\\ &- 2\mathrm{i}\sum_{p=0}^{N_m}\sum_{q=0}^{N_{2m}}\underbrace{I\begin{pmatrix}m & -m & 2m\\ n & p & q\end{pmatrix}b_p^{m\star} b_q^{2m}}_{\text{feedback}} - 2\mathrm{i}\sum_{p=0}^{N_{2m}}\sum_{q=0}^{N_{3m}}\underbrace{\{\cdots\}}_{\text{fdbck}} - \cdots\end{aligned} \tag{9}$$

$$\begin{aligned}\frac{\mathrm{d}b_n^{2m}}{\mathrm{d}t} &= -2\mathrm{i}\sum_{p=0}^{N_0}\sum_{q=0}^{N_{2m}}\overbrace{I\begin{pmatrix}2m & 0 & 2m\\ n & p & q\end{pmatrix}b_p^0 b_q^{2m}}^{\text{advection}} \qquad (n = 0, \cdots, N_{2m})\\ &- 2\mathrm{i}\sum_{p=0}^{N_m}\sum_{q=0}^{N_m}\underbrace{I\begin{pmatrix}2m & m & m\\ n & p & q\end{pmatrix}b_p^m b_q^m}_{\text{first harmonic generation}} - 2\mathrm{i}\sum_{p=0}^{N_{2m}}\sum_{q=0}^{N_{4m}}\underbrace{\{\cdots\}}_{\text{fdbck}} - \cdots\end{aligned} \tag{10}$$

and so on ($\mathrm{Im}(\cdot)$ denotes imaginary part). The appearance of the terms $\mathrm{Im}\left(b_p^m b_q^{m\star}\right)$ in (8) is due to the the anti-symmetry property $I\begin{pmatrix}m & k & l\\ n & p & q\end{pmatrix} = -I\begin{pmatrix}-m & -l & -k\\ n & q & p\end{pmatrix}$, and the fact that $b_p^{-k} = b_p^{k\star}$. The symmetry property $I\begin{pmatrix}m & k & l\\ n & p & q\end{pmatrix} = I\begin{pmatrix}m & l & k\\ n & q & p\end{pmatrix}$, accounts for the presence of the factors 2 on the right-hand sides. Referring to an element φ_n^k in the expansion as a 'mode' and b_n^k as the (complex) amplitude of the mode, the time rate of change of a given mode's amplitude with azimuthal wavenumber m' is determined

Figure 5. Schematic diagram of the interactions in minimal systems using (a) wavenumbers $k = 0, m$ and (b) $k = 0, m, 2m$ with $m = 2$ for tripole formation. With reference to equations (8)–(10) '1'=advection, '2'=feedback, '3' =higher harmonic generation.

by nonlinear interactions of modes with wavenumbers k, l, which satisfy $k + l = m'$. For these interactions we use the notation $k + l \to m'$. In (8)–(10) and Figure 5 'feedback' indicates interactions $k + l \to m'$ where either $|m'| < |k|$ or $|m'| < |l|$. Any interaction between circular and non-circular modes $0 + k \to k$ has been called 'advection'. Harmonics are generated by the quadratic terms that have both $k, l > 0$ or $k, l < 0$. We obtain a system truncated at azimuthal wavenumber $k = 2m$ if the terms that involve $k = 3m, k = 4m, \cdots$ (upper indices $N_{3m}, N_{4m}, \cdots$) are discarded in (8)–(10). We refer to this as the $(0, m, 2m)$ system. If also (10) is discarded plus all terms involving $k = 2m, 3m, \cdots$ in (8) and (9), we get the dynamics for the system truncated at $k = m$. This we call the $(0, m)$ system. Figure 5 shows schematically the $(0, m)$ and $(0, m, 2m)$ systems that with $m = 2$ have been used to study tripole formation. Finally we mention that the systems conserve enstrophy Q at any truncation, where

$$Q = \sum_{k \geq 0} Q_k, \qquad Q_0 = \sum_{p=0}^{N_0} |b_p^0|^2, \qquad Q_{k>0} = 2 \sum_{p=0}^{N_k} |b_p^k|^2. \tag{11}$$

Thus, in the $(0, m)$ system, $Q_0 + Q_m$ is constant and in the $(0, m, 2m)$ system, $Q_0 + Q_m + Q_{2m}$ is constant.

4. Nonlinear dynamics at low-order truncations

We need to consider what minimal truncations are needed so that the main features of tripole formation are well modeled. Clearly instability should occur. The starting point is the vorticity profile (1) which needs to be approximated with a truncated sum of circular modes φ_n^0. They are functions of the non-dimensional variable r. We scale this variable with a scale factor ε and vary it until an optimal approximation is found. Thus we approximate the initial vorticity distribution by

$$\bar{\omega}(r) \approx \bar{\omega}_{N_0}(r; \varepsilon) \equiv \sum_{n=0}^{N_0} \bar{b}_n^0 \varphi_n^0(r/\varepsilon, \theta), \qquad \bar{b}_n^0 = \langle \bar{\omega}, \varphi_n^0(r/\varepsilon, \theta) \rangle, \tag{12}$$

and vary ε until for a given radial truncation N_0 an absolute minimum for the 'distance' $d(\bar{\omega}, \bar{\omega}_{N_0}) \equiv \int_0^\infty (\bar{\omega}(r) - \bar{\omega}_{N_0}(r;\varepsilon))^2 \mathrm{d}r$ is found. For $N_0 = 4$ and $\varepsilon = 0.28$, an almost perfect match was established $[d(\bar{\omega}, \bar{\omega}_{N_0}) = O(10^{-6})]$. For smaller radial truncations poorer approximations resulted. With the corresponding values of the $\bar{b}_n^0$ a linear stability analysis was performed. Perturbations with a given azimuthal wavenumber k are of the form

$$\omega'_{N_k}(r,\theta;t) = \sum_{p=0}^{N_k} \left[b_p^k(t)\varphi_p^k(r/\varepsilon,\theta) + b_p^{k\star}(t)\varphi_p^{k\star}(r/\varepsilon,\theta) \right], \tag{13}$$

with ε equal to value that minimized $d(\bar{\omega}, \bar{\omega}_{N_0})$. In the truncated model the initial condition is $\bar{\omega}_{N_0} + \omega'_{N_k}$. Substitution in (9) and discarding terms quadratic in the perturbation expansion coefficients leads to the linearized dynamics

$$\frac{\mathrm{d}b_p^k(t)}{\mathrm{d}t} = \mathrm{i} \sum_{p'=0}^{N_k} \mathrm{M}_{pp'}^k b_{p'}^k(t), \qquad \mathrm{M}_{pq}^k = -2 \sum_{r=0}^{N_0} I\begin{pmatrix} k & 0 & k \\ p & n & q \end{pmatrix} \bar{b}_n^0, \tag{14}$$

where $p, q = 0, \cdots, N_k$. We write (14) as $\dot{\mathbf{b}}^k = \mathrm{i}\mathbf{M}^k\mathbf{b}^k$ (a dot indicates time derivative), where $\mathbf{M}^k$ is the $(N_k+1) \times (N_k+1)$ real matrix defined above and $\mathbf{b}^k(t)$ is the (N_k+1)-dimensional (complex) vector $\mathbf{b}^k(t) = (b_0^k(t), \cdots, b_{N_k}^k(t))^{\mathrm{T}}$. Assuming exponential time-dependence $\mathbf{b}^k(t) = \mathrm{e}^{\mathrm{i}\lambda t}\mathbf{b}^k$, one gets the matrix eigenvalue problem $\mathbf{M}^k\mathbf{b}^k = \lambda\mathbf{b}^k$. There are unstable modes when at least one eigenvalue λ has $\mathrm{Im}(\lambda) < 0$. For $k = 2$ we found no unstable modes for radial resolutions $N_2 < 3$. For $N_2 \geq 3$ there are unstable modes. The growth rate of the most unstable mode for $N_2 = 3$ was found to deviate roughly 10% from the growth rate determined with a very high-resolution normal modes analysis. For $N_2 = 9$ the difference reduced to about 2%.

4.1. THE $(0, M)$ SYSTEM

The simplest system that *may* mimic tripole formation uses only azimuthal wavenumbers $k = 0, 2$ and must have a radial resolution $N_2 \geq 3$. We used the minimal truncations $N_2 = 3$ and $N_0 = 4$, as sketched in Figure 6(a), in order to keep the system as low-dimensional as possible. The phase space spanned by the expansion coefficients has $(N_0 + 1) + 2(N_2 + 1)$ dimensions, i.e. it is 13-dimensional. The dynamics is a feedback loop as sketched in Figure 5(a). It has advection $0 + m \to m$ and feedback $m + (-m) \to 0$. The initial condition for (6) are the $b_p^0 = \bar{b}_p^0$ determined by the optimal approximation plus small non-zero $b_0^2 = b_0^{2\star}$. This projects on the most unstable mode. Using this as an initial condition we integrated (6) forward

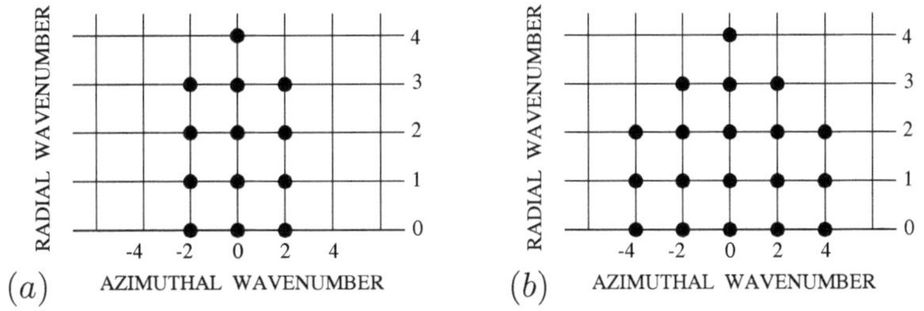

Figure 6. (*a*) The minimal $(0, m)$ system with $m = 2$ and symmetric truncations $N_0 = 4$ and $N_2 = 3$ and (*b*) the $(0, m, 2m)$ system with $m = 2$ and truncations $N_0 = 4$ and $N_2 = 3$ and $N_4 = 2$ used in the simulations described in the text. For (*a*) phase space is 13-dimensional, for (*b*) 19-dimensional.

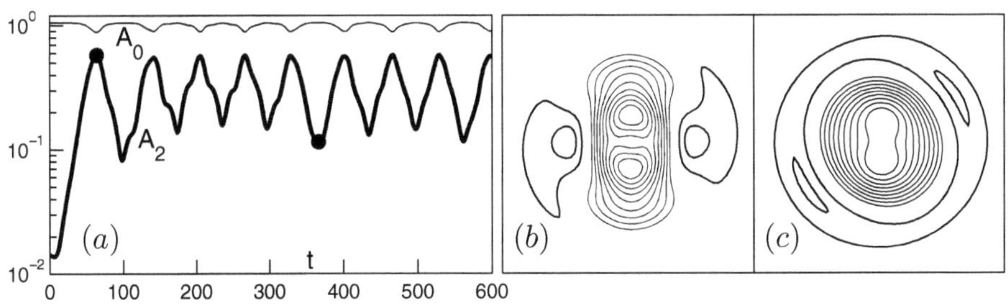

Figure 7. (*a*) Evolution of A_0 and A_2 ($A_k = {Q_k}^{1/2}$) after integration of (6) using $k = 0, 2$ and resolutions $N_0 = 4, N_2 = 3$ with as initial condition the b_p^0 for the optimal approximation ω_{N_0} of (1) plus a small wavenumber 2 perturbation with $b_0^2 = b_0^{2\star} = 10^{-2}$ and all other b_p^2 zero. (*b*) Contours of the resulting field at $t = 65$, (*c*) the field at $t = 366$. These moments are indicated by black dots in panel (*a*). Thick contours indicate positive values, thin ones negative values.

in time with the (0,2) truncation set of expansion modes. In Figure 7(*a*) we show the ensuing evolution of $A_k(t) = (Q_k(t))^{1/2}$ $(k = 0, 2)$, with Q_k given by (11). This is the direct analogue of the $A_k(t)$ defined in (2). Initial exponential growth of A_2 occurs, as in the high-resolution experiments (see Figure 3*a*). This is due to the advection term in (9). Feedback alters the circular components according to (8), and A_0 decreases in amplitude. Since $A_k = {Q_k}^{1/2}$, an amplitude increase in one component is accompanied by a decrease in the other because in the $(0, 2)$ system $Q_0 + Q_2$ is constant. No saturation at finite amplitude occurs, instead both A_0 and A_2 oscillate forever. The fields corresponding to a maximum and a minimum in A_2 are in Figure 7(*b*) and 7(*c*), respectively. A tripole-like pattern is associated with the maxima in A_2, and a distorted circular vortex with the minima. Vacillation between these two patterns continue, no matter how long we integrated the system. In systems with higher radial resolutions N_0, N_2 the

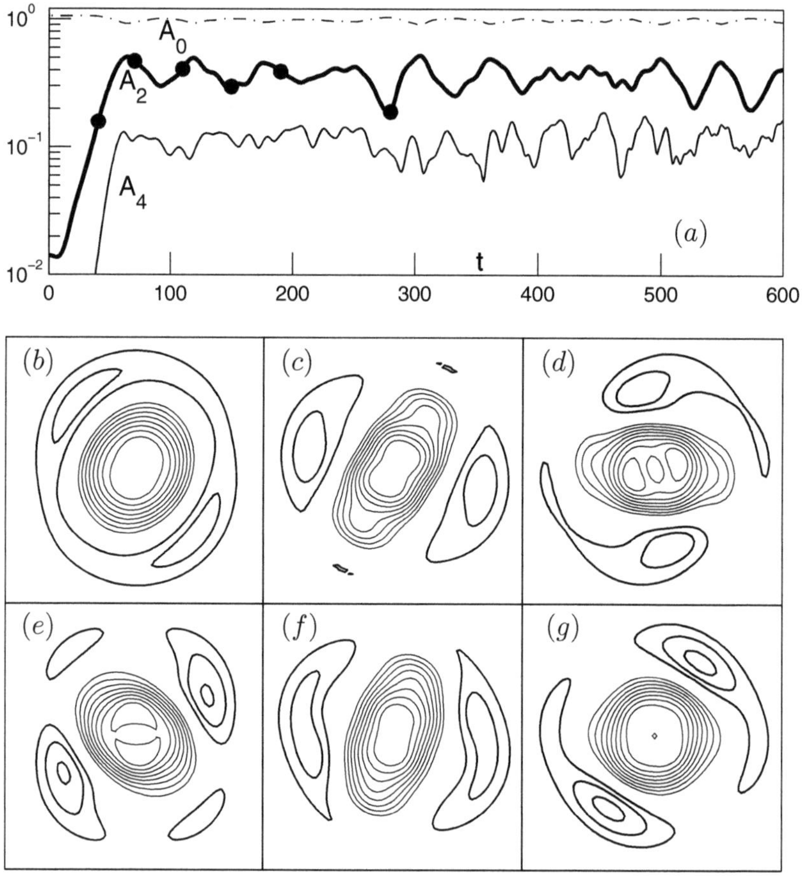

Figure 8. (*a*) Evolution of A_0, A_2 and A_4 after integration of (6) using $k = 0, 2, 4$ and resolutions $N_0 = 4, N_2 = 3, N_4 = 2$ with the same initial condition as for Figure 7 and the fields at (*b*) $t = 40$, (*c*) $t = 70$, (*d*) $t = 110$, (*e*) $t = 150$, (*f*) $t = 190$ and (*g*) $t = 280$.

system exhibited the same behavior. Although it can be shown that phase-space is full of periodic orbits, many of which represent rotating tripoles, the dynamics lacks something needed to keep or put the phase flow in the vicinity of such periodic orbits.

4.2. THE $(0, M, 2M)$ SYSTEM

Next in increasing complexity are systems using $k = 0, m, 2m$. We used the same initial condition as above in the system with radial resolutions $N_0 = 4, N_2 = 3, N_4 = 2$ as shown in Figure 6(*b*). There are now $2(N_4 + 1)$ additional degrees of freedom and phase space is 19-dimensional. The dynamics is far more complex as is seen in Figure 5(*b*): there are three

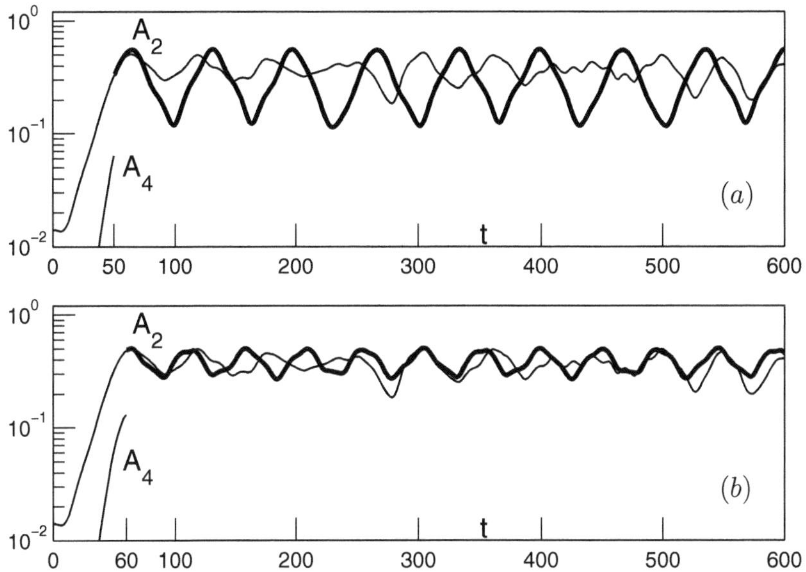

Figure 9. Graphs illustrating the consequences of deleting the first harmonic ω_4 from the dynamics at (a) $t = 50$ and (b) $t = 60$. Initial evolutions are as in Figure 8. Thin lines show A_2 and A_4 from Figure 8(a). Thick lines show the evolution of A_2 starting at (a) $t = 50$ and (b) $t = 60$ in the system using $k = 0, 2$ and truncations $N_0 = 4, N_2 = 3$.

inter-connected feedback loops. In Figure 8(a) we show the evolution of A_0, A_2 and A_4. Again initial exponential growth of A_2 occurs, and A_4, which is initially zero, increases exponentially too. This is due to the higher-harmonic generation in (10). No quasi-periodic behavior occurs, instead there are irregular oscillations in both A_2 and A_4. As in Figure 3(a), A_4 remains at all times smaller than A_2. The amplitude variations in A_2 are smaller than in the simpler dynamics. The six panels in Figure 8 show the field at various representative moments. Even at $t = 280$ when A_2 is relatively small the field is clearly tripole-like with pronounced satellites. The simulation was continued far longer than shown. The irregular oscillations continued but at all times the tripole pattern persisted. Between $t = 100$ and $t = 600$ the major axis of the core made 7 turns which is very close to the number of turns the tripole of Figure 1 made in that time span.

To determine when and how the ω_4-component affects the dynamics we deleted all b_p^4 and $b_p^{4\star}$ from the dynamics at various moments during the evolution and then continued the integration with the system using only $k = 0, 2$. Two examples are shown in Figure 9. In the first case we deleted ω_4 from the dynamics at $t = 50$, in the second at $t = 60$. The evolution of A_2 starting at $t = 50$ for the first case is in Figure 9(a) (thick line). For comparison A_2 from Figure 8(a) in the dynamics including ω_4 at all times is also plotted. The condition at $t = 50$ is such that the ensuing oscillations in A_2 are of the same magnitude as in Figure 7(a). The field vacillates between two extremes similar to those shown in Figure 7(b), (c). In Figure 9(b) the evolution of A_2 for the second case is shown, starting at $t = 60$. Now it stays at levels like seen in Figure 8(a). This was also the case when ω_4 was deleted from the dynamics any time after $t = 60$. At all times tripole-like fields were found, similar to those in Figure 8. Two conclusions are drawn from this. First, once the tripole in the dynamics with $k = 0, 2, 4$ has formed, the ω_4 component is no longer needed to keep ω_2 amplitudes from dropping to low levels with an accompanying disappearance of the tripole. There is mainly a balance due to the feedback loop between ω_0 and ω_m as sketched in Figure 5(a). Secondly, the saturation in the dynamics with $k = 0, 2, 4$ at the levels seen in Figure 8(a) is due to the active role of ω_4 in the period prior to the moment the A_2 and A_4 peak for the first time. It brings about small but important changes in ω_0 and ω_2 so that afterwards the tripole persists even when ω_4 is deleted from the dynamics.

Further information concerning the role of ω_4 was obtained by considering various 'mutilations' of the dynamics. For instance, we cut the feedback from the first harmonic ($k = 4$) to the circular components. This system does not conserve enstrophy. When running the system forward in time we found essentially the same evolution as in the 'uncut' version, i.e. the amplitudes evolved as in Figure 8(a) and only after $t = 200$ small differences became apparent but they remained small. This showed that the feedback $4+(-4) \rightarrow 0$ is not essential. Additionally we also cut the feedback $2 + (-2) \rightarrow 0$ from the dynamics so that only the feedback $4 + (-2) \rightarrow 2$ remained. In this case the ω_0-component does not change, i.e. $\omega_0 = \bar{\omega}_{N_0}$ at all times. Initial exponential growth and saturation of A_2 and A_4 occurred as in Figure 8(a) but at higher levels with $A_4(t) < A_0 < A_2(t)$ and later, i.e. around $t = 100$ instead of around $t = 60$. Nothing resembling a tripole corresponds to this. But it revealed another fact. Had there been no feedback from ω_4 to ω_2, normal modes growth would have continued indefinitely, i.e. the system would have blown up. Analysis showed that in the full dynamics this feedback changes the ω_2 component from the unstable normal mode form towards that of a neutrally stable mode for the altered ω_0 component. The feedback from ω_2 to ω_0 has the added effect that ω_0 is changed to forms for which the growth rate of the most unstable mode is smaller.

The experiments were repeated at various higher resolutions N_2 and N_4. It was found that the minimal model used here is generic and no dramatically different behavior resulted from using higher resolutions. When we increased the resolutions to $N_2 = 5, 6, 7, 8$ with $N_4 = 4, 5, 6, 7$ the saturation of A_2 occurred at higher levels and oscillations were smaller after saturation than in Figure 8(a). Well-defined tripoles formed in all cases.

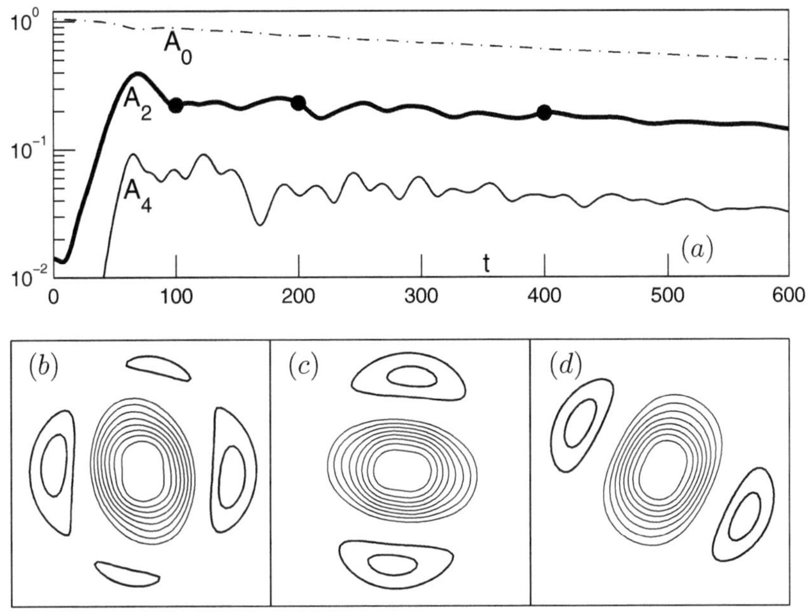

Figure 10. (a) Evolution of A_0, A_2 and A_4 after integration of (15) with $R = 1000$ using $k = 0, 2, 4$, resolutions $N_0 = 4, N_2 = 3, N_4 = 2$ and with the same initial condition as for Figures 7,8 and the fields at (b) $t = 100$, (c) $t = 200$ and (d) $t = 400$.

4.3. VISCOUS DYNAMICS

The truncated systems conserve enstrophy which is not realistic for two reasons. First of all because in reality during the formation process enstrophy cascades to smaller scales. In the truncated dynamics this transfer stops at the highest radial and azimuthal wavenumbers used, and what would have gone to the smaller scales stays in the system. Secondly, in laboratory and numerical experiments there is invariably enstrophy loss. To investigate the effects of enstrophy loss in the dynamical system we added Laplacian diffusion to the dynamics, that is, the right-hand side of (3) was put equal to $\frac{1}{R}\nabla^2\omega$ where R is a Reynolds number. After projection we obtain the system

$$\frac{\mathrm{d}b_n^m}{\mathrm{d}t} + \mathrm{i}\sum_{p=0}\sum_{q=0}\sum_{k+l=m} I\begin{pmatrix} m & k & l \\ n & p & q \end{pmatrix} b_p^k b_q^l = \frac{1}{R}\sum_{i=n-1}^{i=n+1} \mathrm{L}_{in}^m b_i^m . \qquad (15)$$

Note that in the dynamical system the Laplacian is represented by a tri-diagonal matrix. If the basis had consisted of eigenfunctions of the Laplace operator, the right-hand side would have had just one term, proportional to b_n^m. In our case we see that diffusion of a given φ_n^m generates a φ_{n-1}^m

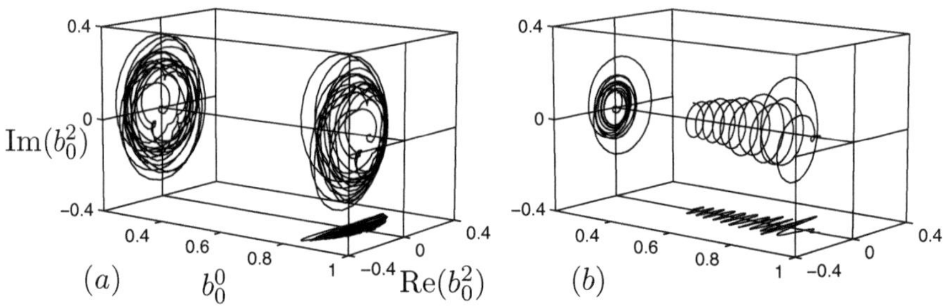

Figure 11. Projections of the phase trajectory on the three-dimensional sub-space spanned by b_0^0, $\mathrm{Re}(b_0^2)$ and $\mathrm{Im}(b_0^2)$ for (a) the inviscid dynamics of Figure 8 and (b) the viscous dynamics of Figure 10. Total time is for both $\triangle T = 600$.

and φ_{n+1}^m component. An example of the evolution for a Reynolds number $R = 1000$ is shown in Figure 10. The same minimal resolutions as before were used, i.e. $N_0 = 4, N_2 = 3, N_4 = 2$, and the same initial condition. The amplitudes in Figure 10(a) evolve along the same lines as in Figure 8(a) until shortly after they first peak. However, after that point the irregular oscillations in the A_k diminish in amplitude while their average values decrease with time which implies enstrophy decay. For higher Reynolds numbers the irregular oscillations diminish to the same extent later in time and the decay is slower. The next three panels show the fields at the times indicated by dots in Figure 10(a). In contrast to the fields of Figure 8, far more symmetric tripoles are found at each instant.

In Figure 11 we compare projections of the phase trajectories in the inviscid and the viscous dynamics. The projections are on the 3-dimensional subspace spanned by b_0^0, $\mathrm{Re}(b_0^2)$ and $\mathrm{Im}(b_0^2)$. 'Shadows' of each trajectory have been drawn at the rear end of the box and on the bottom in order to further facilitate the comparison. These are the trajectories between $t = 0$ and $t = 600$. In the inviscid dynamics (Figure 11a) the trajectory is seen to spiral outwards from the horizontal axis and then to wander around chaotically. With diffusion the trajectory in Figure 11(b) is smooth. In both cases the anti-clockwise loops correspond to clockwise rotation of the tripole (see Figure 4). The inviscid trajectory lies on a hyper-surface of constant Q, given by (11). The projection of this surface fills the entire interior of the ellipsoid $|b_0^0|^2 + 2|b_0^2|^2 = Q$ in the three-dimensional subspace used in Figure 11. The trajectory in Figure 11(a) is at all times within this region. The trajectory in Figure 11(b) cuts hyper-surfaces of progressively smaller Q due to the overall enstrophy decay.

We took the fields shown in Figure 10 as an initial condition for the inviscid dynamics (6) and integrated it forward. In each case amplitudes were scaled so that the total enstrophy Q was the same and equal to the

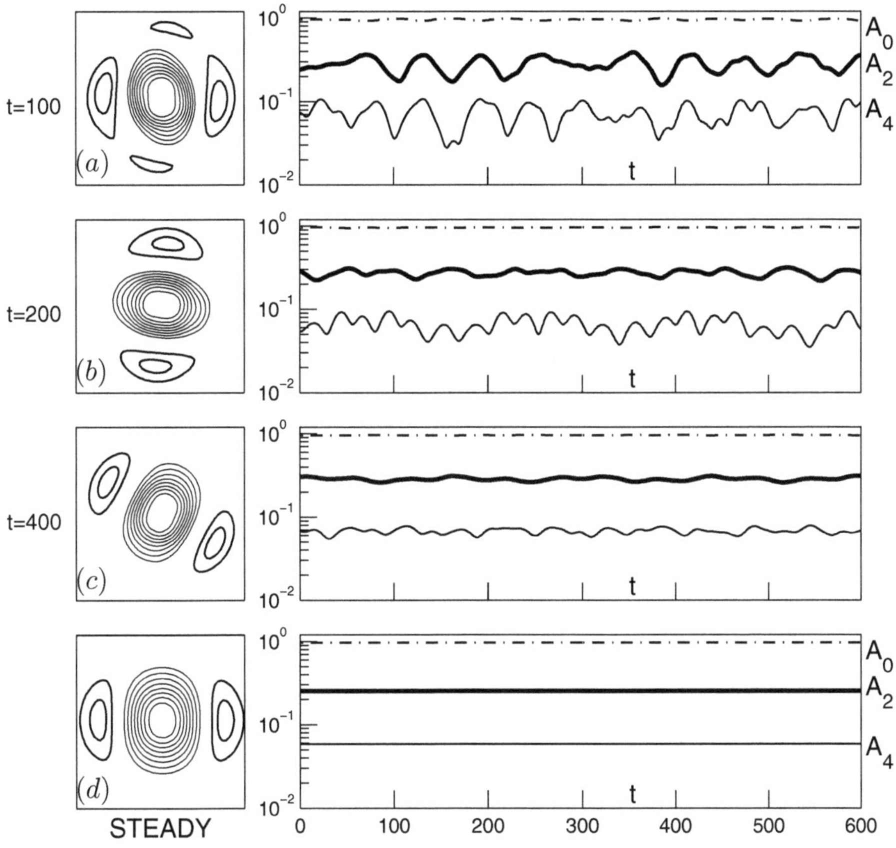

Figure 12. Evolution of A_0, A_2 and A_4 after integration of (6) using $k = 0, 2, 4$, resolutions $N_0 = 4, N_2 = 3, N_4 = 2$ and as initial condition (a) the field from the viscous simulation at $t = 100$, (b) at $t = 200$ and (c) at $t = 400$, each shown in Figure 10, after scaling amplitudes such that Q has the value as in the inviscid simulation of Figure 8. Thin horizontal lines indicate the A_k-levels $(k = 0, 2, 4)$ for the field shown in panel (d) which is a steadily rotating stable solution of (6).

Q-value of the inviscid experiment of Figure 8. The evolution of the A_k in a time span $\triangle T = 600$ is shown in Figure 12 with to the left the fields from Figure 10 that served as initial conditions. It is seen that the ones taken at later times from the viscous experiment lead to smaller amplitude variations with time. In each case the tripole fields of Figure 12(a)–(c) persisted with progressively smaller shape variations and rotated between roughly six to eight times about their axis in the given time span. Thus, as we take tripoles from the viscous experiment at later times, we get closer to steadily rotating solutions for the inviscid dynamics. The thin lines drawn in Figure 12(a)–(c) are the A_k levels of a stable, steadily rotating solution shown in

Figure 12(d). It was found by taking the field from the viscous simulation at $t = 800$ as an initial condition for a numerical search of nearby periodic solutions to (8)–(10), as described in detail by Kloosterziel & Carnevale (1999). It was already very nearly steady in the inviscid dynamics, i.e. amplitude variations were even smaller than in Figure 12(c) for the field at $t = 400$. At $t = 800$ the amplitudes A_k decayed virtually in lockstep at equal ratios. Figure 10(a) shows that at $t = 600$ this is already quite nearly the case. In concise notation, periodic solutions in the $(0, m, 2m)$ system are written as $\mathbf{b}^0 + \mathbf{b}^m \mathrm{e}^{\mathrm{i}m\Omega t} + \mathbf{b}^{2m}\mathrm{e}^{\mathrm{i}2m\Omega t}$, where the vectors $\mathbf{b}^k$ are spanned by fixed expansion coefficients b_p^k $(p = 0, 1, \cdots, N_k)$. Ω is the angular velocity of the rotating pattern. This can be scaled with an arbitrary amplitude γ and more generally solutions are then $\gamma\left(\mathbf{b}^0 + \mathbf{b}^m\mathrm{e}^{\mathrm{i}m\gamma\Omega t} + \mathbf{b}^{2m}\mathrm{e}^{\mathrm{i}2m\gamma\Omega t}\right)$. Without loss of generality we can put $Q = \gamma^2$. What we find is that in the viscous dynamics with increasing time the phase trajectory approaches an orbit determined by fixed $\mathbf{b}^0, \mathbf{b}^m, \mathbf{b}^{2m}, \Omega$ and an amplitude γ for which $Q(t) = \gamma^2$. That is, asymptotically the dynamics converges to a solution $(Q(t))^{1/2}\left(\mathbf{b}^0 + \mathbf{b}^m\mathrm{e}^{\mathrm{i}m(Q(t))^{1/2}\Omega t} + \mathbf{b}^{2m}\mathrm{e}^{\mathrm{i}2m(Q(t))^{1/2}\Omega t}\right)$. The rate of rotation is at each instant $\Omega' = (Q(t))^{1/2}\Omega$. With the decay of Q there is an associated slow-down and with increasing time it takes longer for the orbit in Figure 11(b) to complete a loop. The existence of such solutions to (15) with truncations is not obvious a priori, neither do we know how to predict $\mathbf{b}^0, \mathbf{b}^m, \mathbf{b}^{2m}$ and Ω. The attraction to a particular tripole of fixed structure was noted by Orlandi & van Heijst (1992), who ran high-resolution simulations with Laplacian diffusion and found that besides the overall amplitude decay, at large times the tripole was characterized by a fixed vorticity-streamfunction relation.

5. Summary and discussion

This study of tripole formation had two distinct parts. In the first part (§2) we analyzed data from a high-resolution numerical experiment showing tripole formation (Figure 1). Normal modes growth is followed by the generation of higher harmonics and the formation is completed when nonlinear effects halt the growth and amplitudes level off. Amplitudes A_k of the azimuthal components ω_k have the ordering $A_0 > A_2 > A_4 > \cdots$. In experiments (not shown) with resolutions 64^2 and 128^2 essentially the same tripole formed, rotating at about the same rate. Thin vorticity filaments are not resolved with such low resolutions. They are thus dynamically of little importance and the higher azimuthal components $\omega_6, \omega_8, \cdots$ therefore appeared to play also dynamically a minor role in the adjustment process.

With this in mind we explored in the second part the possibility of reducing the dynamics to as simple as possible a system, that is, a system employing the smallest possible number of azimuthal components ω_k. In §3 a dynamical system was derived by projection of the vorticity equation on the functions (7). In §4 we integrated (6) with severe truncations. The simplest system which uses only $k = 0, 2$ could at sufficiently high radial truncations N_2 mimic normal modes growth, but not saturation with tripole formation as a consequence (Figure 7). Only when also $k = 4$ was included in the dynamics did saturation occur (Figure 8). Substantial amplitude variations occurred with time, but never to the extent that the field lost its clear tripole character. From experiments where the first harmonic $k = 4$ was deleted from the dynamics at various times it was concluded that it is crucial only during a brief period prior to the moment of saturation. In this period ω_4 albeit small in amplitude brings about small but important changes in the ω_0 and ω_2 components. After saturation ω_4 is no longer needed in the dynamics, i.e. from then onward the tripole persists in the system using only $k = 0, 2$. We also explored the consequences of allowing enstrophy to escape from the system through the use of Laplacian diffusion. Saturation with far smaller shape vacillations resulted (Figure 10), as in the high-resolution numerical experiment. Asymptotically the flow converged to a particular tripole with an accompanying uniform amplitude decay. Thus, the main conclusion is that the simplest dynamics that can capture formation process of the compound vortices is that which employs azimuthal wavenumbers $k = 0, 2, 4$. No tendrils like those seen in Figure 1 were resolved by the system, for that higher azimuthal wavenumbers are needed (also with very high radial truncations). But, despite the lack of this feature, tripole formation occurred in the system. The pattern formation process is thus mainly determined by the large scales in the system.

The number of degrees of freedom in minimal systems for tripole formation using only $k = 0, 2, 4$ is still quite high. With the truncations $N_0 = 4, N_2 = 3, N_4 = 2$, phase-space is 19-dimensional. This can be lowered a little by using a poorer approximation to (1), i.e. a smaller N_0 and reducing N_4 to $N_4 = 1$ or $N_4 = 0$. It may be possible to simplify the dynamics further by switching to projections on functions that are more closely related to, say, the dominant components of the tripole. That is, we could for example instead of using the φ_p^k, use the normal modes for either the initial axisymmetric vortex or the final axisymmetric component. By Gram-Schmidt orthogonalization we can then create a new set of orthonormal functions φ'^k_p, and similarly for the circular modes and the $k = 2m$ functions. This is of course similar to 'after the fact' EOF-analysis, but it may lead to the discovery of a truly low-dimensional description of the formation process that is amenable to analysis with the tools of dynamical systems theory. As it is, phase flows are very complex and hard to understand. The phase-space has an infinity of periodic orbits, some of which are stable, others are unstable. In the inviscid dynamics the phase flow wanders chaotically through regions where many nearby periodic orbits reside, but is never asymptotically attracted to one in particular. There can be no domains of attraction surrounding any of the periodic orbits. This is because the inviscid system is time-reversible, i.e. the equations are invariant under sign-changes of vorticity and time. Thus, if there were an attracting orbit which corresponds to, say, a clockwise rotating tripole, then there is another orbit corresponding to an anti-clockwise rotating tripole, with the same structure, which is unstable. Two tripoles with the same structure cannot have different stability properties, and this proves the absence of asymptotically attracting periodic orbits. With diffusion added to the system however, the flow is not reversible and we found that it asymptotically converges to an attracting surface.

Acknowledgements

This work has been supported by Office of Naval Research grants N00014-97-1-0095 and N00014-96-0762 and National Science Foundation grants OCE 97-30843, OCE 97-30843, OCE 01-28991 and OCE 01-29301.

References

Carnevale, G.F. and R. C.Kloosterziel. Emergence and Evolution of Triangular Vortices. *J. Fluid Mech.*, 259:305-331, 1994.

Carton, X., G. R. Flierl, and L. Polvani. The Generation of Tripoles from Unstable Axisymmetric Isolated Vortex Structures. *Europhys. Lett.*, 9:339–344, 1989.

van Heijst, G.J.F., and R. C. Kloosterziel. Tripolar Vortices in a Rotating Fluid. *Nature*, 338:569–571, 1989.

Higgins, J.R. *Completeness and Basic Properties of Sets of Special Functions.* Cambridge University Press, 1977.

Kloosterziel, R.C., and G. F. Carnevale. On the Evolution and Saturation of Instabilities of Two-dimensional Isolated Circular Vortices. *J. Fluid Mech.*, 338:217–257, 1999.

Kloosterziel, R.C., and G. J. F. van Heijst. An Experimental Study of Unstable Barotropic Vortices in a Rotating Fluid. *J. Fluid Mech.*, 223:1–24, 1991.

Legras, B., P. Santangelo and R. Benzi. High-resolution Numerical Experiments for Forced Two-dimensional Turbulence. *Europhys. Lett.*, 5:37–42, 1988.

Lorenz, E.N. Deterministic Non-periodic Flow. *J. Atm. Sci.*, 20:130–141, 1963.

Morel, Y., and X. Carton. Multipolar Vortices in Two-dimensional Incompressible Flow. *J. Fluid Mech.*, 267:23–51, 1994.

Orlandi, P., and G. J. F. van Heijst. Numerical Simulations of Tripolar Vortices in 2D Flow. *Fluid Dyn. Res.*, 9:170–206, 1992.

Patterson, G.S., and S. A. Orszag. Spectral Calculations of Isotropic Turbulence: Efficient Removal of Aliasing Errors. *Phys. Fluids*, 14:2438–2541, 1971.

Index

MIX
Papier aus verantwortungsvollen Quellen
Paper from responsible sources
FSC® C105338

If you have any concerns about our products,
you can contact us on
ProductSafety@springernature.com

In case Publisher is established outside the EU,
the EU authorized representative is:
Springer Nature Customer Service Center GmbH
Europaplatz 3, 69115 Heidelberg, Germany

Printed by Libri Plureos GmbH
in Hamburg, Germany